The Book of the Pig:

ITS SELECTION, BREEDING, FEEDING, AND MANAGEMENT.

By JAMES LONG,

Member of the Council of the Central Chamber of Agriculture;
Formerly Professor of Dairy Farming in the Royal Agricultural
College ; Author of " Modern Dairy Farming," " British Dairy
Farming," "Farming in a Small Way," "The Story of the
Farm," "El ments of Dairy Farming," and of Special
Reports for the Governments cf Canada, New Zealand, and
the United States.

With 16 Plates of Typical Animals and Numerous
Illustrations of Model Piggeries and Appliances.

THIRD EDITION, REVISED.

British Library Cataloguing-in-Publication Data
A catalogue record for this book is available from the
British Library

Pig Farming

A pig is any of the animals in the genus *Sus*, within the Suidae family of even-toed ungulates. Pigs include the domestic pig and its ancestor, the common Eurasian wild boar (*Sus scrofa*), along with other species; related creatures outside the genus include the babirusa and the warthog. Pigs, like all suids, are native to the Eurasian and African continents - they are highly social and intelligent animals. Pigs (hogs in the United States) can be farmed as free range, being allowed to wander around a village, kept in fields, or tethered in a simple house. In developed countries, farming has moved away from traditional pig farming and pigs are now typically intensively farmed.

Almost all of the pig can be used as food, with various parts being made into specialities; sausage, bacon, gammon, ham, skin into pork scratchings, feet into trotters, head into a meat jelly called head cheese (brawn), and consumption of the liver, chitterlings, blood (blood pudding or black pudding) are common. Today, pig farms are significantly larger than in the past, with most large-scale farms housing 5,000 or more pigs in climate-controlled buildings. With 100 million pigs slaughtered each year, these efficiencies deliver affordable meat for consumers and larger profits for producers. This has led to substantial conflict with animal welfare activists as well as those concerned for the environment though. The costs and benefits of large-scale pig farming are still very much up for debate.

Individual farm management focuses on housing facilities, feeding and ventilation systems, temperature and environmental controls and the economic viability of their operations. Procedures and treatments should also carefully weighed, to consider the animals' welfare, health and management in correspondence with accepted husbandry skills. The way in which a stockperson interacts with pigs in agricultural production systems impacts animal welfare, which directly correlates with production levels. Many intensive farms deal with huge numbers of pigs each day, often resulting in handlers becoming complacent with 'positive interactions'. This is not always the case however, and many small farm / 'homesteaders' are turning to raising their own animals, with a view to sustainability and independence.

When handling pigs, there are various methods of handling which can be separated into 'positive' and 'negative' handling techniques, which, in turn, lead to positive and negative animal reactions. These terms are based on how the pigs interpret a handler's behaviour. Negative interactions include heavy tactile interaction, including slaps, kicks, fast movements or even the use of an electric Goad. Negative interactions can result in fear and stress in the animals, which have a variety of negative impacts. For instance, tactile interactions can cause basal cortisol contractions (hypertension). These interactions also lead to animals fearing people, to the point of avoiding human interaction, which can result in injuries to both stock and handlers alike. Fearful animals are also likely to be very stressed, which can result in immunosuppression, leading to an increased susceptibility to disease.

Pigs are very curious animals, and consequently when entering a pen it is good practice for a stockperson to enter with slow and deliberate movements. Such actions minimize the fear for the animal, which reduces stress. Allowing the pigs to approach and smell whilst patting or resting a hand on the pigs back are all examples of positive behaviour. Pigs also respond very positively to verbal interaction - allowing handlers to perform husbandry practices in a much safer and more efficient manner. Pigs are farmed in many countries, though the main consuming countries are in Asia, meaning there is a significant international and even intercontinental trade in live and slaughtered pigs. Despite having the world's largest herd, China is a net importer of pigs, and has been increasing its imports during its economic development. The largest exporters of pigs are the United States, European Union, and Canada. As an example, more than half of Canadian production (22.8 million pigs) in 2008 was exported, going to 143 countries.

Pigs are omnivores, which means they consume both plants and animals. In the wild, they are foraging animals, primarily eating leaves, grasses, roots, fruits and flowers, whereas in confinement, pigs are mostly fed corn and soybean – with a mixture of added vitamins and minerals. However, because pigs are omnivores, they make excellent pasture raised animals, traditionally called 'mortgage lifters' due to their ability to use the excess milk as well as whey from cheese and butter making, combined with pasture. Pigs are incredibly valuable creatures to human beings, utilised for almost all their meat, and can be a relatively easy animal to keep – if cared for correctly. Pig farming will undoubtedly continue for many years to come, and it is hoped that the current reader is inspired by this book to investigate further. Enjoy.

BERKSHIRE BOAR.

Mr. N. Benjafield's "Commander-in-Chief" (Herd Book No. 10,090). Champion at the Bath, and West of England and the Royal Counties Shows, 1904.

PREFACE TO THE THIRD EDITION.

I⊤ is a general custom to preface a book with some remarks of an explanatory or introductory nature. I have, however, written an Introduction at the commencement of the body of the work, and the contents appear to me to need very little explanation. I have endeavoured to write clear, simple, and practical descriptions of the best-known varieties of the domestic pig, and to explain how they should be bred, fed, and managed. What work I have accomplished is not the result merely of reading or of experience in breeding, but of a combination of both. I have expended considerable time in the study of those authors who have previously written—British, American, and French—and such of their views as I have found practical I have recognised in the production of this volume; while the opportunities which have been afforded me of acting as judge at the principal exhibitions in this country have provided facilities for the examination and comparison of the best specimens of British swine. However well or ill the work may have been done, I am satisfied that it will be found practical; and it was an additional inducement to me to undertake it that, in my own study of the subject, I found the greatest difficulty in obtaining information from the scanty pages of the few handbooks, and those non-practical, which were then in existence.

I owe many thanks to a large number of correspondents, both in the New and the Old World. They are too numerous to name here, but I hope that everyone from whom I have received a suggestion or employed a thought will find his name quoted in the work.

JAMES LONG.

CONTENTS.

CHAP.   PAGE

  I. INTRODUCTION 1

 II. PHYSIOLOGY OF THE PIG 14

III. BREEDING AND SELECTION 31

 IV. THE BOAR 55

  V. THE SOW AND HER OFFSPRING 62

 VI. CASTRATION AND SPAYING 86

VII. RINGING PIGS 92

VIII. THE COTTAGER'S PIG 97

 IX. SOME DEBATABLE POINTS 102

  X. THE WHITE BREEDS 109

 XI. THE BERKSHIRE BREEDS 135

XII. THE REMAINING BREEDS 155

XIII. PIGGERIES AND STIES 189

XIV. FOODS AND FEEDING 249

 XV. THE COMMERCE OF THE PIG 296

XVI. HERD BOOKS, RECORDS, AND SHOW CLASSIFICA-
     TION 332

XVII. DISEASES 340

     INDEX 377

LIST OF PLATES.

FACING PAGE

BERKSHIRE BOAR (Mr. N. Benjafield's)—*Frontispiece*

BLACK DORSET SOW 32

SMALL WHITE SOW (Hon. D. P. Bouverie's) 44

POLAND-CHINA BOAR 55

BLACK SUFFOLK SOW 68

TAMWORTH GILT (Mr. R. Ibbotson's) 102

LARGE WHITE SOW (Sir Gilbert Greenall's) 109

MIDDLE WHITE SOWS (Mr. Alfred Brown's) 124

BERKSHIRE SOW (Mr. Alfred Brown's) 135

LARGE BLACK BOAR (Mr. J. Robinson's) 155

BLACK ESSEX BOAR 160

LARGE WHITE SOWS (Mr. Alfred Brown's) 266

BERKSHIRE SOW (Hon. Claud B. Portman's) 284

MIDDLE WHITE SOW (Sir Gilbert Greenall's) 306

LARGE BLACK GILTS (Mr. J. Robinson's) 324

CHESTER WHITE BOAR 344

The Book of the Pig.

CHAPTER I.

INTRODUCTION.

THERE is no animal contributing directly to the food supply of the people which is at the same time so generally kept and so little understood as the pig. Nor is there one which, intimately associated as it is with their welfare, is so generally appreciated by the rural classes. In England, while the farmer breeds and feeds the pig for the twofold purpose of consuming inferior and waste corn and the production of manure, the peasant, where he is thrifty and able, keeps a sow for the profit and satisfaction that she returns him. In Ireland, the pig occupies a high position with farmer and peasant alike, and contributes more to its owner's contentment and prosperity than almost anything to which he devotes attention; indeed, the treatment of the pig by the Irish, as a race, contrasts very favourably with that to which it is subjected, not alone by the Oriental, but by many of our own agriculturists of high degree.

America has long been known as a swine-producing country of surpassing excellence, and its people claim to produce, as well as to consume, more pork than any other nation; indeed, pork forms a large portion of the flesh food of the great majority of the American people. In the year 1900 the exports of pork in one form or another from that vast country amounted in value to 112,159,000 dollars, or about £22,400,000 sterling, while the quantity exported was 15,700,000cwts.; indeed, excepting wheat and cotton, the export of swine products exceeds everything else. It is not easy to understand how immense this industry is, and how worthy the pig is of a separate place in rural literature. In America the pig is a pioneer; it appears contemporaneously with the farmer who breaks up the prairie, and, moreover, provides him with food and furnishes him with the means of obtaining money. Great Britain is America's largest customer, and in 1902, when the total imports of pigmeat reached 7,432,626cwts., we received from the United States 4,954,471cwts.

In France, if the pig is not such a favourite as with us, it is yet largely bred; but although it is an increasingly important item in more northerly dairying districts, especially in Denmark, where it is chiefly kept for the consumption of buttermilk, skim milk, and whey, it is not so extensively bred in the South, no doubt because it is less in keeping with the character of the agricultural industry.

In Germany generally the same remarks apply; but in the North of the Empire, especially in Schleswig-Holstein, pig-breeding is an enormous industry, and there is a growing export trade which is likely to excel that of every other European country, Denmark excepted. Of Germany, however, it may be said that, unlike France, she displays enterprise and effort in this matter, and her large and continued introduction of English blood has been so ably utilised that in truth she is already a formidable rival. If we do not make fresh efforts to preserve the superiority of our own breeds, we shall find, ere long, that Germany, with her energy and perseverance in agricultural affairs, will excel us upon our own particular ground. In Denmark the pig has been immensely improved, and when we remember in how few years her cattle and dairy products came to the front and surpassed those of France and America, both in quality and in value, and how ably her agricultural professors direct all movements of this kind, we shall see that success was a foregone conclusion. Thus it appears that in Europe and America the pig is a conspicuous occupant of the farmyard, as it is also a favourite food of all Christian peoples, although rejected by Jew, Mohammedan, and Brahmin.

"The pig," says Old Markham, "is the husbandman's best scavenger and the housewife's most wholesome sink. By nature greedy, given much to root up grounds and tear down fences, he is very lecherous, and in the act tedious and brutish; he is subject to much anger, and the fights of the boars are exceedingly mortal. They can by no means endure storms, or winds, or foul weather. They are excellent observers of their own homes, and exceeding great lovers one of another, so that they will dye upon any beast that offendeth their fellows. No county in England breedeth naturally better swine one than another, but if any have the preference, then I must prefer Lestershire and some parts of Northamptonshire, and clay counties bordering on Lester, and the reason I take to be their great multiplicity of graine, especially beans and pulse, for the malt countries, tho' they are good breeders, are not large feeders, whence it comes that your wild swine is ever your least swine, but your sweetest bacon."

This history of the pig has never been written, and, perhaps, never will be. This is really a practical age, and those who are interested in the animal desire to know rather how to improve and produce it with profit than to trace its connection with the wild boar, or to learn how that gaunt, grizzly, long-eared and long-snouted beast, so long the object of sport in these islands, and still hunted in many of the forests of Europe, was transformed into the fleshy, fine-boned, symmetrical, and delicate animal of to-day, admired at all the agricultural exhibitions of Europe, and forming an element of many of the most piquant dishes of our tables. We are told by Columella, among other ancient writers, that the swine of his day were large and fatted to excess; and Varro, the Roman historian, who died a few years before the Christian era, speaks of a hog whose "flesh was a foot and three fingers thick." He also relates that a mouse made a nest in the fat of a sow, and produced young ones therein; while another old writer speaks of a pig whose fat was fifteen inches in thickness. Pliny, again, remarks that "the ox, the cow, the hart, and such beasts ask a long time, yet a swine,

which eateth of all sorts of meat, doth very quickly, even
in a month or two, prove worthy of the knife."

There is no doubt that the wild boar of the Middle Ages
is identical with the domestic pig of to-day, and that, whether
or not the swine of the Early English were ever cultivated to
such perfection as those of the Romans, our pigs had little title
to any special qualifications until the introduction of foreign
blood above a century ago. The common pig appears to
flourish in almost every climate, but it has been shown that
in its wild state it is more stupid than, and has not so
complete a development of the senses (smell alone excepted)
as, the cultivated animal. The sense of smell is at all times
acute, and there are records of the use of the pig in the
capacity of a hunter for truffles, which grow below the surface
of the soil. Indeed, this sense is developed in a remarkable
degree, and is only equalled by the ability of the pig to
root up the earth with his powerful snout: this has only
been shortened by the art of the breeder, and by close
confinement to the sty, which does not favour its development.
The sense of taste, which is believed to be imperfect, is at
least remarkable, for while the pig will consume the most
disgusting garbage, it is more particular in its choice of
herbage than almost any of the domestic animals, although
it is omnivorous, thriving upon almost anything.

The pig is called an ugly animal as well as a stupid one;
but however gaunt and inelegant mongrel specimens of the
breed may be, it is impossible to describe the modern races as
anything but symmetrical in form, whether they deserve a
better designation or not. As to its stupidity, a tolerably
extensive experience teaches us that there is no more astute,
we may as well say cunning, animal among those which
afford food for the human race. That he is a glutton will
be admitted by everyone, and that his whole life is divided
between feeding and sleeping is equally true; indeed, the
modern hog, much to the advantage of his owner, will eat—
if he is provided with food—until he is unable to rise upon
his legs.

The pig usually grows until the age of five years, his
natural life reaching from fifteen to thirty years.

There is no food-producing animal which is of more benefit

to mankind; the flesh is most substantial and delicious, and is a favourite dish, both in the fresh and in the salted and dried state—for it can be preserved better than almost any other meat. The offal, except a very small portion, is all consumed as food, even to the blood; indeed, the edible portion of the pig affords a greater variety of dishes than that of any other known beast. The hair is used for brush making, the fat occupies a place of immense importance in commerce, and the skin is largely in demand for the purpose of making leather.

In comparing the habits and appearance of the wild boar with our domestic pig, we may first of all notice that, whereas the snout of the wild animal is extremely long, his tusks are both long and sharp, and his body is remarkably long, narrow, and flat-sided, the domesticated varieties are quite the reverse, being short and broad in the head and snout, much shorter in the leg, considerably finer in the skin, broad across the chine and the loin, and, as it were, composed of a series of beautiful curves. The wild boar during his early life generally follows the sow and her litter, possibly for the purpose of mutual protection; but when he has attained a more mature age, and is in his full strength, he is usually found alone, and does not appear to fear either man or beast, boldly facing his pursuers when attacked. On the contrary, the domestic boar, although herding with his fellows during the earliest period of his life, afterwards roams about alone as he is permitted, but displays no particular affection either for the sow or for her young.

There are various opinions as to when the first importations of an improved variety of swine reached England, and whether they first came from China or Naples; but there are numerous records of hogs of very great weight having been killed during the latter half of the eighteenth century. In the year 1770, for instance, one animal reached 1100lb. in weight, while it measured 9ft. 8in. from nose to tip of tail, and was 4ft. 5½in. in height. Another pig, which was bred and fattened in Surrey, reached a weight of 832lb.; whilst in America, considerably later, a hog was exhibited which reached the marvellous weight of 1325lb. Lastly, we may quote Gilbert White, who describes a sow of enormous size belonging to one of his friends, which was fatted and killed after reaching the age of twenty years; during that time she had had no less than 300 young, having at one time produced a litter of twenty, a feat which it would be difficult to equal even in the present day. Still one instance has come under our own notice in which a mongrel-bred sow of large size produced twelve consecutive litters that averaged eighteen and a half at a birth.

By the art of breeding the varieties of the domestic pig have become decidedly numerous, each race being bred to a particular type as well as colour; but although in the production of these varieties the early breeders resorted to selected animals from the various districts of England—in fact, using good specimens for stock wherever they could find them —yet there is no doubt that we owe the greater portion, if not the whole, of our success to the introduction of Chinese and Neapolitan breeds. The Chinese, in particular, is remarkable for its fertility and for the rapidity with which it lays on flesh without increasing materially in offal or bone. It is a short-headed pig, with small, erect ears, short legs, a high and broad chine, and a wide jowl, and there is no animal that comes to maturity so rapidly. As a general rule, it carries a very small quantity of hair, and the belly hangs close to the ground. The skin is generally dark, while the flesh is especially delicate and white. The Neapolitan breed is entirely black, almost without hair, tolerably short in the face, and with small, erect ears, carried a little more forward than those of the Chinese. It is also short in the legs, long in the body, tolerably wide, and a remarkably easy animal to fatten; but it is not so prolific as the Chinese, and its constitution is much more delicate. It has, however, been of great value in the improvement of some of our English varieties, and it is generally believed that the blacks of the East of England owe much of their quality to the infusion of its blood.

The principal breeds of to-day may be named in a few words. These are the Yorkshire, which is divided into three sub-varieties, sometimes called the Large breed, the Middle breed, and the Small White breed; the black and white pointed Berkshire; the Suffolk, often described as the Small Black breed; the Tamworth, an animal that is sandy-red in colour, and is much appreciated in the Midland counties; the Large Black; the Essex, and the Dorset. The last two varieties, although still recognised by farmers and breeders in their particular districts, are not recognised by experts and the agricultural societies as special breeds; indeed, they are not bred to a fixed type in the same regular and skilful manner as the before-mentioned varieties. There are also the Lincolnshire, the Norfolk, and a Westmoreland breed; but these, again, may be regarded as mere local sub-varieties, for after all it is but the custom of the farmers that regulates. the size, the colour, and the style of animal produced for their particular markets. In Scotland and the North of. England wholly white pigs are generally bred, whereas in the South the black pig will be found in almost every county; and in the Midland counties not only do we find the red Tamworth, but parti-coloured pigs of almost every type, size, and quality. While the Americans would seem to prefer the, Berkshire and the Poland China pig, there is no doubt that. the Germans have set their minds upon the cultivation of the large York breed throughout their country; but the French, up to the present moment, undecided upon the cultivation of one particular foreign variety, have continually introduced specimens of our Large and Middle breeds into their herds, for the purpose of crossing either with the mongrel of their district or with the well-known champion of French races, the Craonnaise, an animal, however, that is decidedly lacking in quality.

We have continually lamented that, notwithstanding the fact of our having by far the best known races of swine, yet except by the few—the very few—intelligent breeders who are scattered throughout the country the pig is not properly cultivated among us. And if we except particular counties, such as Dorset, Suffolk, Wiltshire, Yorkshire, Lancashire, and Cheshire, with one or two others, it is somewhat difficult to find a parish where there is a consistent system of breeding, or where the quality of the animals bred is regular and above the average. This fact is greatly to be lamented, inasmuch as the farmers themselves annually sacrifice a large amount of money by their neglect, while the farm-labourer and the poorer class of people generally, who are unable to introduce fresh blood into their sties, except through the medium or by the assistance of their richer neighbours, continue to this day to breed from animals that are little removed, in appearance and quality, from the wild boar of the early ages. If it were considered how much might be done by the introduction of pedigree swine, even into the herd of one intelligent and liberal-minded man, we feel sure that there are numbers who would, without any further delay, make an effort to improve their own herds, and at the same time to render an important service to all those who may be dependent upon or reside near them. We might quote cases that have come under our notice in which the modern-bred improved pig was until quite recently practically unknown in several parishes—where the animals bred, and very largely bred, were in every instance gaunt and bony, coarse in flesh, long-eared, and long-snouted, and although apparently somewhat precocious in their early growth, still most difficult to fatten with profit. One prominent inhabitant, however, by making judicious purchases and demonstrating the effect of crossing, changed the type and revolutionised the industry, and all in the space of a span of years. In each of the cases to which we refer, and where the introduction of fresh blood occurred only a few years ago, we now find the entire district covered with pigs of an absolutely different type. When it is recognised how prolific the pig is as a breeder, and how rapidly young animals become fit for reproduction, this will not cause surprise, but it shows what can be done by one individual who invites his neighbours to see and to criticise his stock, and permits them to use the boar for stud purposes at a very moderate fee, or, perhaps, as in the case of the labouring classes, without any charge at all. While discussing this subject we would express a hope that those of our readers who happen to reside in a district where the improvement of the pig has not yet been attempted will, without any further delay, introduce a boar of one of our most economical breeds and make some effort to induce those who are engaged in the pig industry, more especially the poor, to use it for stock. Such a course would not only prove immensely serviceable to the inhabitants at large, but would also result in direct and increased profit to the owner himself.

The Influence on Pig-Keeping of Modern Changes.

The improvement which has taken place in the pig, and in the method of production, has during recent years considerably modified the practice of the past so far as relates to its breeding and management, its age, and the condition in which it is sold. In earlier times the pig was much more difficult to fatten; more food was consumed in the production of a pound of pork; and the profit realised was smaller. The farmer, the agricultural labourer, and other country people kept pigs in order to produce pork and bacon for their own tables; but while the farmer was able to obtain what he required at the least cost, his system of management was not only careless, but so badly defined that it became no system at all.

Great efforts have been made to meet the demand for size in the pig. Since the introduction of the Large White, size has been . recognised as an economical factor, and breeders from abroad have sought it in England as much as we have sought it ourselves. Simultaneously with this demand the small breeds have lost their admirers, and one by one the Dorsets, Suffolks, and Small Whites have been abandoned, alike by committees of agricultural shows and by breeders themselves. The small breed is too slow in its growth, and even when it has reached maturity its meat is desirable neither for bacon nor for pork on account of its fatness and cost. The large varieties, on the contrary, are ready for killing at ages which are sometimes astonishing, providing joints of a size to meet the consumers' demand, with less waste of fat, and at a cost of production with which the small breeds could not compete. In a word, the small breeds are unsuitable for economical pig-breeding and feeding, not only because of their smallness and their costing much more to produce, but because their meat is too fat.

To some extent the middle-bred varieties, of which the Middle White and the Berkshire are the only examples, partake of the faults of the small breeds. If they are larger the meat they produce is too fat for the popular taste, and frequently wasteful, although some exception may be made in the case of the Berkshire when it is killed early and not fed with too great liberality Pigs which possess short heads, small ears, and short necks, with collars which carry the largest proportion of fat, as in the Middle

White, are useful for crossing on the long-headed and long-legged
unimproved pig of the country, but these are not what the butcher
requires. For these reasons the large breeds, which have been
immensely improved from the standard of the British pig-keeper,
are more generally used.

Although we have thus dealt with the position of the pig and
the line which the breeder should take in the industry of which it
is the centre, it is advisable to say a few words in defence of the
old system and in opposition to the new, rather, however, in the
interest of the consumer than of the producer. I suppose the
demand for leaner, younger, and smaller meat was created by the
public. If the butcher finds that complaints of over-fatness, the
toughness of the lean, and the large size of the joints are of frequent
occurrence, he tells the breeder the facts, and requires him to
supply what is wanted, if they are to continue their business
relations. The result is that the breeder is forced to make the
required change in his stock and to supply pigs which meet the
butcher's demands. The type of his breeding stock is consequently
changed : young pigs are bred from large dams, which are not of
an excessively fat variety ; they are killed earlier, because they
reach the required weight in less time ; they produce more tender
and less fat meat in consequence, and the question is solved.

When we come to the consumer whose complaints have
been the cause of these changes, we are bound to say that we
believe him to be wrong. The reason is this: young, tender
lean meat is of much less nutritive value than young and tender
fat meat. Of still less value is young lean meat than mature fat
meat. It is possible for a fat joint to provide twice the amount of
nutritive matter that exists in a lean joint of the same weight.
Lean meat contains much more water than fat meat. Thus in
the process of feeding, as the fat increases per cent., the water
decreases, hence the lean of a fat joint, while providing almost all
the nutriment present in the lean of a lean joint, although its
weight may be smaller, is supplemented by a large quantity of
fat, and we must bear in mind that there is no water in fat.

There is another point which is borne out in the practice of the
fat-bacon-eating agricultural labourer. The lean or muscular
tissue of meat is essential in building the muscular tissue of man,
but the adult requires so small a quantity in comparison with the

large quantity of bread and fat which his system demands that it is both wasteful and deleterious to the system to eat much lean, whereas the fat, our chief fuel food, is of much greater value for heating the body and providing the energy which man dissipates in the course of his work. Fat pork or bacon, then, provides the buyer with considerably more nutritive food than lean pork or bacon, and that of a much more valuable character.

In the production of leaner pork the joints are smaller than they used to be where the meat is intended for the trader in large towns and cities. There are still many consumers in the country who prefer large hams and sides for curing, but these do not seriously influence the trade. Consumers of pork are more numerous among the working-classes, and for this among other reasons they prefer joints of a size within their means of purchase. Cold meat is not so agreeable and appetising as hot meat, and the more frequently hot meat is placed upon the table the more people like it.

Again, the best bacon trade of the country is in " sizeable " sides, or flitches, weighing about 56lb. each. These are impossible where the carcase of a pig is large. If, for instance, a pig of a small breed is fed for bacon it must be fed longer ; and this not only increases the toughness of the meat, but it adds considerably to its fatness and the cost of its production. A pig killed very young will produce 1lb. of its live weight upon 4lb. of good meal, or its equivalent in a mixed ration, but as it grows older the quantity of food consumed increases to such an extent that finally the meat produced is worth less than the meal upon which it was fed. One of the most important points, however, in young meat is that it is less fat than mature meat, so that when a pig is killed at from twelve to twenty weeks old the fat will measure from 1in. to 1½in. in thickness on the back, instead of 2in. to 3in., which was usually the case on a mature pig, and still is where the old practice prevails.

The pigs fed for the market in this country may be divided into three types : (1) Those intended for sale as fresh pork ; (2) those intended for curing ; and (3) those fed for the manufacture of sausages, although this industry is comparatively restricted to some parts of the midland counties. Porker pigs should be small, not exceeding 100lb. each in the carcase for the best class of trade, while they sometimes weigh no more than 65lb. It is not always easy to estimate the weight of a carcase as it hangs in the shop of the butcher, but the weight of the best class of carcase has been practically determined for the past three years by the exhibits in the carcase class at the Smithfield Cattle Show in December. Here pigs of various ages are classified, exhibited alive, and subsequently slaughtered, and then exhibited dead, their carcases being so cut that they can be carefully examined. We have made a special point of examining the whole section of the show, and of noting the weights of the winning exhibits and the prices many have realised at the auction at which they are sold. The weights of the prize carcases in the small porker class at the exhibition of 1915 were: 66lb. winner of the champion prize; 71lb., second prize; and 63lb., third prize. The champion carcase, which was only sixteen weeks old, was well fleshed, the substantially thick lean meat being covered with from $\frac{3}{4}$in. to 1in. of fat on the back. Pigs of this type provide small roasting joints; the skin, which most people enjoy when it is scored and crisp, is not cut off as it is in really fat pigs to make a joint fit for roasting, and the result is a delectable dish.

Bacon pigs are of necessity larger, and slightly fatter, than porkers; thus the first prize carcase of a pig weighing between 160lb. and 240lb. scaled 219lb. alive and 178lb. in carcase, showing a proportion of carcase to live weight of 80 per cent. In this pig the fat on the back varied from $1\frac{1}{2}$in. to 2in. in thickness, while the lean meat was thick and the carcase long. It would, however, make sides of bacon which would be too large for the best class of trade, for if we allow 24lb. to account for the loss of weight in curing—and this would be a liberal figure—the sides would scale 77lb. each. In practice a bacon pig to make suitable sides should scale about 170lb. alive. It is needless to say that the streaky parts of the meat intended for rashers should be as perfect as possible, never too fat on the one hand and never "filled" up with lean on the other, while the loin should consist of a rich fillet covered with about 1in. to $1\frac{1}{2}$in. of fat.

Pigs fed for sausage making are usually older, larger, and fatter than bacon pigs, and, as a consequence, they realise smaller prices per stone. Further, it is a curious fact that feeders who would do much better with baconers adhere to a practice which returns them less money and gives them more trouble and risk.

There is little doubt about the fact that those who need cheap pork and bacon the most, and who keep pigs, are agricultural labourers and other small rural breeders. They are accustomed to feed one pig for winter consumption, and the result is that in their own way they make the most of it, and fatten it until it reaches a great weight. They apparently possess no knowledge of the fact that in the later stage of feeding money is lost; nor can it be claimed in defence of this system that where these people grow their own potatoes no loss is sustained. Let us suppose that a pig is usually fed until it weighs 600lb., and is then killed: the hams are sold, the interior organs are consumed, and the remainder is in part eaten fresh and in part pickled and cured. This pig has cost a large sum to feed, and estimated at wholesale market price leaves no profit behind. If, however, three pigs are fed in succession, the cost per pound of dead meat will be less, it will be of superior quality, there will be six hams for sale at substantial prices instead of two, and these of smaller and more saleable size; the pluck and other organs will be conveniently smaller for consumption at home; the provision of meat, and especially of fresh and pickled pork, will be spread over a longer space of time, while more fresh meat can be sold, if necessary.

CHAPTER II.

PHYSIOLOGY OF THE PIG.

THE domestic pig and the wild boar were formerly regarded scientifically as distinct species under the respective names *Sus scrofa* and *S. aper*, while Cuvier even described a third as *S. larvatus;* but the latest classification refers both the domestic varieties and the wild kind to the one species, *S. scrofa.* There is no doubt that we find in the pig of to-day, except for modifications that have been made by modern breeders, the characteristics by which it was known to the ancients. Among English-speaking people, the male pig is known as a boar or a brawn, and the female, after she has had her second litter of pigs, as a sow. Until that period she is respectively called a yelt, hilt, or gilt—terms peculiar to certain districts of this country. There is no special name for small pigs, but the term hog is applied in general to all those of both sexes that have been castrated or spayed for the purpose of rapid fattening.

In France a perfect vocabulary of names is applied to the pig. The general term *verrat* is given to a boar, and that of *truie* to a sow. Small pigs are called *gorets* or *porcelets*, the castrated male a *cochon*, and the spayed female a *coche*. There are also numerous local terms adopted. In Germany a boar is called *eber*, a sow *sau*, and a hog *schwein*. The word *pork*, like the term *porcine*, is undoubtedly derived from *porcus*, the name given to the hog by the Romans.

The pig belongs to the highest class of vertebrate animals, the *Mammalia*, and to the order of *Pachydermata;* the latter formerly included all thick-skinned animals, but this classification has given place to a more convenient arrangement, and the old order *Pachydermata* is obsolete, its representatives being included in the *Ungulata*, in the *Artiodactyla* section of which swine are arranged.

The body of the pig differs in colour according to the race to which it belongs—some being black, others white, red, black-and-white, or red-and-white. Some writers, especially French and American, consider that dark-skinned pigs are more pronounced in numbers in hot countries, and white or light-

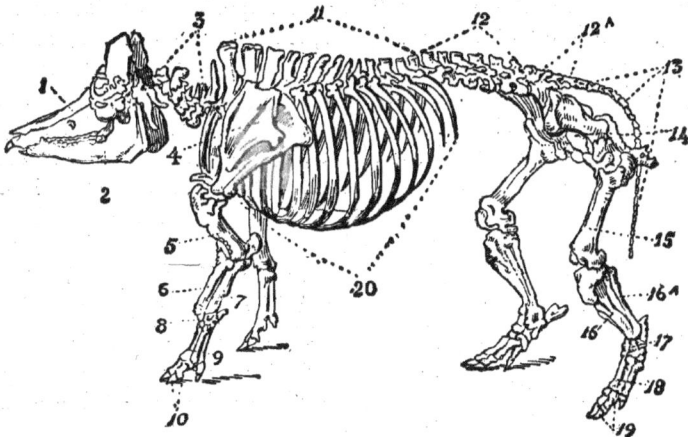

Fig. I.—Skeleton of the Pig.

skinned in cold climates, the former being better able to withstand great heat than the latter. The skin is provided with hair termed bristles, which vary considerably in quantity and texture. Strong hair generally denotes vigour, hence breeders usually endeavour to obtain it; but in-breeding has largely destroyed this point, and high-bred animals are as frequently seen with a small as with a large quantity of hair. In most pure races, the body is of considerable length, with great breadth of loin and chine. The flesh, unlike that of the majority of domestic animals, is not composed of fat nicely mixed or marbled with the lean, but of layers of both. The fat especially can be produced to a considerable extent, and is much softer than that of the sheep or the bullock.

According to Cuvier, the spine is composed of fifty-three vertebræ, fourteen of which are dorsal, five lumbar or loin, seven cervical, twenty-three caudal, or bones of the tail, and four or five bones of the sacrum. The wild boar of the woods, according to Heuzé, has only fifty vertebræ, there being less of the caudal. The tail is usually small and considered to denote breed, but this is principally because high breed means fineness of bone, which is especially noticeable in the tail.

In the skeleton of the pig shown in Fig. 1 the principal bones are numbered as follow: 1, superior maxilla, or upper jaw, composed of several bones; 2, inferior maxilla, or lower jaw (composed of a single bone); 3, cervical vertebræ (the atlas is the first bone, and forms the joint with the head); 4, scapula, or shoulder-blade; 5, humerus; 6 radius, and 7 ulna (these two bones, with the humerus, form the elbow-joint); 8, carpus, or knee-joint (composed of several small bones); 9, metacarpal bones (one large and two small); 10, phalangeal bones, or bones of the foot; 11, dorsal vertebræ; 12, lumbar vertebræ; 12A, sacrum; 13 coccygeal or tail bones; 14, cavity called the pelvis (formed by three bones, the ilium, ischium, and pubis); 15, femur; 16, tibia, and 16A, fibula; 17, tarsus, or hock (formed of several small bones); 18, metatarsal bones (one large and two small); 19, phalangeal, or bones of the foot; 20, ribs.

The head of the pig is pyramidal and coarse, but greatly differs in length, that of the improved kinds being generally short, whereas the common breeds approach their prototype and are often of great length, although in all cases the forehead and brain cavity are very slightly developed. The ears, again, vary as much as the length of face, and their size is commensurate with it, high-bred animals having small ears, generally erect and fine in texture, whereas common pigs have very coarse and usually flopping ears; in other words, the size of the ear is just in proportion to the size of its bone. The eyes are particularly small and the pupils round, and in many of the exhibition pigs of the present day they are almost buried in the creases of the face, formed conjointly by its shape and excessive fat.

The snout, which is extremely powerful, becomes narrower as the point is approached; in the white breed especially it is turned up, thus giving the face that "dished" or pug-like appearance so much sought by pig-breeders. The point of the snout is quite round and resembles a plain flat disc, cartilaginous and flexible, in which there are two small round holes forming the nostrils. The scent of the pig is marvellously developed, and this qualification, combined with the muscular power of the snout, gives great trouble to the farmer. The quickness of the sense of smell enables the pig to scent roots and almost any edible substance beneath the surface of the soil, when it immediately commences to dig: and for this reason it is usually rung. The mouth, again, is large and powerful. The lower jaw is shorter than the upper, and the tongue small and furrowed.

The Teeth.

There is nothing, perhaps, so important in the physiology of the pig, from a practical point of view, as the teeth. The principal reason is that the teeth indicate the age of the animal better than anything else can possibly do. Perhaps, however, this statement should be qualified, for, notwithstanding the acceptance of the test by both scientific men and the pig-breeding community itself, it is not invariably correct, inasmuch as what appear to be natural conditions vary according to race, quality, and system of feeding. In the show yard, however, the dentition test has been adopted for some time, and, valuable as it is, several cases have proved that it is not always accurate. These were doubtless abnormal, for, after all, we must consider that, under ordinary conditions, whether in the horse, the cow, the sheep, or the pig, the teeth will, to all intents and purposes, be found to corroborate the rules which science has laid down.

A knowledge of the age of a pig is absolutely necessary for exhibition purposes; therefore, the breeder and exhibitor should invariably make a careful record of every birth, and, where pigs are forced, additional evidence should always be forthcoming. Under what may be termed artificial conditions as to breeding and feeding, the teeth vary as much in young

C

pigs as in children. In older animals there is little variation,
and in almost every case it is safe to accept the teeth as a
guide to age until that period in their existence when the
mouth is full, and when the scientist leaves it to the breeder
to decide. At this time, judgment and experience alone can
determine what the age of a pig may be, for the test is
comparative; and as the colour and general appearance of
the teeth are the only guides, it will be apparent that the
breeder should understand the variety of the animal, the
manner in which it has been bred, and how it has been fed,
at least since it shed its milk teeth.

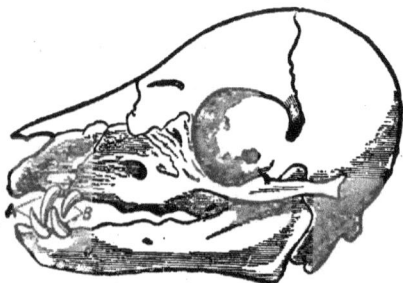

Fig. 2.—Skull of a Newly-born Pig.

A, Corner Incisors ; B, Tusks.

The engravings (Figs. 2 to 10), which are intended to
assist the reader in thoroughly mastering this important
branch of the physiology of the pig, are reproduced from
drawings by the celebrated German scientist, Furstenberg,
and appeared in the *Zeitschrift für Deutsche Landwirthe.*
They are based upon the investigations of Nathusius, Professor
Simonds, and Furstenberg himself, and are the best exem-
plifications of dentition that have yet been prepared.
Unfortunately, perhaps, they represent the teeth and bone
alone, consequently it will be necessary to make allowance for
the absence of the tissue and gums. Dr. Paaren, of Chicago,
who translated Furstenberg's work from the German, remarks
that, in consequence of this peculiarity in the drawings,
it will be requisite, when comparing real teeth with those

represented, to make allowance to the extent of about one-third of the length of the fully-developed teeth, measured from their base to the surface of the jaw-bone. This, however, would not cover the tusks and the undeveloped teeth, more particularly as some of those delineated would not be visible in a living animal at the particular age represented. For example, the intermediate incisors (Fig. 3, c) in the jaw of a pig four weeks old appear to be considerably developed, and, making allowance for the absence of the gums, they are quite correct; but they are never quite visible in the living animal until the third month after birth, while

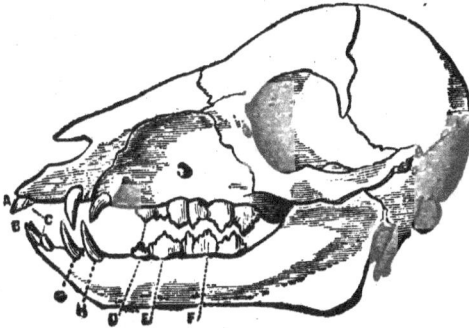

Fig. 3.—Skull of a Young Pig.

A, B, Nippers; C, Intermediary Incisors; D, Third Premolar; E, Second Premolar; F, First Premolar; G, Tusks; H, Wolf Teeth.

the third temporary molars, which, in Fig. 3, D, measure one-twelfth of an inch in length, are only visible after the fifth week from birth.

Furstenberg, whose authority upon this matter is not exceeded by any known writer, says that, according to usage, the teeth are classified into two groups—the incisors and the molars. The full-grown hog has twelve incisors or front teeth—six in the upper and six in the lower jaw—and two canines in each jaw. The incisors in each jaw are divided into two halves, three on each side of the median line, of which the foremost (Fig. 10, E, E) are called the nippers, the next outside of these (Fig. 10, F, F) the intermediary incisors, and the remainder outside of these again (Fig. 10, G, G) the corner incisors.

There are seven molars in each side of the upper and lower jaw, making twenty-eight; and, to facilitate description, each row is divided into three sections. Each of the three hindmost molars in the four rows appears at different periods of a later age, and are permanent teeth (not preceded by milk teeth). The three next in front of these appear soon after birth, one after another. They are called milk teeth (*premolares*), and are in the course of time shed, one after another, in the order in which they appeared, to give place for three permanent molars. These six molars are counted from the hindmost one forward (Nos. 1 to 6 on Fig. 10, which represents a portion of the lower jaw of a full-grown pig).

The seventh molar, or the fourth premolar (Fig. 10, c 4), appears later, in the space between the third premolar and the tusk. This small, apparently supernumerary tooth is sometimes called the wolf tooth, and was formerly considered as an independent one. However, at present it is classed with the molars, to which it undoubtedly belongs. It is a permanent tooth, and sometimes has a very small and crippled appearance, which is accounted for by the near proximity of the large and strong tusk. There are four tusks (*canini*), one on each side of the upper and the lower jaw (Fig. 10, D). Temporary tusks are present at the time of birth.

The full-grown hog has thus forty-four permanent teeth, of which twenty-eight are preceded by milk teeth and sixteen appear without previous shedding and as permanent teeth. The pig is born with eight teeth, which are about one-fourth to three-eighths of an inch in length above the gums. Of these, the two foremost ones in each jaw (Fig. 2, A) are corner incisors, and the other four are tusks (Fig. 2, B). They all have the appearance of tusks, as seen in Fig. 2, which represents the skull of a newly-born pig. On account of their outward direction, these teeth do not hurt the teats of the sow, as is sometimes supposed.

In the course of eight to fourteen days after birth, there appear through the mucous membrane of the gums the second and first of the premolars on each side of the upper and the lower jaws (Fig. 3, E, F). These from the time of birth have been concealed immediately under the gums. Four weeks after birth, the nippers (Fig. 3, A, B) cut their way through the gums in the upper and lower jaw, so that the young animal at this age has eight incisors, four tusks, and eight molars.

Soon after the appearance of the nippers, the third temporary molars break through the gums in both jaws; and at the age of from six to eight weeks these, as well as the nippers, will have so far developed that the young animal is able to subsist independently of the mother. At three

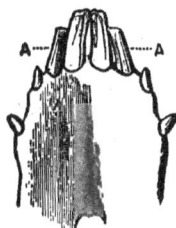

Fig. 4.—Part of the Jaw of a Three-months Pig.

A, A, Intermediary Incisors.

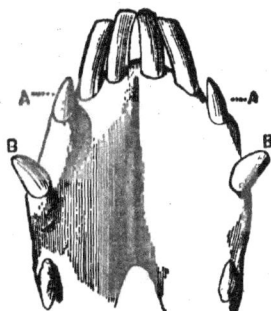

Fig. 5.—Part of the Jaw of a Six-months Pig.

A, A, Corner Incisors ; B, B, Tusks.

months the two intermediary incisors in each jaw (Fig. 4, A, A) appear, and with these all the milk teeth are present.

With advancing age, the size of the teeth increases, so that in a pig of six months the maximum proportions of the existing teeth are attained. As soon as the four middle front teeth of the lower jaw have reached their full length they will present an evenly rounded front (Fig. 6, B); but soon afterwards the edges of their crown begin to wear off, first on the nippers and afterwards on the intermediary incisors. At the sixth month of the pig's age, the so-called wolf teeth (Fig. 6, A) break through the gum. In the lower jaw these grow close behind the tusks, but in the upper they are nearest to the third premolars. At this age appears also in each jaw the first of the permanent molars (Fig. 6, c, and Fig. 10, A 3).

The milk teeth are shed in the order in which they have appeared. The shedding of the corner incisors and the tusks takes place shortly before or during the ninth month, and at the same time appears the second permanent molar (Fig. 8, B 2, and Fig. 10, A 2). The nippers (Fig. 7, A) are shed with the beginning of the twelfth month, and with the end of the first year the three premolars (Fig. 10, B) are shed, exactly in the order in which they first appeared. The cutting

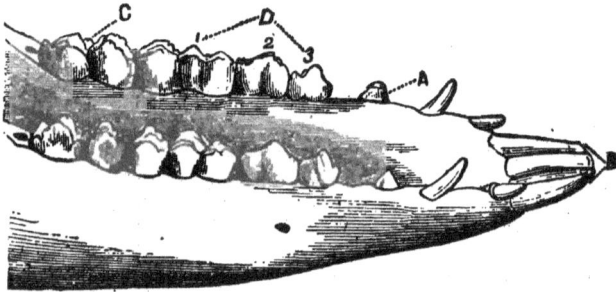

Fig. 6.—Part of the Jaw of a Six-months Pig.

A Wolf Teeth; B, Incisors; C, Third Permanent Molar; D, Premolars.

surface, or crown, of the permanent teeth filling their places (Fig. 8, A) will, at the age of fifteen months, be on a level with that of the permanent molars (Fig. 10, B).

Dr. Paaren remarks that by comparing the temporary intermediary incisors in Fig. 7 (B, B) with the two permanent nippers between them, the difference in shape of a temporary and a permanent incisor will be apparent. In Fig. 10, where the permanent intermediary incisors have succeeded the temporary ones, it will be seen that they are now of the same peculiar form as the permanent nippers between them. The intermediary incisors (Fig. 10, F, F) will, at the end of the eighteenth month, have been succeeded by permanent substitutes. Simultaneously with this change appears the last of the permanent molars (Fig. 10, A 1). This tooth is composed of three principal parts, which again appear to be made up of smaller parts, and its grinding surface is extremely rugged. After a while, when this surface meets that of its mate in the upper jaw, the ruggedness of both wears away, and they become as smooth and level as the other molars. The second and third of the permanent molars (Fig. 10, A 2, 3) are the most worn, because of their constant use

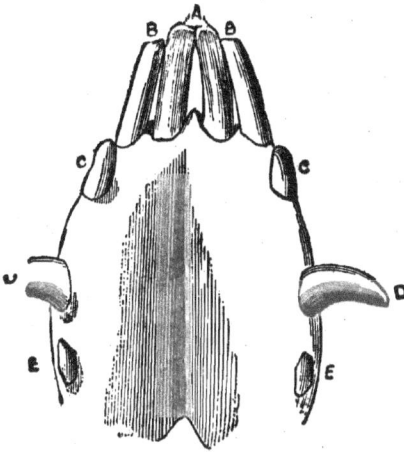

Fig. 7.—Part of the Jaw of a Six-months Pig.

A, Permanent Nippers ;   B, B, Intermediary Incisors ;   C, C, Corner Incisors ;
D, D, Tusks ;   E, E, Wolf Teeth.

between the sixth and ninth months, whereas the premolars only appear after the twelfth month, and the last permanent molars only at the end of the eighteenth month.

After the age of twenty-one months, when the incisors are fully developed, they begin to show the effect of wear, more especially those in the lower jaw, which gradually become shorter, although the extent of wear depends somewhat on the manner in which the animals are kept. Thus, if compelled to hunt on the land for a great part of their existence,

the teeth will wear more quickly than when the animals
are provided with liberal food and kept within limits; consequently, the state of the incisors after the pig is fully
grown is no sure criterion of age. The cutting surfaces or
crowns of the fully-developed incisors not only vary considerably in size and shape, but the incisors of the upper
jaw differ in form from those of the lower, on account of
their different position in the jaw. While the incisors of the
upper jaw take a vertical direction, those of the lower are
slanting, with a tendency to the horizontal.

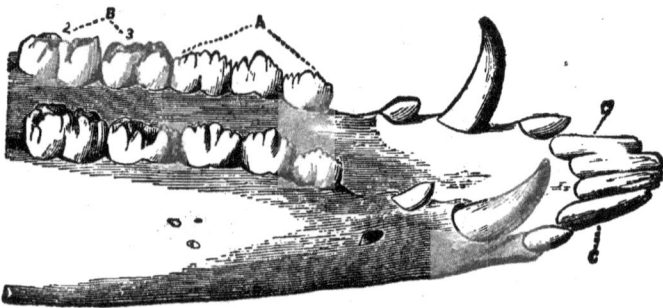

Fig. 8.—Part of the Lower Jaw of a Pig, after Nine Months.

A, Premolars; B, Permanent Molars; C, C, Intermediary Incisors.

It has already been mentioned that the permanent incisors
and molars have, at the end of a year and three-quarters,
fully developed. This is, however, not the case with the
permanent tusks of the boar, which continue to increase in
size up to the age of two and a half to three years. The
tusks of the lower jaw are longer than those of the upper,
and are turned backwards.

Fig. 9 represents the left side of the lower jaw of a ten-
months-old pig. A portion of the bony wall is removed, in
order to show the relative position of the temporary molars
and the dental papillæ, from which the permanent teeth
develop, also the extent of space occupied by the roots of the
other teeth. The dental papillæ, or dental pulps (4, 5, 6),
are at this age of the animal already considerably developed.

The cause of the diminutive size of the wolf tooth (A) will be understood from the position and size of the root of the adjoining tusk (B). The comparatively small size of the corner incisor (c) is likewise due to the limited space allotted to it by being wedged in between the large roots of the tusks and the intermediary incisor (D), which latter, together with the nipper (E), occupies a very slanting position in the jaw. The third permanent molar (3) is fully developed, while the second (2) is just appearing, and the position of the papilla (1) of the last is clearly shown in a far advanced state of development, as are also the papillæ of the permanent nippers and the intermediary incisors, but yet deeply enveloped in the jaw.

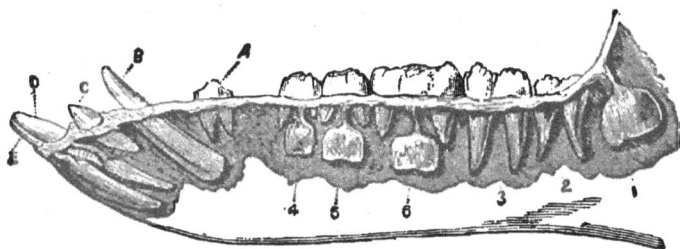

Fig. 9.—Left Side of the Lower Jaw of a Ten-months Pig.

A, Wolf Tooth; B, Tusk; C, Corner Incisor; D, Intermediary Incisor; E, Nipper.
1—3, Permanent Molars in Three Stages of Development; 4—6, Dental Papillæ.

As a rule, the shedding of the teeth proceeds with the same degree of regularity as the cutting, so that irregularity in this respect only occurs in crippled or sickly animals.

Furstenberg has given the following tabular summary as a guide to the determination of age in swine:

The young animal is born with eight teeth — four corner incisors and four tusks (Fig. 2, A, B).

On the eighth or tenth day appear the second and third temporary molars (Fig. 3, E, F).

At four weeks old the four nippers (Fig. 3, A, B) appear— two in the upper and two in the lower jaw.

At the fifth or sixth week the foremost temporary molars are visible in the upper and lower jaw (Fig. 3, D).

At the age of three months the intermediary incisors (Fig. 4, A, A) have appeared above the gums.

At the sixth month (Fig. 5) the so-called wolf teeth (Fig. 6, A) will have appeared, and at the same age appear the third permanent molars (Fig. 6, C).

At the ninth month the following teeth will have appeared: the permanent corner incisors, the permanent tusks, and also the second permanent molars (Fig. 8, B 2).

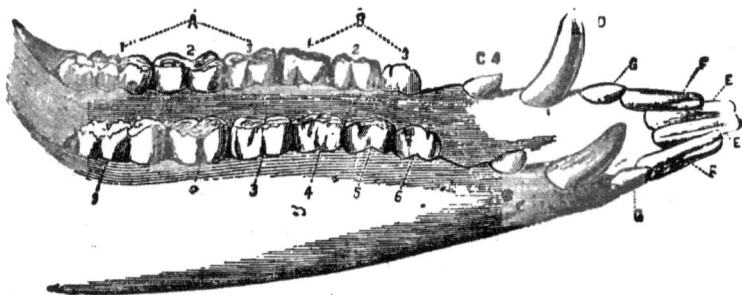

Fig. 10.—Part of the Lower Jaw of a Full-grown Pig.

A, B, Molars;   C 4, Fourth Premolar, sometimes called the Wolf Tooth;   D, Tusks; E, E, Nippers; F, F, Intermediate Incisors; G, G, Corner Incisors.

At the twelfth month, the permanent nippers (Fig. 7, A) will be seen.

With the twelfth and thirteenth months, the three temporary molars will have been shed, and their permanent substitutes, which at fifteen months will have fully appeared (Fig. 10, B), will be just cutting through the gums.

In the eighteenth month the permanent intermediate incisors (Fig. 8, C) and the hindmost permanent molar (Fig. 10, A 1) will have made their appearance, and with the twenty-first month they will be fully developed.

As to the first teeth of the young pig, Professor Simonds has remarked that they resemble small tushes, and being situated at the side of the mouth, do not injure the sow's teat when it is grasped in the act of sucking. The tongue is slightly fringed upon its border, and when doubled in sucking, the fringes overlap the nipple, and, without doubt, protect it against injury from the teeth. Furstenberg, too, points out that the teeth do not hurt the sow, as is supposed, and he thinks that they have no other use than to steady the tongue, and thus, in a manner, assist in holding the teat. Again, the tush or tusk of the boar is often regarded as an especial sign of age, and so it is, but not according to its length, as is commonly supposed. Numerous experts have made comparisons between the tusks of pigs of various races, and have discovered that their development is not the same in each case, but that they differ considerably; the finer the breed the smaller the tusks, and the commoner or coarser the pig —or, to be plainer, the more it resembles the wild boar, either in appearance, habits, or the way in which it is fed— the longer and stronger they are. There are, moreover, other influences which affect their development: a deviation from their normal direction often causes, as it were, a junction between the tusks of the upper and lower jaws. It has been noticed that a bad-tempered boar, or one which is unduly excited by constant proximity to sows, wears his tusks much more quickly than one of a gentle disposition, or one kept entirely away from sows. Many boars of the improved races are so gentle that their mouths can be examined; but, in the majority of cases, it is necessary to ascertain the age of the animal from its tusks and incisors alone. The tusks do not cease to grow until the animal is two and a half years old, and Sidney remarks, very correctly, that in judging a pig for age it is necessary to notice both the state of the incisors and the size of the tusks, although the latter are not, taken by themselves, of sufficient value.

The necessity for a knowledge of the dentition system must be evident to all who have business with pigs. For example, dealers have frequently sold youngsters of three months for those of two, and vice-versâ; but in most instances this can be detected if the mouth is examined for the intermediary incisors, which appear at three months. Again, at the age of six months, which is frequently adopted as a limit by the principal agricultural societies, the wolf teeth and the third permanent molars have appeared, so that at this period there is a double and, we may almost add, absolute guide, which it is possible for even the most uninitiated to follow.

Lastly, as the second and third permanent molars appear at from six to nine months, and the first of them at eighteen months—these being the three hindmost double teeth on each side of both jaws, and thus easily understood—it is apparent that, if they are found in an animal stated to be younger than these particular ages, there is good ground for suspicion, more especially if they are at all worn by constant use. Strangely enough, the mouth is seldom examined by those who breed and feed pigs, but there is no doubt that the imperfect growth of many young ones is caused by something abnormal in the growth of their teeth; and if these "bad doers," as they are called, had their mouths inspected in the first place and their imperfect teeth removed, they would quickly thrive as well as the remainder of the litter.

The Internal System.

The internal arrangement of the pig's system is not difficult to understand. It is stated by Vïborg that the length of the intestinal canal is 16yds. to 19yds., of which 4yds. to 5yds. are taken up by the colon and the rectum, and 5in. to 6in. by the cæcum. Heuzé, however, the able author of "Le Porc" and Inspector-General of Agriculture in France, states that the average length is 21½yds., of which the colon and rectum occupy about 5yds.; and that the intestinal canal of the wild boar is barely 17yds.

The mean capacity of the stomach is about seven quarts, that of the small intestines eight quarts, and that of the large ones seven and a half quarts. The small intestines have an equal diameter throughout the whole of their length, their walls being smooth and furnished with very small and delicate papillæ.

The heart hangs obliquely; the lungs are composed of two quite distinct lobes, while the liver has three lobes and four divisions.

The blood coagulates in from fifteen to thirty minutes, and is composed as follows:

Globules	145,532
Albumen	72,875
Fibrin	3,950
Fat	1,950
Alkaline Phosphate	1,362
Sulphate of Soda	0,089
Alkaline Carbonate	1,198
Chloride of Sodium	4,287
Oxide of Iron	0,782
Lime	0,085
Phosphoric Acid	0,206
Sulphuric Acid	0,041
Water	767.643
	1,000,000

With regard to the flesh of the pig, the late Sir John Bennett Lawes stated that, of 57 parts of the dry matter of the body, 44 are represented by fat; and M. Chevreul has shown that each 100 parts contain 80 of carbon, 11.14 of hydrogen, and 3.75 of oxygen. Upon analysis, the urine is found to contain ten and a half parts of bicarbonate of potash, five of urea, and small proportions of sulphate of potash, chloride of sodium, and carbonate of magnesia. A well-fed pig voids upon the average about seven pounds of urine daily.

We may mention that, at birth, young pigs weigh from 2½lb. to 3½lb. Both boar and sow are able to breed at the early age of five months, and the sow will produce litters every six months afterwards.

The pig has little choice with regard to his food, eating almost everything supplied; but M. Heuzé considers that he is at the same time an egoist and a gourmand, inasmuch as he disputes with almost every other animal the possession of that which he has made up his mind to have.

Few persons are aware how adept the pig is as a swimmer. It may have been noticed by many breeders that their pigs have objected to enter the water, but this is solely because it is strange to them. Once in, they take great delight in it, and it is sometimes difficult to get them out again. In France and Switzerland, where water is frequently provided for pigs to wash in—their bath, in such cases, being a portion of the sty—they regularly indulge in it, and keep themselves exceedingly clean, and their health is in every respect improved by the process.

The memory of the pig is too remarkable to be passed without notice, and although the entire race is considered heavy and stupid, its display of this faculty alone, to say nothing of a marvellous development of other senses, should redeem it from the application of these terms.

It has been stated that there are three modifications in the voice of the pig, each of which has a different meaning; first, the grunt, frequently repeated, which signifies comfort and contentment; secondly, the sharp note by which the pig exhibits its impatience for food, and which, prolonged for a greater length of time, or in a more decided manner, denotes disappointment and grief; and, thirdly, the responding cry to suffering animals, which is often very loud and sometimes threatening. We may, however, certainly add a fourth, which is very quickly uttered directly a sow is approached if she objects to the intruder, or is in fear of her young ones being harmed.

CHAPTER III.

BREEDING AND SELECTION.

IT has been said that Nature provided the various races of swine for the use of man, but that man has perfected them, at the same time preserving those types that have existed for generations. This is partly true, but when the statement is pursued it will appear to be misleading, for whether the Eastern races, whose blood has been borrowed by Western breeders, are really types of their remote ancestors or not, it is certain that British breeds, as we know them to-day, were not among the provisions of Nature. As man has, by his skill and perseverance, moulded, as it were, the various improved races of cattle by a system of selection of animals from other countries as well as from these islands, so has he produced the different families of high-class swine as we now find them, and maintained their types in purity and perfection.

Breeding may be truly described as an art. By skill and judgment our principal races of pigs were produced, and it is only by skill and judgment that they can be maintained. It is not simply necessary to imagine a symmetrical model or type and to attempt to reproduce or to imitate it, for the pig is not alone an animal of form any more than the horse or the cow. If this were so, the critical eye of the connoisseur and the acumen of the judge would be sufficient to enable them to produce the animal that is wanted. But symmetry of form is only one of the qualifications of the perfect pig. The others will be recognised by all who have any claim to a knowledge either of the art of breeding or its requirements, i.e., vigour, prolificness, and quality of flesh. The art of breeding, therefore, can only be sucessfully carried on by men who have practised eyes, considerable experience in the knowledge of the animal, and that judgment which is only attained by years of practice in the study and management of stock. As man has produced our various races of pigs, so we must look to him to maintain the position they have assumed under his guidance, and, indeed, still further to improve them; but without a love for and a study of his subject he will be unable to achieve this result. Let it be our aim to offer some encouraging words upon the subject, and to show that, good as the British races are, they may become better and be still more largely scattered over the face of the globe.

Unfortunately, to a great majority of the people of this country—and we are sorry to include many of the classes who keep and produce live stock—a pig is a pig, whatever may be its quality. It has often been observed that a well-bred animal costs little more to purchase, and no more to feed, than a bad one, and of all animals there are none of which this holds good more clearly than the pig. Indeed, if the truth must be told, a badly-bred pig costs considerably more to keep than a good one, and this is one of the most remarkable features of our improved races. Let us take an example. Presuming, for instance, that a good breeding sow costs £5 a year for feeding, and that she produces two litters, each of ten pigs, during that period, we shall not be accused of overstepping the mark if we say that, whereas she would produce a handsome profit, a bad sow, kept under exactly the same circumstances, would involve a loss of money, as will be seen from the following figures:

Good Sow:

Twenty young pigs (2 litters), at 20s. each .. ∴			£20 0 0
Keep of sow..	£5 0 0		
Extra food	2 0 0		
	—————		7 0 0
Profit (Gross) ., ...			£13 0 0

BLACK DORSET SOW.

The skin of this breed is not really black, but has a bluish tint that is not seen in any other variety, and both skin and hair indicate a lack of vigour. The Black Dorset bears a greater resemblance to the Neapolitan than either the Black Essex or the Black Suffolk. It is easily fatted and matures early, but it does not furnish sufficient lean meat for present taste.

Bad Sow:

One and a half litters, equal to twelve pigs. at 16s. ..		£9 12 0
Cost of feeding, say	£7 0 0	
Extra food	3 0 0	
		10 0 0
Loss		£0 8 0

This is by no means an overdrawn case, and, although we put it roughly, it answers the purpose of our argument without the necessity of showing a minute account. It very often happens that, when an argument of this kind is employed to persuade the owner of inferior stock to improve it, or to change it for a pure race, the question is put, Which is the best breed? If such an inquiry is intended to mean, Do you consider the Berkshire or the Large Black, the York or the Tamworth, the most suitable for my herd? there can be but one answer, for it depends upon circumstances and taste. The circumstances are, whether it is intended to breed for the production of pork or of bacon, or whether it is merely desired to breed young pigs for their sale at weaning time; whereas, with regard to taste, it is a matter for the decision of the breeder himself whether he prefers a white or red pig or a black one, the middle or the large breeds. In truth, however, putting aside the fact that these distinct varieties exist, the most proper answer to make to such a query is, that the breed of a pig depends chiefly upon the breeder, for it must be remembered that the pig is not exempt from the ordinary laws and conditions of breeding.

If an individual takes it into his head to breed for exhibition points or for fat, it is only too possible by such a course to take from the race all its fecundity, and so to destroy a property that is at the same time a necessary qualification of swine, and perhaps the most valuable one of all to the farmer and the breeder. It is perfectly true that the law does not prevent an individual from breeding his stock as he chooses, but, at the same time, there is a moral law that should be exercised to prevent persons who believe they are breeding for the public good from destroying properties which it has taken generations to create.

D

If, then, a breeder, instead of casting about for information
as to the best breed for his own purpose, would make up his
mind, after selecting any suitable type, to produce, nay, to
improve, it for himself, he would be exercising a function of
his own and would be able to ignore the advice of those who
would persuade him to maintain a particular exhibition
variety. It is easy by the pursuit of a certain system
deliberately to convert a most vigorous and prolific race into
one with a delicate constitution, although it may be fattened
far more readily than those from which it came, and this,
too, by a selection of those animals which are the finest in
texture, submitting them to a course of extravagant feeding,
and entirely precluding them from grazing or exercise on the
pastures. On the other hand, it is almost as simple to convert a delicate, fine-grown strain that produces few pigs at
a litter, and these of a very small size, into a coarser, hardy
and prolific family, by breeding from the strongest-haired,
coarsest-typed, longest-eared and longest-snouted beasts, and
allowing them unrestrained liberty in the woods and pastures.

Although these two distinct lines may not be pursued
exactly, yet there are very many breeders of pigs who follow
them in some degree, and thus it is that the various races,
perfect as they may be when they first come into the hands of
individuals, are by degrees either improved in size or quality
or, on the contrary, utterly degenerated as typical or, indeed,
highly-bred animals. What may happen to be entirely successful from every point of view under the hands of one man,
only too often proves quite the reverse in the hands of
another; and this is either because the natural laws of
breeding are not understood or, if understood, carelessly
ignored.

It was formerly supposed that the Berkshire pigs were
superior to any others in prolificacy, but the present system
of exhibition has induced many of the large breeders to treat
them in a manner that, at one time, would have generally
been considered unnatural; and as the animals have become
more refined in quality and more abnormal in obesity, so
have many of them changed in their capacity to produce large
and vigorous litters.

When, however, it is claimed that the improved races rapidly fatten and maintain a refined quality upon a small quantity of food, it is evidently forgotten by those who rate this qualification so highly that these animals cannot be exempt from those natural laws to which we have referred, and that if they gain in one respect they must lose in another, namely, in productiveness. To say that a race is prolific, or that it is not, is manifestly outside the question. The pig is what the breeder and the feeder make it, and although it is not advisable for a person requiring prolificacy to purchase from one who has bred for other qualities, still it is but a question of time and intelligence completely to convert the one qualification into the other. Because, therefore, the Berkshire or the Middle White breed, on account of the influence of the show yard, may be less prolific than of old, it is not right to infer that these breeds, as a whole, are not so valuable as they were.

The Art of Breeding.

Sir John Sebright, who, in his time, was justly celebrated as one of the ablest of that band whose experiments in the breeding of stock are felt to this day, made the following pertinent remarks in a letter addressed to Sir Joseph Banks: "Were I to define what is called the art of breeding, I should say that it consisted in the selection of males and females intended to breed together in reference to both their merits and defects. It is not always by putting the best male to the best female that the best produce will be obtained, for should they both have a tendency to the same defect, although in ever so slight a degree, it will generally preponderate so much in the produce as to render it of little value. A breed of animals may be said to improve when any desired quality has been increased by art beyond what that quality was in the same breed in a state of nature. What has been produced by art must be continued by the same means, for the most improved breeds will soon return to a state of nature, or perhaps defects will arise which did not exist when the breed was in its natural state, unless the greatest attention is paid to the selection of the individuals who are to breed together. We must observe the smallest tendency to imperfection in our

stock the moment it appears, so as to be able to counteract
it before it becomes a defect. The breeder's success will
depend entirely upon the degree in which he may happen to
possess this particular talent. Regard should not only be paid
to the qualities apparent in animals selected for breeding,
but to those which have prevailed in the race from which
they are descended, as they will always show themselves
sooner or later in the progeny."

It will be seen that Sir John Sebright lays great stress upon
the fact that what art has commenced art must perpetuate;
and every breeder should bear in mind that in aiming at a
particular standard of perfection, he will in all probability,
by carelessness or by half measures, do more harm to his
stock than if he allowed them to remain as they were. It
is well also to adopt the advice of Sir John Sebright in another
particular. Remembering how important is the influence of
the sire, he suggests the precaution of, first of all, putting
young males to females of ascertained value as breeders and
of unstained pedigree. Such a trial would prove the capacity
of the boar and enable the breeder to select or reject those
that are to form, as it were, the foundation lines of his
future herd.

It is generally agreed that the likeness or imprint of the
parent is stamped upon the offspring, and, undoubted as the
effect of this is when we are dealing with fixed races, yet
it will be found that the likeness will not always be a strong
or by any means a perfect one. As the perfection of
particular families is only of later growth, and as the blood of
some alien or imperfect sire or dam cannot be very far
distant, it is possible by a cross between two tolerably good
specimens—resembling each other in form and other characteristics, but entirely distinct in blood—to produce offspring
that have very little resemblance to themselves. The litter
may, indeed, revert to a remote ancestor, whose influence,
latent in the blood of one of the parent stock, may have
been developed by this particular cross.

The best breeders of to-day owe much of their success—
indeed, much of the uniformity of their herds—to the fact
that in commencing a pedigree strain of swine they have
endeavoured to form two distinct lines of the same family.

Thus, when crossing for vigour, they are enabled to employ blood which, even if almost as alien as though it came from another country, is practically certain in its effect, because it can be traced back in an unbroken line to the same original and perfect ancestors. If, however, as we have suggested, an animal of the most desirable pedigree, and whose line has for generations been unbroken by the introduction of a single inferior cross, is put to another of equally long pedigree, but which has on one single occasion been stained, either by design or by accident, then it may be considered an established fact that sports are liable to occur, and that the birth of young pigs tainted with the imperfections of the unfortunate ancestor may be expected. Allowing, then, that the influence of a particular ancestor remains long in the blood, as it does when that ancestor was an exceptionally faulty one, it will be seen how important it is that, in breeding to a high standard, the breed of the boar should never be in doubt. Indeed, the breeder should, as far as he knows, take nothing upon trust; and instead of using a sire of doubtful quality, because of some favourable or seemingly opportune circumstance, it will be infinitely better to send sows some distance for service or wait until the use of an undoubtedly pure boar can be obtained.

There are thousands of farmers and breeders of swine who, content to abide by their own judgment, forget that the appearances of the special points of an animal are not an absolute guide to its history or to its pedigree, and that its capacity to breed first-rate stock cannot be guaranteed. This reminds us of the instructions once given to a large American dealer by those who employed him to purchase stock in England, "Do not forget the pedigree." To which he replied, "I shall buy nothing that does not combine the double qualification of pedigree and points." Therefore, just as we should advise breeders not to rely on points alone, but to insist on pedigree as well—for, with pedigree, points may almost infallibly be regarded as fixed—so would we recommend them not to breed from an animal simply because it is backed up by a long and seemingly valuable pedigree, unless it also possesses the points required for the improvement of their stock.

We have referred elsewhere to delicate strains of pigs, and it is as well to remark that many who possess them are under the impression that the only course necessary to reimpart vigour and vitality is to cross with strong specimens of a strain of a similarly valuable family. It must, however, be remembered that, although the offspring of such a union may possess increased stamina and a comparatively vigorous constitution, yet in breeding from their descendants in after years it will be found that the least impulse given to the development of the latent influence of the former delicate ancestors will result in the tendency to throw back that we have noticed in reference to other imperfections.

We may quote a case in point. Some years ago, when one of the most prominent exhibitors of the East Anglian race of swine retired from competition in the show yard, he held a large sale of the pigs which had brought him so many laurels, but which had been, without any doubt, bred to such a state that in many cases the vigour, if not the vitality, of the animals was destroyed. A breeder bought some good-looking specimens, all of which were stated to be in a position to increase their species. They were not required for exhibition, and were consequently not fed, as they had been, up to the exhibition standard. The change in their management, however, immediately proved that, except in one instance, none of them were fertile; the particular animal excepted, which soon afterwards produced a small litter, was put, on different occasions, to different boars, and, without a single exception, brought forth progeny that inherited the characteristics of her ruined constitution, and perpetuated it in many other yards.

Some of our best breeders have studied to produce types of the pig that are not only appreciated in the exhibition yard, but are as vigorous and prolific as they are valuable in other respects. At the same time, there are breeders who have had but one end in view—the winning of prizes—and they have competed with animals which have not only been in-bred, but fatted in an unnatural manner, and which were, consequently, worthless to the purchaser, either for the improvement of stock or for the perpetuation of their own species.

We would here caution the purchaser of pure bred swine intended for the improvement of a herd, against a too common and disastrous practice. If he is inexperienced he will be wise to trust to the judgment of an experienced friend—instead of at all times relying upon the seller—to make him a selection. Breeders with reputations rear almost all they produce, and undoubtedly the majority of their young pigs are sold for stock instead of going to the butcher. First-rate stock animals form only a small percentage of the pigs raised by pedigree breeders, and it practically follows that the average purchaser is supplied with the remainder. The best pigs are in few hands, and this is one of the reasons why there is no plan equal to that of making a selection in the show yard.

The Value of a Herd Book.

Where an owner of stock is breeding up to an ideal, there can be no doubt that his only plan is to start with two distinct families of the same race, and to keep them apart until such time as he finds it necessary to introduce a cross between them. In this way he may be able to continue for generations, and so perpetuate a type, the destruction of which he need never risk by the introduction of alien or, it may be, of still further impoverished blood. In such a case it will be necessary to keep a careful and well-arranged herd book, without which even the most capable breeder will find it difficult to know exactly what he is doing. The necessity for such a work was recognised at the time of the first appearance of this book, with the result that a National Society was formed, and that many volumes of its Herd Book have been issued.

The value of pedigree, however, as we have already indicated, may be exaggerated. Bearing in mind the unprincipled nature of people who, in too many cases, profess to be breeders of pedigree pigs, and who will not scruple to impose upon a herd book any more than they would upon the public at large, we cannot but think that the pedigree of an animal is not alone sufficient to distinguish it as a member of a great race, nor as such a valuable and perfect animal as one that has been brought to the highest state of perfection.

And just as the American buyer properly declined to purchase without both pedigree and points, so do we believe that the only proper course to adopt before entering an animal upon a record so important as a herd book, is either to insist upon its being a prize-winner at one of a number of given exhibitions, or to require it to undergo an inspection by a committee of judges organised for this particular purpose. The latter stipulation would perhaps give rise to a great deal of trouble and cause some dissatisfaction to a number of the less important breeders, if, indeed, it were possible to carry it out; but, if practicable, it would accomplish for British pigs what no other system could—bring them to a state of perfection which they have not yet reached, and, still more, enable them to maintain the superiority that they possess at the present moment. It is not unreasonable to suppose that convenience might be obtained at the hands of the Royal Agricultural Society for the exhibition of pigs specially sent up for the double purpose of competing for its prizes and for entry in the Herd Book. Nor is there any reason why pig breeders should not themselves organise an annual exhibition, when both prize awards and eligibility for entry might be decided by properly appointed judges.

The Importance of the Sire.

A writer in the manual of the American Berkshire Association places this subject before the breeder in a very practical and pithy manner. It is quite true that the sire is half the herd—we might almost add, more than half— and this is shown by the following tables, which will enable the less experienced class of breeders to see as clearly as possible what the result of breeding from good and bad blood really is:

Mongrel }
Full blood } Half blood }
            Full blood } Three-quarter blood }
                        Full blood } Seven-eighths blood.

Full blood }
Mongrel } Half blood }
          Mongrel } Mongrel.

In the first case, it is assumed that the breeder starts with a pedigree boar and common sow; the product of these is naturally a half-bred animal, although, as a matter of fact, the influence of the sire being greater in such a case than that of the dam, the progeny would partake of the character of the pedigree pig to a larger extent than of that of the common pig. This half blood, then, is again mated to a pure-bred boar, the result being a three-quarter blood. Once more mating with a pure-bred boar, we get a seven-eighths, or nearly full blood; and so on, every succeeding sire causing the progeny to approach nearer to his own type.

In the second case, however, how different is the result! The pure-bred boar is put to the common sow, but instead of the continued introduction of pedigree boars the half-bred boar is used, this being regarded as an improvement upon the original or common stock. But what follows? Mated once more with the common herd, the progeny dwindles down to its original state, improved but little by the infusion of the pedigree blood in the first generation. Such, unfortunately, is the practice of a large number of English farmers and breeders, and yet they believe, of course without subjecting the question to any examination, that by the use of one cross of this particular nature they have done a very great thing, and date from it an era of improvement in their herds.

How different, however, would things be if, as in the first table, breeders used nothing but selected pedigree sires! How marked would be the improvement in the stock in every county, how much greater would be the profits, how much grander the general quality of British swine, and how much smaller would be the imports that we receive from the other enlightened stock-breeding nations! It is true that the introduction of a single cross may, in appearance, result in very great improvement. But all who have studied practical breeding will, we think, agree with us that, handsome as such stock may be, it generally fails when it is put to the test, and by its progeny shows how false were the calculations made by those who used it.

Whether a farmer intends to breed to an ideal type or not, he will find that the best way is to follow the praiseworthy example of those who, having attained a real knowledge of breeding by many years of experience, have found that the most reliable system is to use on all occasions a high-bred boar of whatever type may be selected. The most perfect females that are thrown from this cross should be retained and put to a sire of similar perfection to their own. Most persons find that the progeny of a first cross are generally good feeders and possess a valuable growing quality; but, as we have already pointed out, these are not the only desiderata in pig breeding. The proof of value in all-round qualities is only found when the progeny are retained for breeding; for it is then and then only that the results of a proper system of reproduction are understood and appreciated.

What to Avoid in Breeding.

Perhaps of all things in connection with stock breeding there is none which it is more necessary to avoid than in-breeding. This is a question that has for ages been discussed, both by those who are opposed to it and by those who have been compelled to resort to it in order to produce the ideal type that they personally set up. In these days we do not believe it is necessary to cross either parent with off-spring, or even the males and females of that offspring together. To a man of knowledge and judgment it cannot be difficult to procure the type of animal he requires, whereas the little extra trouble and expense of finding and purchasing it can in no way be compared with the trouble and loss that are certain to follow the very pernicious habit of in-breeding. The chief evils created by the practice are debility, or weak-ness of constitution, and infertility; and the longer the system of in-breeding is pursued, the greater will that infertility be. In course of time, the animals themselves will become so difficult to rear that there can be but one result —a clearance of the whole strain and the installation of another.

Besides these evils, in-breeding induces reversion to the particular failings of remote ancestors on either side. If the failing were one of health or vigour, it is certain to be reproduced in a more pronounced form; whereas if it were one of character, colour, or fancy points, it is, upon the same principle, likely to be represented in the offspring in a disagreeably marked degree. Unfortunately, people who do not take the trouble to learn something of the art of breeding too often form a conclusion from a single example, and a person who has used sire and offspring, or paired from the same litter, and has met with a bad result, immediately condemns the entire system; whereas another, who may have adopted the same course with success, concludes that it is a capital idea, and proceeds to recommend it because of the satisfactory result that he had obtained by his one experiment. The objection to in-breeding is, that as both animals are of the same blood it is hardly possible to expect an alteration or improvement in their character; but, as we have already shown, changes do occur from the general nature of the cross, although more with regard to vigour and vitality than to anything else.

What advantage, therefore, is to be gained in pursuing this foolish system it is difficult to imagine. It need hardly be mentioned that when two animals are paired for the reproduction of their species, it is usually for the express purpose of combining in their offspring one-half of the particular qualities of each. If both parents, then, are of the same strain, there can be no alteration in the blood; but if, on the contrary, they are of different strains, the offspring will combine the blood and qualities of each. In the latter case the breeder hopes for a successful result, although it may happen that he is unsuccessful as to the influence of the sire; whereas, in the former case, he cannot expect any marked change in form or character, nor can he hope to avoid a weakening of the vitality and a loss of vigour.

With regard to the system of breeding from sire and daughter, it must be remembered that the case is entirely different, for where the sire and the dam carry in their veins but one portion each of the blood of their parents, the offspring, on the contrary, produced by this close breeding carry three-parts of the blood of one grandparent, and but a quarter of the blood of the other. In this case it is evident that there is great room for variation, and, supposing the influence of the grandsire to be much more potent than that of the granddam, it is reasonable to assume that the former would stamp upon his grand-offspring the entire qualifications of his race rather than those of the family to which he was mated. There is much in a case of this kind to tax the skill of the breeder; and although the practice is one that can only be resorted to in particular cases, it may sometimes prove a valuable one, and should on no account be condemned because of that consanguinity which is apparently so objectionable.

Cases arise where the influence of the pre-potent grandsire is desirable, and where its presence is necessary to the success of a herd, and the intelligent breeder who resorts to it is not to be blamed, provided that he knows what he is about. As in the case of cattle breeders, who very often resort to the practice of close breeding in order to direct as considerable a portion as possible of the blood of some celebrated milking cow into the veins of their heifers, so will it often happen that the swine breeder desiring to accumulate a maximum of particular points—to shorten the head, to obtain a superabundance of hair, to produce quality of flesh, or even to preserve an unusual fecundity—will mate either sire or dam to one of its own litter; but it must not be forgotten that in close breeding for the two last-named qualifications there is a danger of reducing rather than of increasing them.

Points to be Obtained.

We would especially call the attention of all breeders of swine to the necessity of keeping two points in view, for success entirely depends upon these. What is required is prolificness combined with quality. We do not mean by this a capacity to produce fifteen or twenty pigs at every litter, for there is no conceivable value to the breeder in an excessive number. We would call a sow prolific that brought forth ten or a dozen, and that regularly. It must not be forgotten that the value of a sow does not consist in production of numbers alone, but in regularity of production and in a disposition to rear and to protect her offspring. Every breeder of experience knows that animals without " blood " are too often

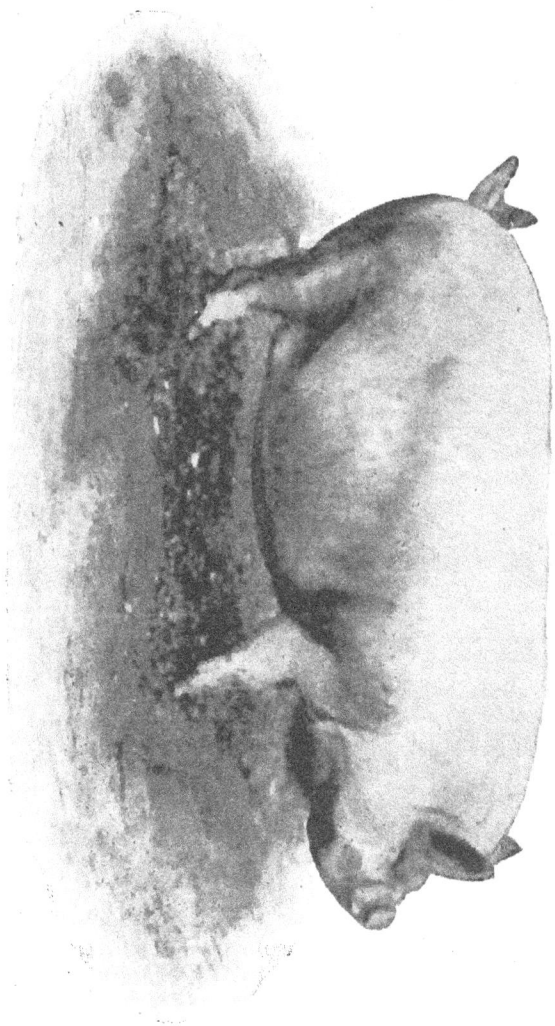

SMALL WHITE SOW.

Bred by the Hon. D. P. Bouverie. Winner of First Prize, Royal Agricultural Society's Show, 1905.

This breed should be pure white, and should have a smooth skin quite free from wrinkles, a fine and silky but abundant coat (which unfortunately is not shown in the photograph), a very short and dished head, a broad, level, and straight back, full and wide shoulders, a thick flank well let down, deep, wide, full, and well-rounded hams, and short legs set well outside the body

prolific breeders, but, on the other hand, they destroy the young either by carelessness or vicious habits as soon as they are farrowed.

Setting points on one side, we regard that sow a good one which produces the number of young we have indicated, which is of a gentle disposition, which can be depended upon to rear them well so long as she is fed well, and which possesses all the quality necessary for early growth and the easy production of meat. It is hardly necessary to say that a race of pigs that is excessively prolific is of little use if it cannot be fattened, for a buyer will soon find out that fact and decline the market in future. The breeding of pigs has but one end, and that the production of meat. The business of the farmer and breeder is to obtain that meat in as large a quantity, of as good a quality, in as short a space of time, and upon as small a quantity of food, as possible. This, then, can only be done by adopting the lines we have indicated and combining fruitfulness with quality.

To answer the breeder who prefers quantity to quality, we need only refer him to the fact that, whereas the supply of inferior-bred pigs is almost always larger than the demand, the demand for pigs of high quality is invariably greater than the supply.

Let us say, further, with regard to breeding for quality, that there is no comparison between a symmetrical fine-bred pig that has been fed upon good sound grain or milk, and the large, coarse, bony beast, devoid of quality, that has been principally fed upon garbage. Why the highly-fed animal produces flesh of better quality is obvious; but why the smaller should be better than the larger is, perhaps, not so apparent. It may therefore be pointed out that, as quality of meat has principally to do with the muscular fibres, or, as they are better known, "the lean," those fibres are not built up in the large pig to a greater extent than in the small; they have not, in fact, been increased in quantity during the growth of the animal, but, on the contrary, as it has increased in size, so have they increased in magnitude, and, consequently, in coarseness. Hence the preference of many breeders and butchers for the smaller over the larger breeds is explained. It is not alone the coarseness of the flesh to which they object, but the increase in bone and in offal, which grows with the development of the body, and constitutes refuse matter out of keeping with it.

The Proportion of Lean to Fat.

One subject constantly creates discussion among those who are interested in the breeding of swine, and that is, the proper proportion of lean to fat. The present system of breeding has materially conduced, not only to the larger production of fat—we may almost call it the excess of fat—but to the reduction of lean, and this must in a measure be considered as a drawback to the value of the pig as a meat-producing animal. It is certainly worthy of the attention of breeders whether it is not possible by care in the selection and by an altered system of feeding to remedy this defect. That the propensity to create fat rather than lean has been induced by a particular method of feeding—or, shall we say, by the use of fat-producing food?—for long generations we cannot doubt; and therefore it is hardly to be supposed that any sudden change can bring about a different result. At the same time, it is evident to those who understand the various races of swine, both of Eastern and Western nations, that the breeds themselves differ in this respect.

If we take the semi-wild hog of the Continent, and compare it with the pig of America, we find that there is an inherent difference in their production of flesh. In the case of the wild hog the food is chiefly nitrogenous, consisting of beech-mast, acorns, &c., while the American pig is fed on maize, which is one of the most successful of fat-forming foods. That a system of pulse feeding, for instance, would bring about a change in the proportion of fat to lean, is quite possible. Not that we believe the muscular fibre, or lean of the meat, would be greater, but it would be more equally distributed among a firmer and coarser-grained fat, thus producing that type of streaked meat which is so much valued by the curers and consumers of bacon.

Indeed, as in the case of other animals, it is quite possible to increase the muscular tissue of the pig by a judicious combination of exercise and nitrogenous food; and, consequently, in a few generations the hog would produce more lean and

less fat than it had hitherto done. Whether such a state of
things would please consumers of pork is, perhaps, a little
doubtful; but we certainly believe the flesh of the pig would
become more valuable as food. The adoption of the plan
advocated would also tend to fertility, and would greatly
minimise the afflictions that at present characterise the live
pig.

At the Smithfield Cattle Show of 1903 classes for pigs in the
carcase were introduced for the first time. These classes we
examined with great interest, and with a result which, as might
have been expected, condemned the existing system of exhibition
of live pigs in no uncertain manner. The exhibits of some of
the most successful showmen were little short of monstrosities,
and at a time when curers and butchers demand a limited
depth of fat, we measured two, three, and nearly four inches
on the back loin or rump. In the class for carcases under
100lb., little fault could be found, although even here we
measured about 1½in.; but in the two classes for carcases up
to 220lb. and above this weight the proportion of fat was so
excessive that, while it must have been difficult for the owners
to find buyers at any current prices—even through the medium
of the auction—the production of such fat must have been
attended with enormous loss.

The Best Colour—Black, White, or Red?

In commencing to breed or improve a stock, one of the first
questions put on appealing for advice is that with regard to
colour. Which is the best, the black pig, the red, or the
white? In the first place, it is a matter of taste; but should
the breeder choose to make his fancy subservient to something
more practical, he has next to consider the fashion of the district. It would in many cases be manifestly unwise for a
farmer feeding in a large way to commence the cultivation of
a strain of black pigs in a district where the people at large
have a decided preference for white, or *vice versâ*. But where
one colour is equally as suitable and as saleable as another,
the general question may be put—Which is the fittest?

In the South of England (a few districts excepted) the
Berkshire pig is regarded as the model for the farmer, and,

indeed, throughout that part of the country, if the question were put to a large number of practical men, we doubt whether the majority would not give their verdict in its favour. It must be admitted, however, that there are very many who, while believing in this race as the most valuable for all practical purposes, are at the same time fanciers and breeders of the Yorkshire, the Tamworth, or the Large Black. If we refer to authorities in America, where the Berkshire pig is a very great favourite, we find this opinion continually expressed in most unqualified terms. The *American Berkshire Record* has quoted a number of records compiled to show that the Berkshire race is gradually superseding the white pig in all parts of the United States.

The superintendent of the Chicago stockyards, who had seen millions upon millions of swine pass through these yards every year, once stated that during a period of three years 90 per cent. of those received were dark; and a former editor of the *American Stockman*, who passed his opinion after several years of close observation and daily attendance in the stock-markets, once stated that not more than 25 per cent. of the pigs sent to the Western market showed any considerable amount of white hair, while only about 5 per cent. were of a defined white breed. The superintendent of the Kansas City stockyard, the next in importance to Chicago, counted 110 trucks of hogs, which came from all quarters, and of those but 6.66 per cent. were white, the remainder, 93.34, being black, although ten years previously the colours were about equally divided.

It must not, however, be supposed that these dark pigs were all Berkshires. The Americans are rather proud of their great native manufacture—the Poland China—which is chiefly black, and this, with the Berkshire breed, occupies a superior position to the whites, which, however, have secured greater attention since the above facts were observed. We speak, of course, of the improved whites. • On the other hand, the white pig is more general on the north of the Thames, while in the North of England the black races are comparatively unknown and unappreciated, and believed by those who understand swine to be less economical than the whites. It is just as easy to quote evidence from America in support of the white breeds, for, as in England, different States have different preferences. For example, the State Breeders' Association of Indiana, where large numbers of swine are bred, decided that the white pig was the best in the market, and obtained by far the largest sale.

The best pig to keep is that which commands the best sale in the district. The difference between good whites and good blacks is so imperceptible that this practical method of deciding the matter is after all the soundest.

The Selection of Stock.

There is no more important element in the commencement of a breeding herd than the selection of animals for stock purposes. The remarks already made have an important bearing upon this question; but, reducing them to practice, we may say that, having determined upon the type to be produced, the first object should be to procure boars as near to that type as it is possible to obtain them, taking care to reject any which do not approach it sufficiently close, or which have one single failing in their composition. The animal whose pedigree and breeding are not thoroughly satisfactory should not be looked at a second time, and, without delineating the numerous points of the pig in this chapter, we may suggest that the breeder should insist upon strength of hair, fineness in the quality of the flesh, comparative gentleness of disposition, uniform breadth, not only across the shoulders, but across the loins, and as much length as can possibly be obtained consistent with size.

A breeder may be pardoned for purchasing a sow whose looks are tempting and whose disposition is kind, although her pedigree may be imperfect and her breeding capacity less valuable than it ought to be; but there is no excuse for a man who, intending to breed largely himself and to influence the stock of his neighbours, wilfully selects a boar with a glaring fault, a savage disposition, or a lineage only removed from that of a mongrel by his good looks.

We should certainly select a sow with a strong and unbroken pedigree, a temper of the gentlest, a vigorous constitution, undeniable quality, an udder furnished with

E

twelve well-developed and well-placed teats, of a strain tolerably famous for its prolificacy, and well known for the ability of the females to rear their young successfully; and all this, too, in addition to the requisite points of symmetry. Where opinions differ as to the class of sow of a particular strain to be used for stock, it is easy to make an experiment, and certainly not costly, as would be the case if similar experiments were made with a boar. Provided the stock boar is thoroughly reliable, it is easy enough to mate him with sows of different types, and to be guided by the progeny that they throw. The inferior animals, which it is not intended to keep for stock purposes, should all be converted into pork, and the progeny of the most successful cross retained for the continuation of the strain. It will, perhaps, be found by such an experiment that one or more of the families thus tested will produce litters of uneven quality, and however good the selected youngsters from these may be, there should be no hesitation about what to do with them. If any are retained for breeding, it will be found that they have the same tendency to produce uneven litters, just as they would have to perpetuate any other fault inherent in their ancestors. Undoubtedly it is possible, by continual crossing, to breed out imperfections of this kind; but an individual who is able to begin with stock of a better class will be foolish indeed to waste a considerable amount of time in the perfecting of imperfect animals.

When we speak of uniformity in a breed, we do not refer to the one miserable little animal which is usually called the "Harry," and which is sometimes found in litters of the majority of the very best sows, while some sows continually throw two, three, and four of these wasters in every litter they produce. Uniformity in a breed we take to be a thorough and general development throughout the whole of the body, general breadth and depth, whether it be in the chine or the loin, the rump or the ham—not so much the uniformity in colour or marking, the form of the head and the ear, valuable characteristic and indicative of feeding powers as these points may be, for they are but fancy points, and, although necessary in a herd bred for the purposes of exhibition or the production of exhibition pigs, they have little weight in the yard of a breeder whose principal object is something of a more substantial nature.

"In the choice of your swine let them be long and large of body, deep-sided and deep-bellied, thick thighs and short legs, for tho' the long-legged swine appear a goodly beast, yet he but consulteth the eye, and is not so profitable to the butcher; high, clean, thick neck, short and strong groin, and a good thick chine, well set with strong bristles; the colour is best which is all of one piece, as all white or all sanded— the pide are the worst, and apt to take meazels; the black is tolerable, but our kingdom, thro' the coldnesse, breedeth them seldome." Such was Markham's opinion some 200 years ago. Let us add a word or two. A short head, wide jowl, and thick neck are usually allied to excessive fatness, and a long head, narrow between the eyes, to defective constitution. Well sprung ribs give room for the heart and lungs, and both need freedom from contraction in order to ensure vigour and vitality. A good loin supports a good rib; short legs provide plump hams; while strong hair, not too coarse, indicates good constitution. Unless there are ample and normal teats the weakest youngsters go to the wall, and this defect often accounts for the number of weaklings in a litter.

Age, Size, and Vigour.

With regard to the age for breeding, again do persons differ. Where some breeders are content to wait until a gilt has reached the age of fifteen months, the majority prefer to commence when she is nine months old; indeed, we have met many thoroughly practical men who believe that she cannot be too young, and who regularly put her to stock at six months, provided she evinces a desire at that early age. The last-named persons have frequently stated to us their belief that a young sow grows quite as well when she is with young as when she is not, and they consider it a waste of time, and consequently a loss of money, to wait until she is of more mature age, and when, in fact, under their system, she is ready for a second litter. On the other hand, those who prefer to wait until maturity believe that early breeding invariably checks the pig's growth and prevents a proper

E 2

development of the frame; and for this reason they continue
to feed their young stock well until they are at least fifteen
months old.   Our own observation has shown us many
instances in which the systems of both classes have proved
advantageous, and, on the contrary, equally disadvantageous;
but we have also known instances where unusually precocious
gilts have evinced a desire to be mated to the boar at a
very early age, but not having been permitted for many
months, were with the greatest difficulty induced to breed
at all.

While on no account advocating the system of such early
breeding, we can also refer to instances in which sows have
had litters at a very early age, have afterwards continued
to produce with great regularity, and have grown, indeed,
large, prolific, and handsome animals. If the question of the
size of the breed is allowed to enter into the argument,
there can be no doubt that it is more advantageous to leave
larger pigs to reach a more mature age before they are put to
the boar than the smaller and more precocious ones. Some
believe that by breeding from very young pigs of a rather
coarse strain at an early age they are able to reduce them
to a finer quality; but, if this be so, there is no doubt that
such fineness brings with it impaired vigour and a liability
to become less prolific than the animals might have been.
Others, again, believe that early breeding prevents a proper
development of milk; but this is, after all, a matter of
conjecture, for almost all experienced breeders, at any rate
those acquainted with the management of very young mothers,
can give instances in which the sows have had a full development of milk as well as the reverse, just, in fact, as is the
case with older animals.   We therefore think that age
has little or nothing to do with this qualification.

In districts where young pigs are bred for sale at certain
seasons of the year, it is necessary that they should farrow
during particular months—January or February, or August or
September, for instance. In such cases it is a very common
practice for breeders to put the spring-born gilts to the boar
at the age of eight months, in order that they may produce
their litters in the natural course of things in the following
spring, for, if this plan is not followed, it is necessary to keep

them until they are at least fourteen months old, and this is regarded as late by many breeders, who, as a rule, prefer to err rather on the side of youth than of maturity. The young stock which is to be retained for breeding should, if possible, always be selected from the litters born in spring; and the reason is obvious, for where the autumn-bred pigs are generally confined to the sty for many months, and are fed chiefly upon artificial food, the spring youngsters, except during the first few weeks of their existence, are continually in the air, feeding upon a large amount of natural food, and enjoying that form of life which is by far the best for constitution and physical development.

Where the very highest type of animal is desired, we believe it is a better plan to commence to breed from those born, for instance, in February, so that their first litters may drop about April of the succeeding year. This cannot be considered too early, and, on the other hand, it is by no means late, for the sows will then enjoy the whole of the summer on the pasture with their young, and will develop into mature and handsome swine.

The influence of age on the size and weight of litters has been well tested at the Agricultural Experiment Station of Wisconsin, where so much splendid work has been done in relation to the swine industry, and the results may be briefly stated. The performances of sows of various ages were recorded, and the average number of pigs in each litter— together with the average weight of each litter—was ascertained as follows :

Age of Sows.	Number of Pigs.	Weight of Litters in Pounds.
Sows four to five years old ..	9.0	26.0
Sows two and three years old ..	7.5	19.7
Sows one year old	7.8	14.2

If we accept the lesson that these figures teach we are bound to believe that the older sows are better mothers than young sows, and that the too common practice of reserving the young for stock and selling the older animals is not to be commended. The figures, however, but confirm the experience of many careful observers of the past.

With regard to the boar himself, it is the custom with the majority of the best breeders to use him at an early age, and to sell him before he has become too heavy or too clumsy —in some instances, indeed, before he becomes too savage. The very best strains, however, while they are carefully managed and the stock animals kindly treated, seldom produce a savage beast. The young boar, if he is perfectly developed and of high quality, may be used without the slightest hesitation from the age of nine months; and we are not alone in our opinion that better stock is got by relying upon a comparatively young than upon a very old boar, if his quality is of the highest.

Respecting size and stamina, it must be remembered that, in the small breeds as well as the large ones, these points are required by the duties of maternity and by the demand that is certain to be made upon the sow when she is contributing to the growth of a strong and numerous litter. The fact that she is being robbed of her vitality day by day, while her power is being continually reduced, is a sufficient argument in this respect, for it is absolutely essential to her future that she should have something more than the flesh that she has accumulated within her to enable her to submit to the tremendous strain to which she is subjected, and to enable her to regain the physical condition that she had previously enjoyed. If, then, the dam is endowed with the vigour to which we have referred, it is evident that the progeny will inherit it, and will continue to improve in a manner which they would not be likely to do without the possession of such a valuable qualification. Unfortunately, the most vigorous breeding sows are too often fattened at an age when their vitality is at its highest, and this because of the prevailing but mistaken notion that it is time to sell or kill a sow when she has attained the age of from four to five years. For the production and maintenance of stock of the most vigorous description, there can be no doubt that, provided all other qualifications are perfect, the most profitable breeding animal is the sow of from two to three years.

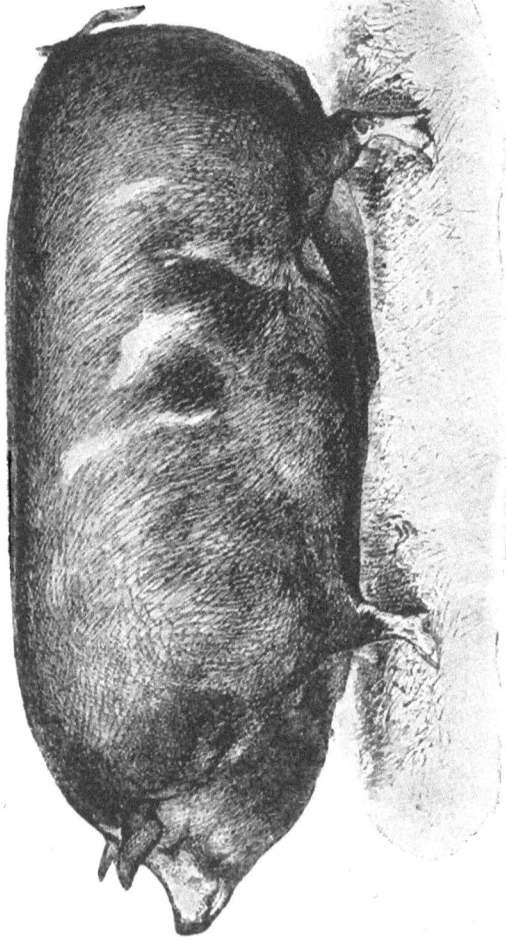

POLAND-CHINA BOAR.

In colour this breed is dark, usually black, with small, irregularly-placed white marks, and its coat is fine and thick. It is strong and hardy, an excellent breeder and suckler, can be fattened readily at any age, and is considered by American authorities to make more pounds of pork per bushel of corn than any other pig known in the States. Although a purely American pig, it is also largely bred in Germany.

CHAPTER IV.

THE BOAR.

An American writer has stated that a first-class boar, properly used, would improve a herd more rapidly than thirty equally good sows. For instance, if this number of sows had each a litter averaging six pigs, they would produce 180 young ones. If sired by a badly-bred boar, more than half of the progeny would be inferior to their dams, in consequence of the prepotency of the sire. The old truism that the boar is half the herd is to-day more palpable than ever. A person intending to breed pigs should make it his chief business to select a really good sire. He should not only visit reputed herds for the purpose, but take care to notice if the quality of the animals composing them is sufficiently high and uniform.

It is customary to obtain the sows before the boar is considered, and many people take very little trouble with regard to their selection. Buyers are only too willing to listen to advice which will save them any extra expense. A moderately well-bred, or even a half-bred, sow is oftentimes a desirable animal: she can always be put to a pure-bred boar, and the quality of her progeny will be greatly improved. On the other hand, the slightest blemish or cross in the blood of the boar not only prevents his regular use by other people, but materially diminishes the value of his progeny.

It is admitted by many shrewd breeders that, in breeding, the boar furnishes the form, quality, and fancy points of the herd, while the sow supplies the frame and the internal

structure.   Others have remarked that sows take after the
sire and boars after the dam in their general characteristics.
These ideas cannot both be correct; still, it may be taken
for granted that in all pure breeds the boar has the greatest
influence upon the quality and the general appearance of his
offspring.

Where quality of flesh or a propensity to fatten rapidly is
demanded, there is nothing like a boar of one of the improved
races.   If, however, large, roomy-framed, hardy, growing
pigs are required, then there is no better plan than to introduce sows of a breed like the Large White.   In some districts
in England large-framed youngsters are in request, and these
alone; but the buyers have not arrived at such a climax of
porcine education as to be able to understand that a meaty
pig of medium size carries more flesh than a large bony one.
Most pig-keepers still believe the fallacy that they gain more
profit by the production of quantity than of quality, of bone
and offal rather than of fine meat in the smallest possible
compass.

In selecting a boar—whether he be of the black or the
white breeds—he should be taken from a litter remarkable for
its uniformity and bred from parents exhibiting those qualifications which the buyer wishes to reproduce in his own herd.
Absolute health and vigour the animal must have, and, above
all things, a good temper.

In rearing a young boar it should be observed that his
future temperament depends very much upon the treatment he
receives from his feeder.   This is a very important consideration; an owner should never leave the animal so completely
in charge of a servant that it may be spoiled either by bad
feeding or by rough usage.   As a rule, the better the breed
of the boar, the more gentle will be his disposition; and it
is quite true that general appearance is a good guide in
breeding.

A highly-bred boar of almost every British breed has a
dished face, a broad jowl, a fine snout, small, almost delicate
ears, and exceedingly fine bone; indeed, the extremities are all
remarkable in this respect, not omitting the tail.   The chine
is broad and thick, the back is level, the loins are wide, the
ribs well sprung, and the hams fully developed.   The whole

animal is square in build, standing firmly upon his legs, which, both back and front, are wide apart. The hair should be plentiful, but, contrary to the common opinion, fine and silky, coarse hair being accompanied by coarse flesh, and sometimes by bad temper.

A boar should always be considered from four points of view—namely, his disposition, his quality, his capacity for stock-getting, and his ability to put on flesh. The last-named qualification enables his owner to keep him at much less cost.

Management of the Boar.

The management of the boar at the present day is most imperfect. It is impossible to deprecate any system more distinctly than that of confining him to a small house. This is frequently without a court, and here, as a rule, the animal passes his existence, except when he is required for service. There is no animal so prone to fatten when it has no exercise as the pig, and there is nothing more objectionable in a breeding animal than excess of fat; nevertheless, our farmers and breeders generally confine the boar to a few square feet in an undrained sty, and feed him profusely upon fattening food. To maintain health and stock-getting condition a boar must have exercise, and it is often the lack of this that causes imperfect and small litters.

The boar's house should be strongly built, furnished with an asphalted floor, and provided with a dry and strong wooden bench. The house should lead into a large yard, court, or paddock, where he can enjoy himself and graze as freely as other animals in the herd. If the boar is really so important a factor in breeding as we all believe, it is astonishing that more consideration is not shown him in this respect. He need not necessarily be allowed to run loose among a number of his species, but where there is any system of breeding at all, or any pretence to produce high-bred stock, a run or yard is certainly a *sine quâ non*. It may be urged that there is always some danger to be apprehended from male adult pigs which are at liberty, but there is no difficulty in " drawing the fangs " of the most savage, and thus rendering him comparatively harmless.

A boar may be used with advantage from nine months to five years old; indeed, this is the best period of his life, although in all parts of the country farmers and others have no scruples whatever in breeding from much younger animals, sometimes even under six months old. This is a most unwise practice, as it prevents them from properly maturing. Some persons, however, who have watched with great keenness the effects of breeding, maintain that an old boar is not only the most prolific, but produces the most vigorous and perfect offspring, especially when he is mated with sows which approach him in age.

There is, again, quite as much carelessness displayed by many pig-keepers in their use of good boars as in the age at which they put them to the sow. For instance, it will be admitted by all those who understand the conditions of breeding that a male should not be too frequently used for service. This question, however, never enters the mind of many breeders, who use the boar at any time or season. In addition to this, they follow the old and ignorant custom of serving the sows a second time, and are compelled to grant the same privilege to those sent to the lord of the harem by their neighbours. Second service is not necessary—nay, more, it does considerable harm to a boar to be in such constant use. The generality of people, however, may be excused their ignorance on this point as many authors refrain from all reference to subjects of this kind, either from reprehensible prudishness or from ignorance.

The practice should be to keep the boar as far as possible from the sties of the breeding sows; otherwise, during the periods of heat, he may be troublesome, and, perhaps, on some occasions, break through or jump over the door of his sty. Old breeders know what this means. There is probably no animal more difficult to catch, to hold, or to manage, especially at such times, than a strong, lusty boar, although there is a great deal of difference in individual cases. We have known many animals of moderate breed which, although not positively savage, have been dangerously uncertain, and at times feared by their keepers. On the other hand, there are plenty of gigantic beasts of the Large York breed which are comparatively gentle, which are handled, even to the mouth, by their owners or feeders, and which allow entire strangers to approach and even caress them.

When a boar has developed his tusks, and is considered dangerous, an early opportunity should be taken to remove them. This may be done in one or two ways. Although it sometimes requires a bold man to take the initiative, the best plan is as follows: The attendant enters the sty with a noose, which it is his object to slip over the upper jaw, behind the tusks. If this is impracticable, as is often the case, the animal should be quietly fed. While the boar is eating from his trough, he may be led unsuspectingly to put his nose over the noose, which is laid in position among his food, and is then dexterously slipped over the jaw. This noose is attached to a long rope, one end of which is run through a stout staple some five feet from the ground, in any convenient spot near at hand. Immediately the noose is fixed and tightened, a couple of men should haul the pig up slightly, and hold him; he will not attempt to commit damage, but will immediately struggle to gain his freedom, by pulling at the noose. He can then be approached, and the tusks removed, one by one, with a small file saw.

Sometimes the boar cannot be noosed over the jaw; in this case, it will not be found difficult to throw a larger noose over the head, and to haul him up in the same way, afterwards passing the small noose over the jaw as directed. Even then, he frequently refuses to open his mouth, which has to be forced and held open. It is at all times better to compel a boar to pass through this operation than to risk the possibility of damage from his tusks.

Boars are sometimes as difficult to catch as they are to drive into a sty to which they object. Some breeders prefer to run them down, as, if fat, they soon give up; but if in lusty, store condition this often takes a long time. If, however, more than usual trouble is given, the first opportunity of getting a cord round a hind leg should be embraced, and as in the hunt the boar will frequently get into a corner, this will be easily managed. The leg can then be lifted, when another man will be able to slip a noose over the jaw.

A boar should never be beaten on the body or legs, although in these cases he is frequently chased with forks and broom handles. A short, fine ash stick is much the best, and this, used upon his snout, will have far greater effect than anything else; he will fear it more, and suffer no actual harm from it.

The Boar on the Farm.

Generally, farmers of the ordinary class decline to keep a boar for their own use, their objection being chiefly upon the ground of expense. It should not be forgotten that, if a person only keeps five or six breeding sows, and produces from these offspring of an inferior kind, he is throwing away considerably more money than would suffice to purchase and keep one of the best boars in the kingdom. Putting this on one side, however, he would find it by no means difficult to arrange with some of his neighbours, either to share the expense of a well-bred animal or to place one at their service at such a fee as would amply compensate him for his trouble and expense. Upon a farm, it does not, as a general rule, cost more than half-a-crown per week to keep a large boar, and we question very much whether he costs more than three or four shillings at any time, even when the whole of his food has to be purchased. In the case of small boars, such as those of the Small White or Black breeds, half this sum will suffice if they have the advantage of a grass run. Both breeds will exist, and can be kept in capital condition, upon an exceedingly small quantity of food.

When the boar is no longer useful for service, he may be put up to fatten for bacon, after submitting to the following method of castration. This operation is described by our friend, Mr. Coburn, author of the popular American work on "Swine Husbandry." After drawing up one hind leg, and fastening it securely to a post or stake, run another rope round the upper jaw, back to the tusks, draw it tightly, and fasten it to another stake. In this position the animal can offer no serious resistance. The cut should be low down, and as small as possible; the low cut will afford a ready means of escape for all extraneous matter, and allow the wound to keep itself clean, there being no sac or pocket left to hold the pus formed during the healing process. This operation should not be performed when the boar is very fat, or the weather too warm, as the risk is much greater. If castrated early in the season, and kept on grass during the summer, the flesh, when fat, will be but little more rank than that of other pigs. If kept with others, there is danger, if the boar is quarrelsome, of his doing them injury with his tusks. Hence, it is desirable to fatten a stag hog by itself. It is at this period that the old boar's true proportions will show themselves, as he will put on fat very rapidly, and present a greatly improved appearance. When about to sell, the buyer will probably insist upon paying for the "stag" only two-thirds the price of other hogs, and, in many cases, we consider the deduction is far too much.

The castration of old boars is often most successfully performed by dealers who are practised in the work, and as they usually buy at a low price, the profits are considerable. The castration of young pigs is described in another chapter.

CHAPTER V.

THE SOW AND HER OFFSPRING.

In dealing with a subject of this practical nature, it will be found most convenient to commence with "the gilt," or, as she is called in different localities, "the hilt" or "yelt." By these names the young sow is known until she has had a litter, although in some parts of the country she retains her "maiden name" until the second litter, when she becomes a sow. There are, however, breeding and feeding gilts, and the latter, like castrated boars, are termed hogs. These feeding gilts are spayed for fattening purposes. The operation (which is described further on) is analogous to that performed upon the male pig when he is castrated; in other words, the ovaries are removed, the animal is no longer in a condition to breed, and therefore fattens more rapidly.

Breeding.

A breeding gilt that is not pregnant or with young is commonly termed an "open" gilt, just as a more mature female is called an "open" sow. It has now become a very general custom for persons commencing to breed pigs to purchase gilts before they are of a breeding age, and when they are only in moderate store condition; consequently, they are less costly than if either tolerably fat or, as it is called, "in pig," or in farrow. There are, however, numbers of amateur pig-keepers who prefer to pay more money in order to obtain an animal actually in farrow. They do this for two reasons: first, because they consider they get infinitely more for their money; and, secondly, because they do not thoroughly understand the system of management necessary to bring about the desired end.

It is a strange thing that, in almost every book dealing with domestic animals, there is an entire absence of practical information with regard to mating, and the signs which are apparent in the female when she should be put to the male. When a gilt or a sow arrives at the period of heat—or, if we may put it in another way, is in season and desires service—she will exhibit considerable restlessness, the sexual organs will be enlarged and perhaps inflamed, and in many cases she will be continually jumping on any pig near at hand. There can scarcely be any mistake made in the recognition of these signs by even a novice in pig-keeping. At the same time, we may remark that there are, now and then, animals which do not, to the practised eye, exhibit any signs at all, and which would never breed if left in their sty and kept for the period of heat to manifest itself. Sows of this description, which have passed two or three months without requiring the boar, are far better turned loose in a large yard or pasture, where he can have constant access to them. Indeed, in some important herds, it is the custom of breeders to allow the boar to run with the sows regularly; and, if the animal is a tried one, there is seldom a miss, or even a noticeable delay.

When, then, the period of heat arrives, whatever may be the time or day, the sow should be at once driven to the nearest boar whose quality is such as to commend him. If, as is often the case, she declines, she had better be taken home again (if the distance is not too far), and brought back on the following morning. Sometimes it is necessary to take sows several times to a boar before service is effected, and even then it is often necessary to leave them with him for four-and-twenty hours.

Owners of boars, in some cases, decline to permit the introduction of sows from strange yards, especially when swine fever or any other contagious diseases are prevalent. At such times they are certainly wise in adopting this course, but there is often considerable selfishness exhibited by breeders

who decline to permit the use of their stock when fair
payment is offered them.

Feeding the Farrow Gilt.

It is at all times a good plan to allow gilts, whether they are
in farrow or not, to have their liberty, for the regular
exercise that they get in the open air makes them vigorous,
and undoubtedly assists in imparting a prolificacy not found
in pigs which are always kept in the sty.

It is now the rule among the best breeders in this country
to provide a certain area of pasture for their pigs, just as
they do for their cows; while in France, under the new
system of pig-keeping, a grass run, termed a "park," is
provided to almost every set of sties. Where there is plenty
of good grass, as well as ample room for exercise, the
animals thrive much better, produce far hardier stock, and
cost much less for food. Indeed, it has been stated over and
over again by breeders of the Small White and the Small
Black, that during the few best months of the year, when
grass is plentiful, the sows require no other food.

As a general rule, although there are some differences of
opinion upon this point, gilts should be rung, otherwise they
will turn up the pasture with their snouts in search of roots,
and commit considerable damage, as well as make the fields
most unsightly. The ringing, however, should be done before
they are in farrow, or at all events immediately afterwards,
to prevent the possibility of abortion, which is sometimes
the result of over-excitement.

With regard to food, there is nothing better than pollard
or sharps, or even bran may be used with good results, if of
good quality. If these meals are mixed with a quantity of
wash and vegetable refuse from the garden, it will be found
that the animals will thrive upon a comparatively small
quantity, for they require but little purchased food indeed
for three months out of the four they are pregnant. During
the fourth month, however, the feeding should be increased,
as heavier claims are made upon their systems for the
support of the coming family, and in order that they may not

lose condition and strength, both of which will be very necessary after parturition.

It is an unwise plan to feed in-pig or farrow sows upon dry meal or grain, as they put on too much flesh, which is most undesirable when they litter, although we have bred very respectable litters from exceedingly fat sows. Highly fed animals, too, are more liable to become bad tempered at this important period, and are more easily excited.

During the summer season gilts in farrow, like sows, may be largely fed upon succulent grasses. As a general rule, where they can obtain a sufficient quantity of lucerne, vetches, sainfoin, clover, ryegrass, or a mixture of clover and grass, they will require little or no corn until shortly before parturition. In the same way, they may be largely fed upon turnips, mangels, and potatoes in winter. Experience has shown that as regards roots, good as they are, they should not be largely given, unless in conjunction with meal.

Gilts in farrow, like sows, should never be confined in sties with strange or older animals. It may happen that the rough treatment they receive will cause them to abort; and this is not only to be avoided on account of its immediate consequences, but because of the future liability of the sow to repeat the misfortune.

Gestation.

The period of gestation in the pig varies from 110 to 116 days, although in many instances we have found sows with their litters upon the 112th day, at which time certain animals have invariably brought them forth. There is, in our judgment, no domestic animal so true to time in this respect as the sow. The French breeders have a rule, in which they implicitly believe, that a breeding sow goes three months, three weeks, and three days with her young; but, as will immediately be seen, this may mean almost anything, for a month is a term with a somewhat wide margin.

The indications shown by a sow about to farrow are numerous, but there are three sufficiently prominent for the guidance of those who are quite unused to stock. In the first place the udder develops largely, every teat being

F

apparently filled with milk. The gradual development can, indeed, be seen for some days by the accomplished eye. In the next place, the vagina also develops in a most unusual manner, and is a certain sign, even to the unpractised, that parturition is near. In the third place, the sow generally makes her bed, and will be seen carrying mouthfuls of straw to the spot she has selected in which to deposit her young.

Unfortunately, persons who have no knowledge of stock-breeding give themselves trouble and excite themselves unnecessarily when they expect an addition to their stock; but if they have taken note of the day of service, they may rest assured that, accidents excepted, all will go well, and that on the 112th or the 113th day they will, if they let the sow alone, find her safe and happy with her newly-born litter. The excitement, however, which these events create in the minds of novices in pig-breeding does not stop here, for they insist upon their stockman sitting up all night with the sow—sometimes even for two or three nights—watching her every action, and feeding her liberally with most improper food, so that their excitement is communicated to the animal herself, and she finishes either by lying upon and crushing a portion of her litter, or, what is worse still, by killing and perhaps devouring two or three in her frenzy.

It has long been a maxim among old breeders that sows are best attended to at farrowing time when they are not attended at all; for, if they have a warm and dry sty, it matters very little if they are not fed for twelve hours, and they are much better left alone to pass through the operation without any assistance whatever. Naturally enough, there are cases in which assistance is necessary; but, of all animals, the sow is the one that requires it the least. There are some stockmen who insist upon the necessity of their presence, and who not uncommonly make a bungle, such as cutting the cord too closely, thereby causing excessive hæmorrhage.

Visitors should be rigidly excluded from the sty of the farrowing sow, and even the owner himself should make a practice of keeping away and leaving the management to the man who usually feeds the animal, and whose presence is known and recognised by her. If it is requisite to watch the sow at all, a hole should be made in some convenient place in the sty, so that she can be seen without discovering the presence of the observer.

The following table shows the date on which a sow is due to farrow, allowing sixteen weeks. The dates in the first column are dates of service: thus, a sow served on January 1 is due to farrow on April 23.

January, due April		February, due May		March, due June		April, due July		May, due August		June, due September		July, due October		August, due November		September, due December		October, due January		November, due February		December, due March	
1	23	1	24	1	21	1	22	1	21	1	21	1	21	1	21	1	22	1	21	1	21	1	23
2	24	2	25	2	22	2	23	2	22	2	22	2	22	2	22	2	23	2	22	2	22	2	24
3	25	3	26	3	23	3	24	3	23	3	23	3	23	3	23	3	24	3	23	3	23	3	25
4	26	4	27	4	24	4	25	4	24	4	24	4	24	4	24	4	25	4	24	4	24	4	26
5	27	5	28	5	25	5	26	5	25	5	25	5	25	5	25	5	26	5	25	5	25	5	27
6	28	6	29	6	26	6	27	6	26	6	26	6	26	6	26	6	27	6	26	6	26	6	28
7	29	7	30	7	27	7	28	7	27	7	27	7	27	7	27	7	28	7	27	7	27	7	29
8	30 May	8	31 Jun.	8	28	8	29	8	28	8	28	8	28	8	28	8	29	8	28	8	28 Mar	8	30
9	1	9	1	9	29	9	30	9	29	9	29	9	29	9	29	9	30	9	29	9	1	9	31 Apl.
10	2	10	2	10	30 July	10	31 Au.	10	30	10	30 Oct.	10	30	10	30 Dec.	10	31 Jan.	10	30	10	2	10	1
11	3	11	3	11	1	11	1	11	31 Sep	11	1	11	31 Nov.	11	1	11	1	11	31 Feb.	11	3	11	2
12	4	12	4	12	2	12	2	12	1	12	2	12	1	12	2	12	2	12	1	12	4	12	3
13	5	13	5	13	3	13	3	13	2	13	3	13	2	13	3	13	3	13	2	13	5	13	4
14	6	14	6	14	4	14	4	14	3	14	4	14	3	14	4	14	4	14	3	14	6	14	5
15	7	15	7	15	5	15	5	15	4	15	5	15	4	15	5	15	5	15	4	15	7	15	6
16	8	16	8	16	6	16	6	16	5	16	6	16	5	16	6	16	6	16	5	16	8	16	7
17	9	17	9	17	7	17	7	17	6	17	7	17	6	17	7	17	7	17	6	17	9	17	8
18	10	18	10	18	8	18	8	18	7	18	8	18	7	18	8	18	8	18	7	18	10	18	9
19	11	19	11	19	9	19	9	19	8	19	9	19	8	19	9	19	9	19	8	19	11	19	10
20	12	20	12	20	10	20	10	20	9	20	10	20	9	20	10	20	10	20	9	20	12	20	11
21	13	21	13	21	11	21	11	21	10	21	11	21	10	21	11	21	11	21	10	21	13	21	12
22	14	22	14	22	12	22	12	22	11	22	12	22	11	22	12	22	12	22	11	22	14	22	13
23	15	23	15	23	13	23	13	23	12	23	13	23	12	23	13	23	13	23	12	23	15	23	14
24	16	24	16	24	14	24	14	24	13	24	14	24	13	24	14	24	14	24	13	24	16	24	15
25	17	25	17	25	15	25	15	25	14	25	15	25	14	25	15	25	15	25	14	25	17	25	16
26	18	26	18	26	16	26	16	26	15	26	16	26	15	26	16	26	16	26	15	26	18	26	17
27	19	27	19	27	17	27	17	27	16	27	17	27	16	27	17	27	17	27	16	27	19	27	18
28	20	28	20	28	18	28	18	28	17	28	18	28	17	28	18	28	18	28	17	28	20	28	19
29	21			29	19	29	19	29	18	29	19	29	18	29	19	29	19	29	18	29	21	29	20
30	22			30	20	30	20	30	19	30	20	30	19	30	20	30	20	30	19	30	22	30	21
31	23			31	21			31	20			31	20	31	21			31	20			31	22

Parturition, or Farrowing.

It is a good plan to move the farrow sow or gilt into her particular sty at least a month before parturition. Some

animals are a long time in becoming accustomed to their new
home, and are especially restless if they happen to farrow
in a strange place. If careful note be taken, it will be
found that, generally speaking, the sow brings her litter
almost to the day, seldom farrowing more than one, or at
the outside two, days from the date of service. In most cases,
a sign of the coming event will be the making of her bed, if
she is provided with straw, for she will take large quantities
in her mouth and lay it carefully in the selected spot.
Although this is the custom with pigs in general, there are
many that never attempt it, for reasons which are generally
found to be very simple.

Parturition, or farrowing, with the pig most frequently
takes place during the night. In the majority of cases, where
the animals are not specially watched, the sow will be found
lying quietly with her family when the stockman arrives in
the morning. It is, however, desirable to attend them, for
although not always necessary, yet there are occasions when
one or more pigs may be saved. Some sows are particularly
vicious, seizing each pig as it is born, and killing it with
one nip of the teeth. Therefore, if anything of this kind is
apprehended, a sow should be partitioned off in a corner of
the sty, by a hurdle or gate; and as each little pig is born it
should be taken by the attendant and laid in a warm corner
of straw, or placed in a basket, if the weather is cold, and
removed to the fireside. When a litter is complete, and the
sow calm and quiet, her pigs may be returned to her without
fear, and, as a general rule, she will prove gentle and kind
to them, and in a very short time will allow the attendant
to put them to her udder.

There are other pigs that make a practice of devouring
their young, and it is difficult indeed to suggest any plan to
prevent a recurrence of the practice. We would rather
recommend those who are owners of animals of this kind to
take the first opportunity of converting them into bacon.

If the young pigs squeal, and the sow is of an excitable
nature, it is at all times advisable to carry them off as
they are born, as there is nothing more calculated to make
the dam troublesome than the squealing of her young.

Some sows are exceedingly clumsy at the time of farrowing,

BLACK SUFFOLK SOW.

The Small Black breed known as the Black Suffolk was probably produced by one or two crosses on the original Essex pig. It closely resembles the Small White in size and symmetry, although a perfect contrast in colour, and the hair is seldom so curly or so woolly as in the White Breed. The Suffolk is a good grazer, and when properly managed is very prolific.

and unless the stockman is present, they are almost certain to lie down upon two or three of their young, or, if they do not actually do this, to suffocate them as the underlying little ones are sucking at the lower row of teats. These sows should be watched, and not be left until they, with their young, are settled for the night in comfortable and safe positions.

Treatment of the Sow and Young.

Sows that have farrowed should not, as is popularly supposed, be immediately fed upon rich foods, given in large quantities, for few things are more calculated to do them harm. Having had pollard, middlings, or bran for a few weeks before, the same food should be continued and mixed with cold water, although it may be given a little oftener. Perhaps on the whole good sound middlings form the most serviceable ration for a sow when suckling her young. Rich, heating foods are not so much required as what we may term cooling and nourishing drinks. As the young ones increase in size, however, common sense will dictate that the food should be augmented both in quantity and quality; but at no time is it advisable to give the sow, while with young, such a food as barley-meal, which is too heating. Some kinds of food diminish the quantity of her milk, and others (especially changes) are liable to cause scour, or diarrhœa, in the litter; they should consequently be avoided.

At the end of the first month, no matter how well the sow may have been fed, it will be noticed that her family have considerably pulled down her condition, and that the little pigs will now require some food to themselves. They will, however, frequently be seen feeding with their dam when her trough is filled, and this is as it should be, for, while sucking, little pigs should learn to feed by taking only what is given to the mother.

At all times when the weather is fine, and it is possible to do so, the sow should be let out of the sty, after her morning's meal, for a couple of hours (when she should be driven to her young to suckle them, and to feed), and again let out for the afternoon. In this case, she will be feeding

on the pasture most of the day, while the litter will be
dividing their time, as young pigs should, between feeding
and sleeping. Professor Henry found that while the periods
of suckling were irregular up to the fourth week after farrowing, after this date the sows nursed their young about two
hours during the day and four hours during the night.

At the end of a month, it is a good plan to commence to
feed young pigs with a handful of wheat or peas, scattered
among the straw in the sty, the mother also getting a little
at feeding time. There are no foods so good as these, although
oats are frequently and strongly recommended. If the
quantity given is increased as the pigs grow, until weaning
time, it will be found that they wean well, and continue
growing, just as though they had not been deprived of the
mother's milk.

If milk can be regularly obtained, more particularly skim
milk, it should be given to the sow when she is suckling,
and especially when the young pigs can" help themselves to
a little at the same time; but on no account should it be
given to sucking pigs by themselves.

In cases where the sows are not let out of the sty, it will
be found advisable to feed the little pigs separately, unless
their food can be placed so as to be out of the sow's reach.
To this end, some breeders have two sties, side by side, a
small opening in the partition being made for the little pigs
to run through, so as to get at the food provided for them.

In the splendid series of experiments conducted at the
Wisconsin Station under Professor Henry, it was found that
1.19lb. of grain and 2.4lb. of skim milk per 100lb. live
weight were required to maintain a brood sow after her
young were weaned. The common Razorback sows of the
country consumed less than the pure-bred sow; but they
ate more in proportion to their weight, although they made
a slight gain in weight, which the purchased sows did not.
Three Berkshires, averaging 395lb. in weight, consumed an
average of 3.6lb. of grain and 7.5lb. of milk; two Poland
chines, averaging 362lb., consumed 3.49lb. of grain and 6.8lb.
of milk; three crossbred sows—the pure on the mongrel—
averaging 320lb., consumed 3.65lb. of grain and 7.2lb. of
milk; while four common sows ate 3.19lb. of grain and 6.3lb.

of milk, their weights averaging 226lb. In each case the older sows ate the most food, and the food consumed was always in proportion to their weight. Data of this kind enabled the experimenters to ascertain the cost of maintenance, and therefore how much of the food consumed by the sows, when suckling their young, was utilised in the production of milk.

There are many opinions with regard to the use of straw for pigs, but it is quite safe to say that the only straw absolutely fit for a litter of young pigs is wheat straw. Whatever may be the cost, it will be found far cheaper and more profitable to use than any other. Upon wheat straw, if supplied in sufficient quantity, pigs are always clean, and much less liable to become infested with vermin. With barley straw, however, they are certain to be troubled with the latter, while they will not thrive so well, and never at any time look, as healthy little pigs should, pink and clean.

There is one important point in the gilt that should never be omitted when inquiry or purchase is made, i.e., the number of teats with which she is provided. It is scarcely fair to expect a sow to bring up more than twelve good pigs. Even this is above the average, as few animals, whatever be their quality, do justice to this number. Every gilt, however, should have twelve teats; otherwise, if she happen to produce a litter of thirteen, for instance, it will be necessary to destroy those over and above the number of teats that she carries. Of course, if another sow is at hand, which has farrowed about the same time, and will take the surplus young, or if any person has the patience and ability to rear them by hand, then let it be done.

When the teats are sufficiently numerous, it often happens that one or two strong youngsters prefer particular ones, and continually fight for them. When this is the case, one or two of the weakest in the litter usually suffer, and grow up small and comparatively worthless pigs.

It has frequently been recommended — among others by the late Mr. James Howard, M.P.—that the teeth of new-born pigs should be nipped off with a pair of pliers. When the teeth are unusually sharp, the young ones may irritate the dam, and cause damage by continually battling with their little messmates for certain teats; but in practice it is seldom found that the pliers are necessary, and unless it is found so we would distinctly recommend breeders to have nothing to do with them.

Bedding or Benches.

There is much difference of opinion as to the necessity for a large quantity of straw in the sty at the time of farrowing. Some persons urge the use of battened floors, with a little straw on the top, or of wooden benches. Their objection to the latter is that, during the first few days, the young pigs are liable to roll off, and, being unable to get back, to die of cold. Others conclude that it is necessary to provide every breeding sty with a small rail, which is erected a few inches from the walls, to prevent the sow from lying upon her young or squeezing them against the sides of the sty.

We have a particular objection to manure of any kind in the sty in which the litter of young ones is placed, but there is no doubt that the nest usually formed by the sow for her young in farmyard piggeries—which are generally primitive enough, and a foot or two deep in dung—prevents the destruction of young pigs through the clumsiness of their dams. These beds yield to pressure, and, although buried in the straw, the little ones are still able to breathe, and are consequently not so easily suffocated as in sties in which straw is less abundantly provided. We would, therefore, advise the use of plenty of straw for the bedding of the sow, although we would distinctly urge the necessity of keeping the sty entirely free from manure.

In addition to a thoroughly clean floor, the sties should be limewashed before the appearance of the litters that are to occupy them. There is nothing so liable to cause disease on the one hand, and harbour vermin on the other, as are filthy walls and dirty floors. It is an old custom among farmers to make fresh beds for fatting stock upon the dung they have made, and not to clear this out (although it sometimes rises to a height of three and even four feet) until the end of every season. And yet the animals are expected to thrive and to keep entirely free from disease! Every good pigsty should be provided with a solid floor. It will be found in practice that the sow is a particularly clean animal, although this is somewhat contradictory to her supposed nature. She will generally make her bed in one corner of the apartment, and void manure in another. This fact is well known, and hence there is no possible excuse for preferring muck to a sty as clean as it can be made.

With regard to the habit of sows lying upon and crushing or suffocating their young, we think it will be found that this is somewhat of an inherent fault. Some animals are most careful in all their actions among their newly-born offspring: others are invariably so stupid or careless that they can scarcely lie down without inflicting some damage to them. Well-bred (not necessarily pedigree) sows are generally most gentle, and it is not at all uncommon for breeders to allow two, and even three, to live together in the same sty quite up to the day of farrowing, or permit two to farrow together.

If persons would select their stock from gentle sows, and rear the young kindly, they would soon abolish a costly and obnoxious failing. It is not, however, always the case that the best mothers are the best-looking pigs. Cross-bred animals are frequently very gentle in disposition, especially those that have been bred from sows without pedigree and a boar of an improved race. Their nature, too, is generally much kinder if they have been bred and reared with others rather than in a sty by themselves. Solitary confinement too often develops a bad temper in the best bred animals.

Unequal Litters.

As there is no animal that produces her young so closely to time as the pig, it is a capital plan to arrange, as far as possible, for two or three to be served at the same time. We have noticed that gilts from the same litter will often require their first service together, and will continue to do so, and to farrow almost together, for years. This practice should be encouraged, for in case of one having an unusually large, or the other an undesirably small, litter, it is possible to equalise them by removing a few of the young pigs from their own mother to a foster-mother. This should be carefully managed at the earliest possible moment, and, as far as

possible, during the night. If those to be added have been
born before their foster-mother produces her litter, they
should be given to her with her own pigs; but if, on the
contrary, they are not born so soon, it is a good plan to
place them in her bed at night, or if possible, to put them
to some of her teats when the rest are sucking, and to leave
them in this position in the dark.

A gilt generally produces from two to eight pigs, and
although the former number is absurdly small, yet we have
found in practice that, when weaned and ready to sell, they
usually realise as much as four pigs from an ordinary gilt's
litter. A gilt should not rear more than eight, unless she
is unusually large, has plenty of milk, and is more mature
in age than is customary.

Sows generally produce from seven to twenty, and we know
a case in which a common-bred animal produced a dozen
litters, none of which were less than seventeen, while one
reached the enormous number of twenty-one. It has been
stated by some authorities that sows average twelve, but we
very much doubt this, and should think ten is nearer the
mark. At all events, twelve should not in all cases be allowed
them, for they do far more justice to ten. A larger number
pulls them down considerably, while the young are frequently
not satisfied before they are able to eat. With large and
strong litters, it is a capital plan to teach the young to eat
as soon as possible, and to feed them well between times
upon similar foods to those supplied to the sow.

It has been noted by Mr. Springer, the well-known
American authority, that persons are too prone to boast of
the extraordinary prolificacy of their sows, but that reports
of results from these large litters are very seldom seen. There
is much truth in this statement, and we believe that, after
all, a sow regularly producing moderate litters, to which
she does justice, is much more valuable than one producing
unusually large litters, some of which she kills or does not
properly nourish.

The Growing Litter.

Some breeders make a practice of selling to private
customers the majority of pigs in each litter as sucking pigs.

It must be confessed that the system has much to recommend it when pork is low in price. The breeder frequently obtains almost as much per head at the age of three weeks as he would do at eight weeks, when they had been weaned, in which case he would not only save a certain outlay for food, &c., but he would gain time with his sow, and in the labour necessary in her management.

It has often been remarked that young pigs require less attention than other young animals: and, provided the management in the yard is really good, this is no exaggeration; but if it is irregular and careless, trouble is not far off. Litters of pigs kept in a dirty state quickly become covered with lice, which are troublesome to remove, and can be better dealt with by the free use of carbolic soap than anything else.

If, from careless feeding or any other simple cause, scour supervenes, there is no home-made remedy better to check it in its incipient form than a little soot placed in the food. If the sty is dry, clean, healthily situated, and facing the sun, and if the pigs are regularly fed upon good and appropriate food, there is, as a rule, little difficulty. If a meal is occasionally missed, if the foods are given irrespective of their quality and fitness, or if the sty is frequently left untouched, the little pigs may stand the change for a few days, but they are certain, ultimately, to go wrong. On the other hand, there need be no petting or over-feeding. The attendant should be gentle, and should occasionally handle them so that they may become accustomed to him, and then, as a general rule, they will be fit for weaning at the age of about eight weeks. In some cases, however, it may be nine or ten weeks before little pigs can be taken from the sow.

Sometimes weaners are worth from £1 to £1 5s. per head, unless very inferior, but prices fluctuate remarkably, and as buyers make the great mistake of preferring size and bone to quality—which really means more meat in proportion to bone—it has become necessary to use the large races, and to breed pigs that are very much larger at weaning than they used to be, in order to command the higher prices. This is one reason why the improved small breeds, such as the Small

White and the Small Black, have not remained popular. To
our certain knowledge, dealers resolutely decline to purchase
small-bred weaners at anything like the price that they will
give for large-bred mongrels, although they know perfectly well
that the one is superior to the other in quality. The time
has, however, arrived when all concerned unite in preferring a
combination of size and quality.

Some persons prefer to let out their litters with the sows
at a very early age, if the weather is fine. There is no doubt
that this plan assists in maintaining their health and vigour,
and, if they are equally well fed, in developing their growth.
Pigs at liberty, however, and more particularly young pigs,
are excessively troublesome. If there be the smallest possible
chance of their making their way on to a neighbour's ground,
they will quickly do so, destroying his crops and causing
considerable annoyance, and sometimes expense, both to him
and to their owner. They also commence to root up grass
very early, and should be rung immediately this propensity is
noticed, if they have not been rung before. Previous to
turning out pigs, therefore, at whatever season of the year,
the fences should be thoroughly examined, as they must be
both perfect and strong. The rails or wires should be sufficiently close together to keep in pigs of any age. Hedges,
unless unusually thick, will not prevent them from gradually
making their way through. The ringing of young pigs is very
simple, and there are two or three systems, which we shall
describe later on, that may be easily managed by the owner
or stockman.

Cutting, however—i.e., castrating or spaying—is generally
performed by a practical man. The young boars, or at all
events those not required for stock, should be castrated at
from seven to eight weeks, if sufficiently strong. Most
breeders prefer to perform the operation before the young
leave the sow. It is absolutely necessary to cut boars not
intended for stock, but we question very much whether the
advantages of spaying gilts are sufficiently numerous to
warrant the operation. It is much more severe than castration, and usually stops the growth, although perhaps not in
a very great degree. There is, moreover, an element of
danger that seldom exists in the operation upon the boar.

If the gilts are to be fed for bacon, there is certainly some little trouble at the periods of heat if they have not been spayed; but this is not serious, and, for porkers, it is more than questionable whether any commensurate advantage is gained by the operation.

Weaning.

When pigs are weaned, it is better to remove them to a fresh sty, in which case they will miss their dams far less than if they remain where they have been brought up. If the sow is let out daily, they should be fed in their sty until she has thoroughly dried her udder. If they have become used to feeding with her, and by themselves in her absence, they will generally thrive well upon a slight increase of the same kind of food after having been taken from her.

Gluttons as pigs are, few animals so readily become disordered by the use of improper or variable food, more especially at this time. It is therefore imperative that changes should not be made, unless by degrees, otherwise the growth of the pigs may be retarded or their stomachs deranged. It will also be found better to feed them five times a day at first, upon a small quantity of food each time, rather than to give them three heavy meals. Most people think that anything is good enough for the pig, and expect it to consume whatever is provided, and to thrive equally as well as though it had its regular meals of ground corn. This is not the case; and although it refuses very little, yet it will be found useful to pay some attention to its likes and dislikes, and to give only what is found to suit it, rather than to feed upon cheaper foods, which leave no appreciable results. In other words, the feeding and management of the pig demand the continual thought of the breeder, just as is the case with other, although higher, classes of stock.

Some consideration is necessary at the time of weaning as to the purpose to which the young pigs are to be devoted. For instance, if they are to be sold as weaners, they should be kept in a good growing fleshy condition, and as clean as possible. If it is intended to keep them as porkers, and, consequently, to put on flesh rapidly, their feed should be increased by degrees, as well as improved in quality, until, a few weeks after weaning, they will consume as much as is given to them. In this case, too, they should be confined to the sty and furnished with plenty of clean straw, to encourage them to divide their time as much as possible between eating and sleeping. If, however, they are to be grown for stock purposes, they should be still fed thoroughly well—not necessarily upon fattening food—and be let out upon the pasture whenever the weather is favourable.

Great care, too, must be taken as to dividing a weaned litter, for pigs, like other young animals, are troublesome when strange to each other; and they quickly become so if separated for a time. Litters that are bred and kept together always do much better than those that have been mixed. In the latter case, in addition to regular fights, some youngsters invariably fail to get their proper share of food, and fall off in growth, and possibly in health.

If one portion of a litter is intended for fattening, and another for breeding, it is wise to divide them at once, and at the end of a week to commence to treat each lot according to the special circumstances of their case. The picked gilts of a litter, for example, that are intended for breeding will do much better if taken from the boars and kept together until they have become far advanced in farrow. At the same time, there should be no overcrowding. Although pigs can be kept in a marvellously small space when well managed, yet overcrowding is one of the worst faults that can be found upon a stock farm.

Drying the Sow.

To return to the sow. It may be remarked that some persons do nothing more at weaning time than to separate her from her family, and allow her to dry without assistance; others prefer to change the system of feeding by giving barley meal instead of the pollard with which she has hitherto been supplied. The meal is considerably more heating, and assists her to dry her milk, as well as to bring on the period of heat, which generally follows weaning in from three to ten days. It may be remarked, however, that the sow will

exhibit fewer signs of that condition now than at any other
time. Some persons prefer to wean gradually by allowing
her to suckle her pigs only every other day for a week, but
this trouble is scarcely necessary. In the ordinary way, her
milk soon goes, she is rapidly brought into service, and, if
served, success is generally certain, after which the meal
feeding may be reduced. For some time she will thrive upon
wash or grass, slightly supplemented with pollard or sharps.

It has been stated by some persons that a sow should
yield five litters in two years; she cannot, however, be expected to produce more than two litters per annum, and if
she rears these, each litter embracing a fair number, she will
do remarkably well. A breeder attempting to obtain five
litters can have no possible system : his pigs would come in
at all periods of the years instead of, as they should do,
at particular seasons.

Feeding by Hand.

In some cases—it is to be hoped few—it will be found
necessary to rear a litter of pigs by hand. This can be done
successfully if care be taken. For three or four weeks the
trouble involved will be considerable, and unless the work
be persevered in to the end, the whole time may be lost.

The little ones should have the richest cows' milk that can
be obtained, sweetened with coarse sugar, and given them
at first from a spoon, unless they can be induced to take it
from a teat fixed upon the end of a bottle. With a little
perseverance, they will soon learn to drink. In a few days,
they will not only look for it, but take it readily when poured
into a basin or any similarly convenient vessel. They will
require a small quantity of milk at least eight times a day
for at least a month. For an average litter from a sow of
medium size, at least a gallon will be required, and at the
end of this period, and as the little ones commence to feed,
the quantity supplied may be very gradually increased, until
at weaning they are consuming at least two gallons of new
milk, with as much skim milk and meal as they require.
Subsequently the new milk may be diminished, until the pigs

are receiving the normal ration for a store or a porker,
which, if possible, should still include skimmed milk.

At the end of a month, some feeders provide a small quantity of boiled wheat, which is really good for the young, and
this food may be increased until they are able to feed like
weaners, at the end of seven or eight weeks.

Purchasing at Shows.

It must not be forgotten that in stock pigs fat is not
required. If a novice in breeding were to visit the principal
agricultural shows of this country, and to take especial note
of the exhibits of swine, he would not be favourably impressed
with the truth of this remark. It is a lamentable fact that,
in order to obtain prizes, pigs, whether for stock purposes
or not, are necessarily fed and fattened to a standard in this
respect that is often disgusting. The system largely prevents
the introduction of high-class stock into the ordinary farmyard. The farmer or stock-breeder visiting an exhibition for
the purpose of selecting animals for the improvement of his
herd, is sometimes unable to do so without, on the one hand,
purchasing those whose breeding capacity has been partially
or wholly spoiled, and, on the other, paying a long price for
excessive obesity. It may be observed, however, that of late
years considerable improvement has taken place in this respect,
especially among young stock.

Setting aside the fancy value of high-bred stock, the
exhibitor of an unnecessarily fat pig naturally remarks, in
answer to a question as to price, that his pig is worth £8 or
£10, as the case may be, to the butcher, and that it should
therefore be worth this sum to the breeder. Now, in order
to prepare an animal for stock the superfluous fat must be
reduced, inasmuch as it is of no possible use to the buyer.
We therefore advise every breeder to keep the young pigs he
intends for stock purposes in the very best store condition,
and on no consideration to fatten them. The principal
qualification in sows as well as in boars is growth. This is best
encouraged by sound feeding upon nitrogenous rather than
fattening food, with the addition of plenty of exercise in a
meadow in which there is good grazing.

It is wise to practise kindness and gentleness in dealing with stock pigs. Few inexperienced persons know how troublesome badly-bred pigs are, and what losses they frequently cause. Some of our prominent breeders have urged the necessity of grooming stock pigs, as it results in the conversion of a most sullen animal into an unusually gentle one. It will be seen that there is considerable advantage to be derived from this practice, although we can hardly expect farmers in general to undertake it.

Management of the Young Stock.

We have already stated that gilts intended for breeding should be kept together. It may be added that on no consideration should they be kept in the same sties with older sows. In addition to their being continually beaten, they are generally deprived of most of their food, and sometimes would obtain none at all unless the attendant remained during their feeding-time to see that they had fair play.

Gilts that are kept in good condition often reach maturity very early: the first period of heat sometimes arrives before they are half the age at which they should be put to the boar. As a general rule, it will be found that, if they are then served, the service will be successful; but, if not, they may be sometimes served afterwards without becoming pregnant. There is no doubt that the service of gilts of the small breeds may take place at an earlier period than those of the large breeds, if they have fairly developed, and are in good order. The larger the breed, the more advisable it is to postpone the date of service, for it is quite possible to reduce the size of a large variety by early breeding, and thus, as it were, diminish a valuable quality.

With regard to young boars, they should be allowed the same privileges as to exercise and grazing as the gilts; but immediately they begin to show signs of work, they should be kept in their sties, and only allowed their liberty when the sows are shut in. Every breeder should keep his own boar, for the reasons given in previous pages; and, if the animal is a thoroughly good one, the receipts from persons who are unable to keep their own will generally pay the expense of food and management. In every country district where pigs are kept in any number, there are always some persons who are glad to send their sows to a good stock boar, for the use of which they generally pay from 2s. to 2s. 6d.

The greatest care should be exercised in the reception of sows during the prevalence of fever and other contagious diseases. If there are many gilts in the herd which have to be served, and which, being sisters, are very much alike, it is a good plan to mark them after service, either with tar or paint. These substances, however, are quickly removed, and must be renewed if further identification is necessary. The pigs may with greater advantage be marked in the ear with a stock-marking button, numbered and removable (although occasionally it may be torn out by accident or design).

A Registration Book.

As it is very necessary to know the date upon which a sow is due to farrow, great care should be exercised by the breeder or his stockman in noting the day on which she is served: hence, in all large herds, some system of registration is absolutely necessary. A book should be kept in which all these little matters may be entered, so that note can be taken of the particular day three weeks after service, in order that the animal may be watched, and again served if the first service has proved unsuccessful, as she will then come in season again.

There is another point to which both owner and herdsman should rigidly adhere where pigs are bred for exhibition. It is well known that at the principal exhibitions competing pigs are required to comply with conditions as to age, and that they are frequently disqualified owing to the fact of their teeth not agreeing with the age, as shown by the entry. As it is to some extent a reflection upon the veracity of an exhibitor to be "disqualified," he should be at all times prepared with proof; and this can only be shown by an entry in his herd book, properly authenticated by two or more witnesses of undeniable respectability.

Treatment after Farrowing.

After farrowing, it is the custom with some people to give the sow a little treacle, either mixed with water or poured into her ordinary wash; but this is rarely necessary, and if, in a large breeding yard, a person makes a practice of adopting any one of the nostrums so frequently recommended, he will find that the additional expense is an item of some consequence, to say nothing of the trouble occasioned by their composition, more especially upon a farm where drugs are seldom at hand when they are wanted, and when they can only be obtained by sending to the nearest town. If a sow, after farrowing, shows any signs of constipation, it is then time to administer an aperient or a purgative, according to the severity of the attack; for the complaint is one that must not be neglected, resulting, as it may in serious cases, in inflammation of the bladder and sometimes in death.

Sometimes it happens that a sow is provided with very little milk and is physically incapable of feeding her young. In such a case, they will rapidly lose condition, unless something is done to supplement the small quantity of food they obtain. If possible, the sow should be given warm milk or whey, in which some bran, pollard, or fine oatmeal is mixed. The little pigs should be encouraged to take some of this food. If they are very young they may have, in preference, warm new milk, thickened with a little ground rice or corn meal. A very small quantity will suffice to keep them going until they feed better by themselves; but this they must have if they are to be saved: for, if a young pig is not kept well up to the mark during the first three weeks of its existence, it scarcely ever thrives.

Weak Pigs Bad Breeders.

It occasionally happens that two or three pigs in a litter are unusually weak, and it is almost invariably the case that there is one much smaller and more delicate than the rest. We would scarcely advise that a sow should be retained in milk after her healthy pigs are weaned; but where, for example, there are three or four youngsters that stand in need of further nursing, it is a good plan to let them remain with their dam for another fortnight, when they will derive considerable benefit from the extra nursing that she will afford them.

In addition to the suggestions that have been made in this chapter, we would advise the breeder, if he wishes to maintain a high character for his stock, to dispose of every breeding sow that brings him less than eight pigs at a litter; indeed, we question very much whether this number should be admitted. As previously remarked, we should prefer an average of ten, regarding a less number as unsatisfactory; whereas a higher number demands considerably more nourishment from almost all sows, and can only be appreciated when they are farrowed by those which are thoroughly able to rear them. It is a very common mistake for persons to suppose that large litters are more profitable than such moderate ones as we have suggested. In the majority of cases they are less profitable, for nothing feels the advantage of a slight addition to its normal food more than the pig; and, on the contrary, no animal so quickly loses health and condition when that food at all falls short. This it generally does when the litter is a large one, for few sows secrete a sufficient quantity of milk to satisfy from a dozen to fifteen determined little youngsters continually tugging at the udder.

Useful Hints.

Another point to remember—and this has much to do with the general health of the pig—is to see that a small quantity of cinders or small coal, with a little lime rubbish, is put into one corner of the sty, at least once a week. The pig has a great partiality for anything in the shape of coals, and, what is more, these greatly conduce to the maintenance of a healthy condition. The same may be said of salt, which should be used in the food, and which gives it a relish, although it must be admitted that it is seldom necessary to add anything to pigs' food of any kind. Water should always be provided, and, where pigs are allowed to run loose in the yard, or in a field, an iron vessel of some sort, filled with water, should invariably be within their reach.

Breeders frequently complain that their young sows do not thrive or attain such a respectable size as their boars; but it often happens that these persons, either from ignorance or from want of room, rear both sexes together. This is quite contrary to rule and experience. Unless young pigs are running loose in a farmyard, where all can shift for themselves, and where each has an opportunity of obtaining its proper share of food, they should be separated after weaning: otherwise, the boars, being stronger and generally larger, will obtain a greater share of the food than the gilts, and the latter will consequently lose condition in a serious degree. The pig is such a rapid eater that, where a number are feeding from a single trough, and one or two fail to get their heads into it during the first minute or two, they stand a very bad chance of getting any food at all, for the whole is usually consumed in a very little more than this space of time. The farmyard system has its advantages in this respect, for the feeder, as a rule, scatters upon the ground, or among a quantity of straw, some whole grain, peas, or beans, when every youngster obtains his proper share.

Whenever a young pig dies whilst the litter is with their dam, it should be removed and buried deeply, for the sow will eat almost anything, even to her offspring; and, once having tasted the flesh, she is liable to acquire a relish that may prove dangerous to the lives of the other pigs or to her future young ones.

It is the custom of some pig-breeders to pay the pig-man sixpence a head for all pigs properly reared and weaned. This is certainly some incentive to the man to do his best; and, although the system may be said by some to have its faults, especially by farmers, who consider that a servant should do his best for his weekly pay, yet it is, nevertheless, a wise one; for, in practice, it is found that very few men give the careful attention that stock require unless they have a special interest in them. The cost, even to a large breeder, is almost immaterial, and the experiment is almost sure to have a good result; at all events, it is worth the trial.

CHAPTER VI.

CASTRATION AND SPAYING.

Castration of Young Pigs.

THAT castration is a necessary operation amongst pigs there can be no doubt whatever, for the young boars thrive much better when they have been cut, whereas, if left entire, they are exceedingly troublesome when kept in any number together. The best time to perform the operation is a week or ten days before they leave the sow, in which case they recover from the effects while upon their mother's milk. There is very little difficulty in performing it upon the pig at this age; and, where large numbers are bred, the owner or the stockman should take an early opportunity of learning and practising the process, for it will save a considerable sum of money in the course of a year, even though the local genius is only paid at the rate of 1s. or 1s. 6d. a litter, which is usually the price. In almost every village there is a farrier or blacksmith, or an old labourer, skilled in this work, and any sharp man, once having assisted, could very well perform it himself, if possessed of a sufficient amount of confidence.

An attendant, who should be strong if the pigs are large for their age, holds them by the hind legs, with their backs to himself, and their heads and fore feet between his knees. The scrotum is taken between the thumb and finger by the operator, who, having pressed forward one of the testicles, makes an incision from back to front, and immediately presses it through, partly cutting or tearing it off. The other

is withdrawn in the same way, and the pig put back in the
sty, when, with ordinary care, and if no bungle has been
made, he will do just as well as if nothing had occurred. Pigs
should be fasted for at least twelve hours before castration.

The best time of the year for this operation is in spring,
as during both hot and cold weather there is greater liability
for the part to go wrong.

Sometimes it happens that there is a rupture in the young
pig, and if this is the case, although local advisers may
recommend the immediate use of a needle and thread, we
would advise the owner of such an animal to have it attended
to at once by a skilled person. It also occasionally happens
that the part which has been cut swells. In this case, a little
goose oil, applied with a feather, will be found beneficial.
Accidents and deaths through castration occur very seldom;
indeed, young pigs generally manage to get through it with
great ease.

Castration of the Adult Boar.

Adult boars are often cut by men who are practised in the
work, and as most breeders in these days prefer to use a boar,
they ought certainly to have some knowledge of the operation
of castrating them, for by this means the animals are, by good
feeding, considerably increased in value, growing better, and
producing meat of much finer flavour and quality than they
would have done had they remained entire.

In castrating an old boar, it is first of all necessary to catch
and hold him. This should be done as described in the
chapter on the boar, or by means of one of the devices shown
at the end of this chapter. If his head be thoroughly secured,
little difficulty will be experienced in removing the testicles,
which should be cut and drawn out as in the case of the
young pig, although greater care must be exercised in
removing them entirely. There are two plans usually adopted:
one is to separate the testicles from the ligaments to which
they are held with a red-hot iron; and the other is, first to
tie the ligament with a thin cord, leaving some inches of
the latter remaining, and then to sever the connection by
tearing. The object of leaving a portion of the cord

remaining is that the ligament may be drawn again if necessary; for its muscular nature causes it to spring back into the body, so that it cannot again be otherwise reached: thus, if, through an imperfect operation, hæmorrhage takes place, the divided part can be drawn out and the bleeding stopped by searing or tightening the ligature.

Spaying.

The operation of spaying is performed upon the female with the same object that is held in view in the castration of the male animal—the greater rapidity and perfection of fattening. It is questionable whether the benefits derived from the process are at all equal to the risks run. Young pigs are usually spayed at ten weeks old, care being taken that they have not been fed for at least twelve hours before the operation takes place, and that they are very sparingly fed for some days afterwards. Sows that are to be spayed must not be in season at the time, but with them the operation is much more severe than with young pigs.

Spaying consists in the removal of the ovaries, and, in the case of the young pig, the womb (which is very slightly developed) is also taken away through a cut made in the left flank. Sometimes it happens that a small portion of one of the ovaries is allowed to remain, through imperfect manipulation; and, in this event, notwithstanding the fact that the entire womb has been removed, the animal will frequently come into season when it has arrived at a mature age, although, of course, it will be unfruitful. When the operation is over, the wound is sewn up, and the pig kept very quiet, and fed sparingly until it has thoroughly healed.

Some persons prefer to have young gilts served instead of cut when they intend to convert them into porkers; this gives an impetus to the feeding propensity of the animals, and to the laying on of flesh, and if they are killed in five or six weeks, no ill effect will result, for the immature state of the young will have caused little or no drain upon the system

We do not advise the operation of spaying, which, in our judgment, is not only disagreeable, but, to some extent, revolting. However, if it is to be performed, it should be undertaken by a thoroughly skilled person, in order that the animals may suffer as little pain as possible.

M. Gourdon, the author of "Traité de la Castration des Animaux Domestiques," has dealt with this subject very distinctly. He says that, to perform the operation properly,

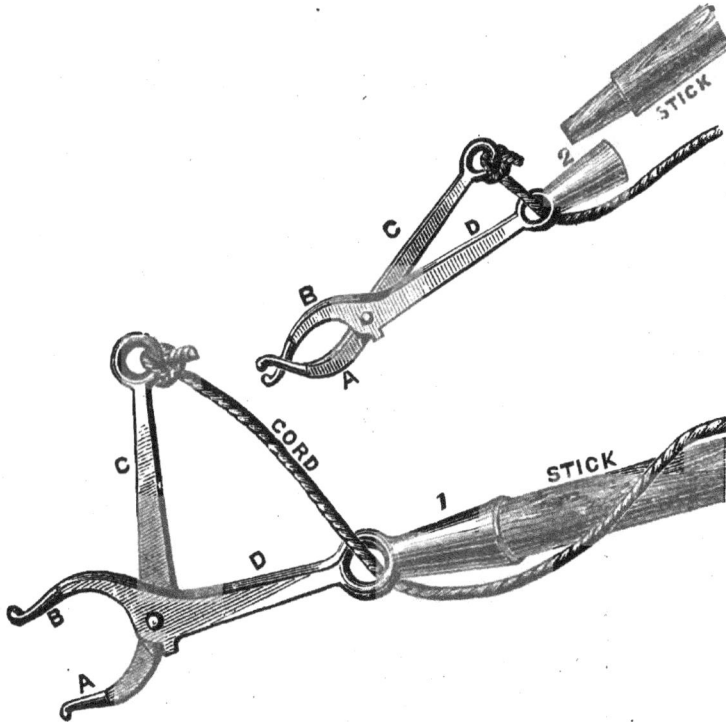

Fig. II.—American Pig-catcher.

A, B, Jaws for gripping the leg; C, Movable Lever, with cord attached, which passes through the eye in D.

two persons should assist, the one holding the head and forequarters, and the other the hind-quarters. When the incision has been made, the index finger should be introduced into the abdomen to search for the left ovary, which will be found in the lumbar region, and will be recognised by its size. In the adult sow this is equal to that of a small

walnut, and can be recognised by the granulous inequalities
of its surface. Having seized it with the finger, it is drawn
outside the incision, and detached by tearing it from the
ligaments to which it adheres. In the case of young sows,
the two portions of the womb are then removed; but this is
not so with the adult, for, as this organ is largely developed,
such an operation would be extremely violent and dangerous.
The finger is again introduced, and, reaching the womb, finds
that the left portion, which had been moved in detaching
the left ovary, is back in its place. Pursuing its course with
great care and gentleness, the finger next reaches the other
corner of the womb, when the right ovary is found and removed in a similar manner, unless—as sometimes happens
—the ovary is found in the centre of numerous circumvolu-
tions of the womb, in which case it is previously disengaged.

Pig-catchers.

The illustration, Fig. 11, shows a pig-catcher which is both
ingenious and useful, and which is made by Messrs. Chambers,

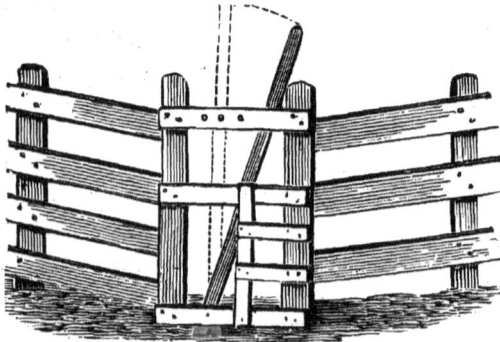

Fig. 12.—Pig-catching Gate. This can be used either in a straight fence or
in a cul-de-sac. The Pigs are held by the movable lever.

Bering, and Quinlan, of Dectur, Illinois, U.S.A., manufacturers
of the Champion Pig Ring. The implement is fastened at the
end of a stick, and the pig caught by one of its hind legs,

when the string is immediately tightened to, hold it fast, and the stick simultaneously withdrawn. The larger figure (1) shows the instrument attached to the stick, and the smaller figure (2) the method of fixing the stick within the socket of the catcher. A and B are the jaws which grip the leg of the pig; C is the movable lever to which the cord is attached, and which holds the leg tight against B; D is provided with an eye, through which the cord passes, enabling the holder to keep a perfect mastery over the animal.

The catching gate shown at Fig. 12 is simple but ingenious. It can be used either in a straight fence or a *cul-de-sac* made for the purpose. When open, as in the engraving, the pig, being driven, attempts to pass through, finding no other outlet, and the attendant at once pulls back the bar, which, falling behind his jaw, holds him secure.

CHAPTER VII.

RINGING PIGS.

THE necessity for ringing pigs arises at some time in every yard. Some people think the custom a cruel one, but these are generally non-practical men, and little difficulty would be found in convincing them if they owned half-a-dozen sows which they desired to turn into a meadow during the months of spring. At some seasons of the year, pigs may be turned out into the fields by the dozen, and will not attempt to dig. At others, and more especially in spring, the instinct is so strong that one, and even two rings, are sometimes useless in preventing them from indulging in the habit, and searching for roots which lie two and three inches below the surface of the soil.

In our opinion, all pigs should be rung from the time they are weaned, and at all seasons of the year, although some breeders maintain that it is advisable to remove the rings in autumn and to insert them again in the spring. There is certainly a difference in the amount of energy, as applied to digging, that various pigs possess; but, rather than risk the damage that one determined sow can accomplish in a single day, we should prefer to keep them all under control by one of the several kinds of rings in the market.

There are other reasons why ringing is necessary. The operation tames the animal, more particularly the boar, and if it is found that one ring has little effect, two, three, and even four may be inserted in his snout, with the certainty of a good result. Pigs that are determined fence-breakers and gate-lifters are all checked by the operation—and we have seen a boar lift an exceedingly heavy yard gate with the greatest ease.

Some persons prefer to slit the ligaments and cartilage of the nose in the centre when the pig is very young, and, as these do not unite again very easily, the animal is left powerless so far as his rooting and similar undesirable propensities are concerned. Another plan is to thrust the point of a penknife through the rim of the snout, and to draw it half an inch to the right and left, so that, as the pigs roots, the piece of loose flesh will give sufficient pain to cause him to desist. These practices, however, are disagreeable ones, because we believe them to be unnecessarily cruel, while they certainly disfigure the pig.

Rings in Use.

Let us now see what kind of ring is generally used. In every country village, where the blacksmith is entrusted with the operation of ringing the litters of the neighbouring farmer, an ordinary horse-shoe nail is employed. A hole is bored through

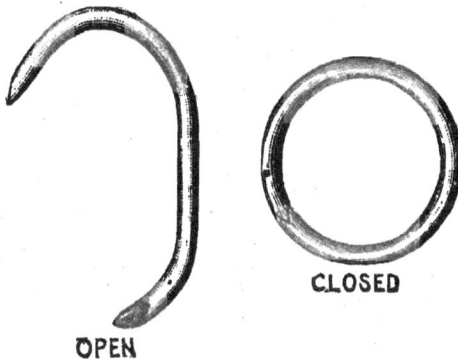

CLOSED

OPEN

Fig. 13.—Brown's American Pig Ring, shown both open and closed.

the snout with a brad-awl, the nail inserted, and curled round with a pair of pliers. The operation is simple and effective, but is neither so rapid nor so humane as when Brown's

American ring is used. This is a round copper ring (as shown
in Fig. 13, both open and closed). The open ring is inserted

Fig. 14.—Method of Inserting Brown's Ring by means of the special
Pincers-like Implement.

within the jaws of an instrument resembling a pair of nippers
or pincers, and the pointed ends are forced through the snout
by the action of closing the instrument, the work being done
in an instant; indeed, if the pigs are caught and held, it is
as easy to ring a dozen by this plan as a single pig by the
common system. Fig. 14 shows the method.

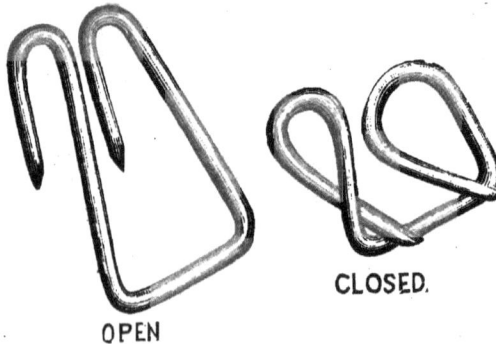

CLOSED.

OPEN

Fig. 15.—The Champion Double Ring.

There are two kinds of round rings, in one of which the
joint remains within the snout, and which is not well

finished; in the other, the joint is outside, the two points
finishing off the ring in a close and satisfactory manner: the
latter only should be used. Another pig ring, however, called
the Champion, is the one that we prefer to any other, as it

Fig. 16.—Champion Pig-holder, by means of which the strongest Boar can
be held for ringing.

is still more efficacious than the round ring referred to. It
is shown at Fig. 15 (both open and closed), the two points
being thrust simultaneously through the snout by means of
a similar instrument to that previously mentioned, when
they are brought to the front and lapped under the front
portion of the ring.

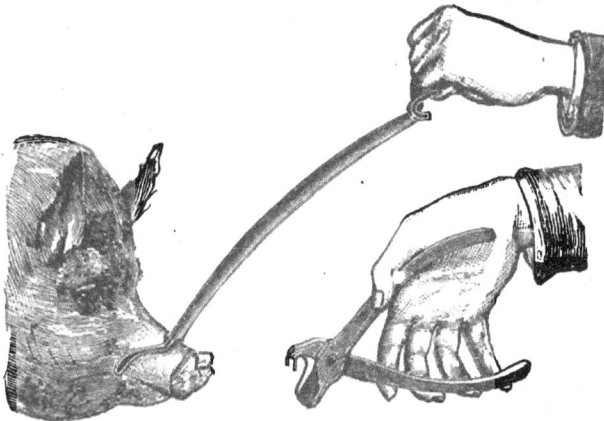

Fig. 17.—Method of Using the Champion Pig-holder and Inserting the Ring.

Fig. 16 shows an instrument used in America, by means
of which the strongest boar can be held when it is placed
behind his tusks. The manner of using it is shown in Fig. 17,
where the Champion ring is about to be inserted.

The best time to ring young pigs is after weaning and when they have thoroughly recovered from castration. The ring should not be forced too far back or too near the nasal bone, nor should it come too close to the front of the snout. If badly rung, the pigs will either get rid of the ring altogether or tear the snout.

It is a good plan to examine the whole of the pigs at regular intervals, and to re-ring those that have lost one or more of their rings, which will generally be found to be the case after every few months. Breeding sows should be noticed before they are sent to the boar, and, if necessary, rung then, as the practice is dangerous when they are pregnant, for it may result in abortion.

CHAPTER VIII.

THE COTTAGER'S PIG.

In spite of the objections made by the farming community in many parts of the country—although, we are glad to believe, not in all—we are of the opinion that every encouragement should be given to the labourer to keep a pig. Not only is the animal a useful assistant in the economy of his household, providing him with valuable food during that part of the year when he is so much in need of it, but it also assists him, as a labourer's wife once remarked to us, "in keeping the money together"—a matter all-important to the cottager, whose opportunities of collecting together a few pounds are small in the extreme.

There are differences of opinion as to how the cottager should go about pig-keeping. Some authorities consider that his most profitable course is to buy a well-bred young pig about twenty weeks old when he takes his harvest money, to fatten it as rapidly as he possibly can, and to kill it at Christmas, by which time the pork will be most acceptable, especially in severe weather, when labourers are so frequently out of work. There can be no doubt that if fifty shillings or sixty shillings are well laid out in the purchase of a pig, with the extra money earned at harvest time, and the gleanings, beech-nuts, and acorns collected by the family, there is every opportunity for the cottager to fatten the animal well at a small cost. It must be remembered, however, that the experiment will bear but one result, and, if that fails, the poor man had better not have seen the pig at all. In other words, to

H

make the matter worth his while, his pig must be thoroughly fed from the commencement, gradually giving it as much good food as it will eat and fattening it as completely as possible.

Cheap Food.

Now that the system of preserving green stuff in the silo has proved so successful, the labourer has one more advantage, which he should not neglect, in the management of his piggery. There are few persons who would deny him the privilege of cutting the coarse herbage that grows by the side of hedges and ditches on the high road; and this, if he is a wise man, and properly educated into a knowledge of the subject, he will collect as largely as possible, pack away in old casks, or, still better, in a pit in his garden, and preserve for winter use.

When the silo system was first introduced, it was esteemed only as a system for the rich, or for those persons who could get a pit built for them; but that idea has been dispelled. Nothing is simpler than to erect a partition of wood across an outhouse, or to line a pit, dug in a light, chalky, or stony soil, with wood, so as to keep out the wet. This, like the casks, should be as well filled as possible, treading the whole thoroughly, and weighting it with eighty to ninety pounds to the square foot for a pit, say, 12ft. deep; indeed, so simple is this matter, and it costs so little beyond labour, that the working-man is quite able to undertake it for himself upon a scale commensurate with his requirements.

If the labourer has no garden, and is unable to obtain either gleanings, acorns, or any other suitable material, but is compelled to buy the whole of the food that he uses in fattening a pig, then we distinctly say that we believe this plan will not pay him. We have seen it tried in numbers of instances, in the great majority of which no gain had been made, while, in the few in which a skilful man could show a profit, this has scarcely risen above two shillings and sixpence to five shillings. It is quite true that, at the moment at which we write, feeding material of all kinds is abnormally dear; we are therefore inclined to the opinion that the cottager's best plan is to keep pigs that he can feed chiefly upon the produce of his own land, and especially upon the potato.

Rearing a Young Sow.

We regard the persevering labourer who has a garden of good dimensions as a man who, by means of thrifty pig-keeping, is in a position to raise himself above his fellows, and, in the end, to obtain a little farm of his own and to enjoy greater profit and happiness than has hitherto fallen to his lot. Many men have begun with a pig and have reached positions of eminence and wealth; but, if a labourer can reach the climax to which we have referred, he will not have much to complain of in this world.

In normal times it is quite possible for a labourer to purchase a well-bred gilt pig after weaning at from fifteen to twenty shillings, a sum which it ought not to be difficult for him to save. He should make every effort to induce the best breeder or the owner of the best pigs within reach to sell him a vigorous specimen from a spring litter. With this he should commence, and he would find that, through the summer season, he might be able to keep it almost for nothing, wash, garden refuse, grass, and, in fact, weeds of many kinds, forming its chief food. At the same time, it would be more advantageous if he gave the animal a shillingsworth of corn or corn-meal weekly. The principal part of its winter food should be provided by the garden (chiefly through the potato) and by the system to which we have above referred; and, as fattening will not in any sense be necessary, the pig being merely kept in store and growing condition, it can then be kept in a thoroughly warm, dry sty (and warmth, it must be remembered, is as important as food), for a very trifling sum, until the following March, when, if well grown, it might produce its first litter.

The labourer should have been careful to watch for a season of heat in November, and, if necessary, to force it so far as possible, should he fear losing service during that month. He should take it to a boar of the very best breed in the district, even though it cost him an extra shilling and a

journey of an extra mile or two, for upon the cross depend
the character of the young and the future profit.   In the
North of England it is not at all uncommon for labourers
to keep pedigree pigs, which they sell at unusually high prices.
We were once asked ten guineas for a young sow of wonderful
quality from a person in a very humble position.

What to do with the Litter.

When the young pigs arrive, the expenditure of a few
shillings per week will be necessary for food; at the same
time, the whole summer will be before them, and careful cultivation in the garden, with the herbage taken from the road-
side, should do very much to minimise the miller's bill. If they
have thriven unusually well, the pigs may, perhaps, realise
eighteen to twenty-two shillings each at ten or eleven weeks
old, although possibly this is too high a figure for a gilt's
first litter, which are generally smaller and less precocious
in growth.

The sale of the litter and the gilt's second visit to the boar
should take place as close together as possible, that her
second litter may come in in the autumn and have grown
well before winter has approached.   During the summer,
she ought not to cost more than a shilling a week for food;
whereas for winter the same system should be adopted as in
the previous year, the labourer getting, if possible, still more
food together for her and her young family.   The money
that he obtained for his first litter should be carefully laid out
in food for the second, so that the pigs should have every
possible chance to grow and to sell at better prices.   In this
case, it may be found advisable to keep the picked gilt of the
litter for growth into a sow, according to the system that
was adopted with the mother.   Thus the labourer may be
continually, if slowly, adding to his stock—feeling his way,
as it were—taking care not to keep more than he can feed
properly, and so progressing until such time as, by means of
his piggery, he has arrived at the still more desirable position
of owner of a cow.

There are times, both in spring and autumn, when little
pigs are remarkably cheap and scarcely pay for the purchase

of food. In such a case, the labourer would do well to sell the whole of his young pigs at three weeks old for roasting. In the past there have been many arguments against this system; but in particular seasons it may be depended upon in many cases as a most advisable one to adopt. If ten shillings each can be obtained at this period, the labouring man will certainly be in pocket; for he will escape the many risks of the next few weeks—often the loss of a pig by death —he will save a pound or two in food, and, at the worst, the young pigs will yield him very little less than they might do if he sent them to market, considering the then state of trade.

Moreover, a cottager should never attempt to fatten a pig unless he knows something about its breed, for it might cost him much more than it would be worth when killed. He should never purchase a badly-bred pig, however cheap he can obtain it; and he should never attempt to breed young ones unless he is quite satisfied that he can feed them well. A starved sow entirely ruins her young; and if, at three weeks old, they have lost condition, they will never regain it, but become a certain loss to him.

We have referred to weeds as a food for pigs, although they have, from time immemorial, been considered the bane of the garden as well as of the farm. If a labourer is compelled to use weeds in summer time, he should take the trouble to note which the sow prefers and which she rejects, if she rejects any. In this way, he may save himself much trouble, especially if he collect them by the road or hedge-side. Indeed, nothing should be too much trouble in the management of stock, which can only be made to succeed by the greatest care and attention.

CHAPTER IX.

SOME DEBATABLE POINTS.

EVERY person who has had extended experience in the management of pigs knows that there are several points—to all of which we shall refer in due course—upon which there are differences of opinion. We therefore addressed a number of questions to some of the most experienced breeders, with a view to giving the benefit of their practical experience to the readers of this volume; and we cannot thank these gentlemen too cordially for their replies. We value them the more highly, as the opinions are those of trustworthy men, and, in the present day, the importance of such a point cannot be too greatly estimated. The amateur and the beginner are both too liable to be carried away by appearances, and by the advice of men who are neither legitimate breeders nor feeders, but who can, perhaps, point to a certain number of honours which their pigs may have obtained at some of the principal exhibitions. Just as there are many first-rate breeders who are comparatively unknown in the show ring, so are there exhibitors who, although frequently beaten, and oftentimes by purchased animals, are yet breeders of the very highest class.

We shall divide the questions put to, and replies received from, the gentlemen named below into two series; the first being more particularly applicable to the chapters with which we have already dealt, while the second will be referred to when we arrive at the chapter upon feeding. For the sake of convenience to the reader, we give the questions and answers in the following order:

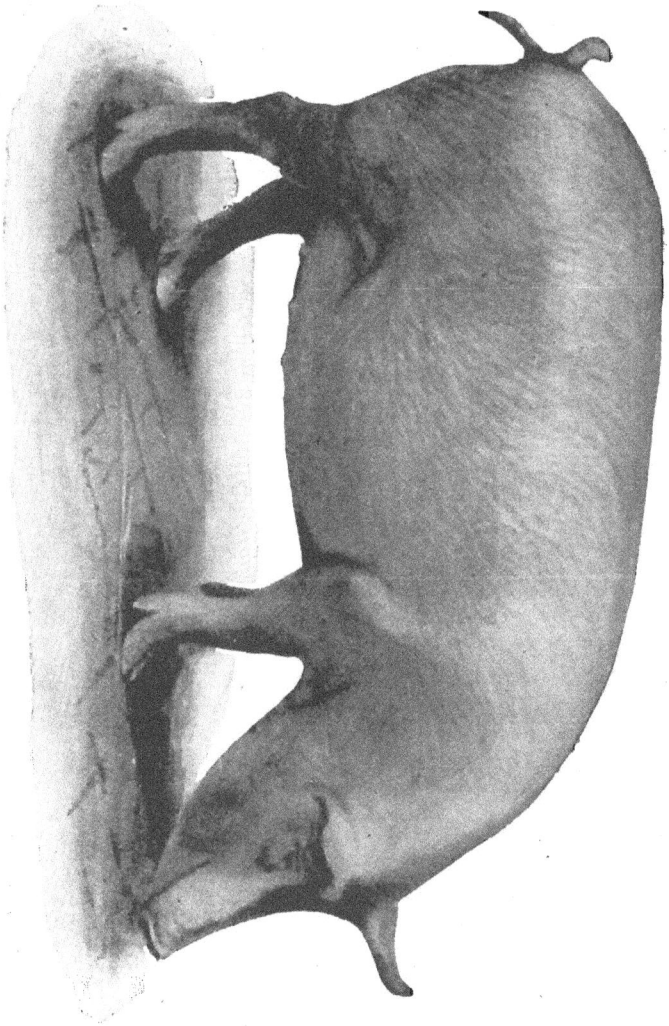

TAMWORTH GILT.

Mr. R. Ibbotson's "Knowle Favourite," winner of First Prize and Silver Plate, Lancashire; First and Gold Medal, Warwickshire; Second, Royal Agricultural Society's Show; all in 1905.

The Tamworth Pig is an ancient and pure race, quite distinct from any other English variety. It should have an abundant coat of long, straight golden-red hair on a flesh-coloured skin, and should be quite free from black hair, very light or ginger hair, or black spots on the skin. This breed will stand more forcing than any other.

The Age for Breeding Gilts.

(1) What do you consider to be the proper age at which a gilt should be used for breeding?

The late Mr. James Howard, M.P., said: "From nine to twelve months, according to the growth. If bred from before they are well grown, they do not attain their natural size."

Colonel Platt says: "Twelve months."

Mr. James Robertson, Steward to the Earl of Radnor, thinks that from seven to nine months is the proper time.

Colonel Walker-Jones says: "From eight to nine months."

Mr. Nathaniel Benjafield, Motcombe, Shaftesbury, says: "I put my gilts to the boar at from six to eight months old, as a rule."

Mr. Frederick Coate, Newtown House, Sturminster Newton, Dorset, says: "Six months is the age at which I first put my young sows with the boar, and I have always found the young pigs come strong and healthy. Some people commence at a still younger age; but I do not approve of this. I think six to nine months is the best age."

Mr. Thomas Bennett, of Rossville, Vermilion County, Illinois, says: "I prefer not to breed from a gilt before she is eight months old, so that the pigs come at one year old. My experience is, however, that if gilts are not bred from before they are eighteen months old, they make finer sows."

The somewhat variable nature of the replies given may, perhaps, be explained when we state that Mr. James Howard chiefly bred the Large White Pigs; Colonel Platt the white varieties; the Earl of Radnor, the Small White variety; Colonel Walker-Jones, the Middle White; Mr. Benjafield, the Berkshire; Mr. Coate, the Black Dorset; and Mr. Bennett, Tamworths and Duroc-Jerseys.

The Best Floor for a Sty.

(2) Do you prefer brick or concrete floors and straw beds, or wooden benches?

Mr. James Howard preferred concrete floors *with* wooden benches and plenty of straw; but said that if the floor was made with hard bricks or clinkers, they should be laid in and grouted with cement.

Colonel Platt prefers brick floors and plenty of clean straw.

Mr. Robertson says: "I prefer brick floors, and good straw for litter."

Colonel Walker-Jones says: "I prefer brick floors, and wooden frames for pigs to lie down upon. The litter should be raised about three inches from the floor, in order to allow the water to run through, and they will be found warm, and to save straw. I also use pine-wood sawdust, which is a disinfectant, keeping the sty very sweet. In 1883 I had forty-eight cows affected with foot-and-mouth disease. The pig-sties were all adjacent to the cow-sheds in which the diseased cattle were stalled. I gave up straw, and used pine-wood sawdust freely upon the benches, and among the pigs themselves, and had not a single case amongst them. I do not like concrete: it is dangerously slippery in very cold weather."

Mr. Benjafield says: "I like brick floors, with a wooden bench for the pigs to sleep upon, and plenty of bedding of some kind."

Mr. Coate says: "Pigs should always be kept clean, moderately warm, and dry. I like brick floors, with a nice heap of straw in the corner, and I almost always find the pigs buried amongst it. In summer time, straw is not so necessary, as the pigs can lie anywhere, although, in spite of their dirty habits, they prefer spending the night in a dry place, and this keeps them thoroughly healthy."

Mr. Bennett says: "I prefer concrete floors, with plenty of straw."

Mr. Mechi was a great advocate for sparred floors, which many breeders have used with great advantage, as the bedding is never wet, nor is the wood so cold as brick or concrete. We remember seeing a large number of pigs upon sparred floors at a large dairy farm near Caen, in France. The whole of the manure went through, and was carried away to a tank.

Vicious Sows.

(3) What, in your opinion, is the cause of, and remedy for, sows lying upon and eating their young?

Mr. James Howard said: "Pigs may be overlaid when too much straw is used as bedding, or when no protection is given in the shape of rails round the sides of the sty. If a sty is level, or nearly so, the sow will lie down as readily in one part as in another, often crushing her young in doing so. I would therefore make a sty slope from the outside to the centre, where the drain might be placed, and where the sow would naturally make her bed, as being the most comfortable position in the sty. Less litter would be used in such a case, as the urine would immediately pass away; whereas the young ones could escape to the sides of the sty. I use iron tubes, fixed round the sides of the sty, to the height of ten inches, which prevent all danger from this source.

"The cause of sows eating their young is often attributed to the fact of little pigs having teeth, and biting the teats. All pigs when first born have teeth, which are very sharp, and should be broken off, especially in the case of large litters. When there are only sufficient teats for the number of pigs in a litter, a scramble for a position at feeding-time often results in one or other of them biting a teat, which causes the sow to snap at the offender, and sometimes to hurt it severely. When blood is once drawn, the sow will often eat the youngster; and when sows and their pigs are continually shut up, they get into an unhealthy condition for want of earth and other mineral substances: the little ones dwindle away, and the smell of them becomes more and more offensive to the sow, until she ultimately finishes them off. It will, therefore, be seen that exercise and liberty on a grass field will go far in preventing sows from contracting this disgusting habit, and, at the same time, tend to keep them in good health."

Colonel Platt says: "Scarcity of room is often a cause of pigs lying on their young, when a heavy pig is apt to do more damage than a lighter one. Sometimes, a rail is fixed round the sty, nine inches from the floor and sides, thus giving the young pigs room to escape." The latter part of the question he cannot account for.

Mr. Robertson believes that death is sometimes caused by the entanglement of lengthy litter, when the dam is in an inactive condition, and suggests as a remedy the use of chopped

straw, or "cavings" (the material between straw and chaff
which appears after threshing), or of a rail fixed round the
inside of the sty, nine inches from the wall, and six inches
from the ground. A pig that eats her young he would send
to the butcher as soon as possible.

Mr. Spencer believes that if sows are kept in a healthy
store condition, and are not exhausted by breeding, they will
not lie upon their young. Of sows eating their young, he
has had but little experience, and he considers that the only
cure is the butcher's knife.

Colonel Walker-Jones considers that want of care is the
cause of young pigs being crushed by their dams. He has a
rail, similar to that referred to above, fixed in his sties, to
allow the young pigs to pass between the wall and their
mother, who naturally prefers to lean against something when
lying down : if without a barrier of the kind, she would
certainly crush them. Sows get careless with age, if of great
weight, and do not rise so soon, even when their young
scream on being crushed. Colonel Walker-Jones adds :
Attention, when farrowing, is the best preventive for sows
eating their young. I sometimes take out the sharp needle-
like teeth of the litter, as the bite of the youngsters is at
times apt to make a sow snap at them, when, if she catches
the delinquent and draws blood, she finishes by eating him.
The butcher's knife is the only remedy if a sow becomes
addicted to this habit.

Mr. Benjafield says : "All sties in which sows farrow should
have a rail fixed about six inches from the wall and eight
inches from the ground, so that the little pigs may have a
chance of escape if the sow is clumsy. I believe that, when
sows eat their young, they are, as a rule, in ill-health."

Mr. Coate says : "The reason why so many pigs are laid
upon by the sow is, that they have too much litter or too
small a house, when the sow is large and fat. I rarely lose
any in this way. My house in which the sows farrow is about
twelve feet by eight feet. I use 'cavings' for litter for several
days, and, as a double security, have a small rail fixed all
round the house, a foot high, and a foot from the wall; and
when the sow lies down, the little ones run under the barrier,
and are generally safe. I have never had a pig eat her young,

and I think it would be a rare occurrence, if the sow—for it generally happens with the first litter—were kept very quiet and tended only by the man she had been accustomed to. I fancy excitability is the cause, but should, nevertheless, fear to breed a second litter from a sow which had once practised this bad habit."

Mr. Bennett says: "In nine cases out of ten, the cause of a sow lying upon her pigs is that she is too fat, and consequently lazy, and not in a proper condition for breeding. If a sow is in proper condition, is allowed to go to a straw stack and to make her own bed, she will raise every pig she has, providing there is nothing to interrupt her. If she is supplied with food daily, and not disturbed, the pigs will be found running about the yard in two or three weeks, although they have never been seen or the sow moved. I know this to be a fact, as I have experienced it myself.

"With regard to sows eating their pigs, I would remark that, if a sow is out of condition, her system calls for something she does not and cannot obtain. It must be remembered that the pig is an omnivorous animal, and that its system requires something of an animal nature; and if a sow is not able to obtain this, she will help herself upon the first opportunity, even though it be upon her own pigs. I think I have reasons for this statement. I have observed that, as a general rule, those farmers who complain of sows eating their pigs are 'kid glove' farmers, who are anxious to keep their pigs so nice and clean that they may be better than those of other people; in other words, they attempt to change the nature of the pig, and they reap the consequences. I raise from 500 to 700 hogs every year, and I have never yet had a sow eat her pigs. I believe it is a bad plan to ring sows, for the practice robs them of Nature's remedy. Their noses were made to root, and when they are deprived of this capacity, they lose something which the breeder is unable to replace. I have noticed for many years that sows which are placed in my barn for farrowing during winter, say March and April, invariably eat the placenta, or after-birth, and the dead pigs, if there be any; but this practice I do not object to. Again, I have noticed that sows which farrow in May or June, when grass is plentiful, and they are able to roam about at liberty, and when they make their beds where they please, seldom or never eat either the placenta or a dead ·pig. To me the difference is clear : in the one case, the ground is frozen, and the pigs are unable to procure anything beyond what is given to them ; while, in the other, they obtain grass, and are able to dig and root, and thus supply the wants of nature."

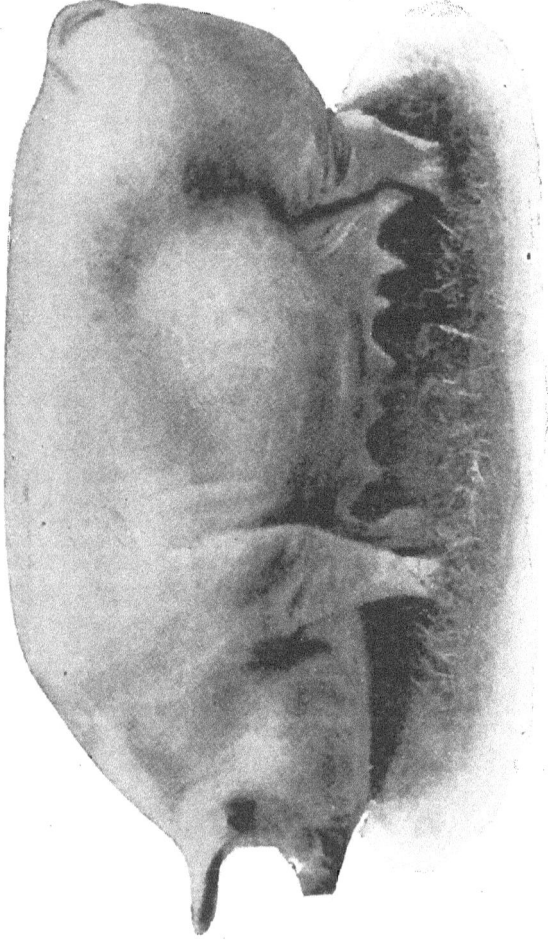

LARGE WHITE SOW

Sir Gilbert Greenall's "Lindsey A" (Herd Book No. 14,186). Winner of Second Prize, Bath and West; First and Medal, Royal Counties; First, Highland; First, Lincolnshire; First, Gold Medal, and Championship, Royal Lancashire; First, Great Yorkshire; all in 1905.

The Large White Variety is not only highly popular both as an exhibition and as a utility pig in this country, but is largely used by Continental breeders to improve their strains. It should have a long and moderately fine coat without any black hairs and a skin as free from blue spots as possible.

CHAPTER X.

THE WHITE BREEDS.

The Large White Breed.

THERE is no doubt that the Large White breed has contributed more to the popularity of English pigs than any other, for it is largely sought by Continental as well as by home breeders, for the improvement of local varieties. It has, moreover, if we consider its quality as well as its size, been the most useful of any race, and its wonderful improvement has largely contributed to the present high position that the pig holds as a domestic animal. It is true that large pigs have existed for generations, more especially in the North of England; but they have been chiefly remarkable for their large quantity of bone, their entire want of symmetry, their narrow backs, and their flat sides. Crossed with the Chinese, they were improved in their fattening qualities, as well as in pre-cocity of growth, but to a great extent their size was reduced. The quality of the hair was, however, improved, and the white colour retained, while they were, generally, both vigorous and prolific, and reared their young with marked success. Sometimes, they were crossed with the Neapolitan race, but the result was never so successful as with the Chinese, although it increased their prolificacy; for, in addition to the loss of a self-colour, the already small quantity of hair was reduced, their constitution was impaired, and their young were more difficult to rear.

It has been suggested by a competent authority that the

Berkshire pig was produced from a cross between the
original Large White and the imported Black Chinese; and it
is strange that, up to the present day, a cross between a
white race and the Berkshire invariably results in the production of a white pig. The fact that crossing a pure-bred
white pig with a pure-bred Black Berkshire brings a white
progeny was first pointed out to us by Lord Moreton, and
we have often observed that this is the case. All experienced breeders are able to substantiate the fact that with
almost all pure black breeds the result is the same; but, like
most rules, it has an exception, and, for the first time,
perhaps, since pure races became fixed, a pen of pigs exhibited
at Smithfield, in 1884, by Mr. Tom Coate, of Sturminster
Newton, Dorset, were entirely black, although bred between an
improved Dorset boar and a white sow. They were marvels of
perfection, and obtained the champion cup; but even in this
case there is some explanation in the fact that the Dorset
type is not so distinct as the Berkshire, and does not, therefore, so completely affect the well-known principle to which
we have referred. When crossed with the Berkshire, the old
white breed produced a slow-feeding pig, which almost attained
its second year before being fit for bacon. It was, moreover,
coarse, but extremely hardy.

How THE LARGE WHITE BREED WAS MANUFACTURED.—It was in
the year 1851, when the exhibition of the Royal Agricultural
Society was held at Windsor, that the improved Large White
pig first became famous. It was exhibited by one Joseph
Tuley, a weaver, of Keighley, in Yorkshire, and at once
arrested the attention of everyone interested in the live stock
of the farm. This individual had, by considerable skill and
judgment, produced the most wonderful pigs ever seen in this
or perhaps any country; and his famous strain proved to be
the foundation of the entire race of our modern Large White
breed, which was for many years called the "Yorkshire
breed."

Joseph Tuley sold his pigs at high prices, and the late Mr.
John Fisher stated that Tuley built his Gothic cottage, which
he named "Matchless House," after his famous sow, "Matchless," out of the proceeds of the sale of one of her litters.
To this sow, and to his well-known boar, "Sampson," some

of our large breeders can still trace the origin of their strains, although they owe the fact principally to the enterprise of Mr. Wainman, of Carhead, Yorkshire, who was for some years the most celebrated breeder and exhibitor in the kingdom; and to his manager, Mr. John Fisher, whose skill was never sufficiently recognised out of his immediate district. Mr. Wainman purchased largely of the Tuley blood, and from his herd pigs were sent to almost all parts of the world, their merits being recognised, not only by the farming community, but by royal and imperial owners and by the representatives of foreign governments.

Mr. Fisher has remarked that, in the terrible year of 1860, a struggling tenant of Mr. Wainman's was enabled to clear off the whole of his rent by the production of pigs from a young breeding sow which his landlord had lent him. She was sent to the best Carhead boar, and Mr. Wainman purchased the produce, for which there was then a great demand. In all, the returns from this one sow, in young pigs alone, amounted to nearly £1000; one of her litters of thirteen breeding stores actually realised £116 10s. at the age of twelve weeks, being delivered at the Royal Show yard at Salisbury. Another sow of the same family, after having reared 153 pigs in thirteen litters, weighed 1203lb. (live weight) when despatched to the same exhibiton; and a daughter of this animal reared thirty-three pigs in three litters before she had reached the age of twenty-two months, in addition to which she had won nine prizes.

It was found in Mr. Wainman's herd that, with judicious feeding, choice pigs of the large breed would reach 35st. at twelve months old; but perhaps their greatest merit was in their capacity to make baconers of considerable weight at the end of the year in which they were born, March litters reaching, as a rule, 20st. to 24st. by Christmas, if fed for the purpose. It was therefore claimed that the necessity of keeping a pig into the second year to make bacon was entirely avoided.

STANDARD OF EXCELLENCE.—The following is the standard of excellence of the Large White breed formulated by the National Pig Breeders' Association and the British Berkshire Society:

Colour white, free from black hairs, and as far as possible
from blue spots on the skin; head moderately long, face
slightly dished, snout broad, not too much turned up, jowl not
too heavy, wide between ears; ears long, thin, slightly inclined forward, and fringed with fine hair; neck long, and
proportionately full to the shoulders; chest wide and deep;
shoulders level across the top, not too wide, free from coarse-.
ness; legs straight and well set, level with the outside of the
body, with flat bone; pasterns short and springy; feet strong,
even, and wide; back long, level, and wide from neck to
rump; loin broad; tail set high, stout and long, but not
coarse, with tassel of fine hair; sides deep; ribs well sprung;
belly full, but not flabby, with straight under line; flank
thick, and well let down; quarters long and wide; hams broad,
full, and deep to hocks; coat long and moderately fine; action
firm and free; skin not too thick, quite free from wrinkles.
Large bred pigs do not develop their points until some months
old, the pig at five months often proving at a year or fifteen
months a much better animal than could be anticipated at
the earlier age and *vice versâ;* but size and quality are most
important. Objections: black hairs, black spots, a curly coat,
a coarse mane, short snout, inbent knees, hollowness at back
of shoulders.

MR. JAMES HOWARD'S OPINION.—We are indebted to the
late Mr. James Howard, M.P., for the following remarks upon
the Large White breed:

"I have pleasure in acceding to your request that I should
give my views upon the points which male and female animals
of this popular breed should possess. I would preface my
remarks by observing that my herd was founded in 1863 by
the purchase of two sows and a boar from the most celebrated
breeder and exhibitor in the kingdom at that period, Mr.
W. B. Wainman, Carhead, Yorkshire, whose animals have
never been surpassed. From Mr. Wainman's shrewd and
successful manager, Mr. Fisher, I received my early lessons in
the selection of animals and the art of pig-breeding. It was
not, therefore, without a feeling akin to regret that, within
three years after my commencement, against my early tutor,
I carried off the first prize for the best breeding sow at the
Royal Agricultural Show.

"Thirty to forty years ago, many of the Large White pigs exhibited in our show-yards were coarse mammoth specimens of the genus *Sus*, with large, wide, lop ears, and often weighing upwards of half a ton. It was evident that they possessed hardy, strong constitutions; and, as they were chiefly found in Yorkshire and adjacent counties, there can be little doubt that they formed the foundation of the choice herds of Mr. Wainman and other Yorkshire breeders, boars of a smaller race and the highest quality having been introduced from time to time. I have occasionally had a pig 'throw back' to this mammoth variety, and which, judiciously used, I may say, has proved most useful in maintaining the size and constitution of my herd.

"The Large White pigs, if well descended, are good breeders and mothers; the young ones grow fast, and attain great weight at an early age. I have sold many at ten months old which have weighed over fifteen score, and I have realised ten guineas each for fat pigs under nine months, and fourteen guineas under fourteen months old. In addition to their being fast-growing pigs, they have thick bellies, and a good proportion of lean meat—a point of great importance to the bacon-curer.

"Some years ago, with a view to improving the Irish race of pigs, the great bacon firm of Messrs. Denny, Waterford and London, bought specimen pigs of the various English breeds, and from different herds; the result was that, upon being slaughtered, the Large White breed were found to possess the thickest bellies and most streaky flesh. Messrs. Denny subsequently introduced into Ireland a considerable number of Large Whites, not a few of which were selected from my herd.

"In the selection of a boar for breeding purposes, as the external form and locomotive organs are unquestionably chiefly contributed by the male parent, I attach the first importance to these features. The fore legs should be straight and well outside, giving width to the chest; the hoofs short and straight, the fetlock joints strong. The hind legs should be strong, and not cow-hocked; the hams wide and deep. I always prefer a boar with testicles set beneath the thickest part of the hams, and not too large; when they protrude immediately beneath

I

the tail, I should castrate him. Above all, I should never use a boar which does not show quality, both in flesh and hair.

"With respect to the selection of sows, I attach more importance to size than in the boar, and less importance to quality. Of course, it is preferable to have quality on both sides; but my experience has taught me that, if size is to be maintained, the sow must be large and roomy. The head should be lengthy: nothing is more objectionable than the head of a small-bred pig upon a large-bred sow or boar. The head should naturally correspond in length and width with the size of the pig: a short snout on a large pig is an indication of slow growth. The jaws should be of great width, and the forehead broad, giving a good space between the eyes as well as ears. The ears should be fine, and of good length and fair width, also erect, or slightly pointing forward. The 'collar,' or neck, should be wide and well filled up; the skin fine and clean, denoting thinness; and the hair should be abundant, long and silky, which is an indication of quality and constitution as well as of lean flesh. The legs should be straight and short; the shoulders well outside; the chest thick and deep. The body must be of good length, and the back slightly arched, so as to sustain weight without drooping. There should be great width throughout, and the ribs should be well sprung, giving rotundity; the loins wide, flank well filled up; the hams reaching as near down to the hocks as possible. The tail should be long, but not coarse, with a good tassel at the end, and set on nearly in a line with the back.

"A sow should have twelve teats, but if she proved a bad milker, I should fat her off after weaning, inasmuch as milk to the young during the first few weeks is all-important; hence, in selecting young sows for breeding, it is highly desirable that they should be the produce of dams possessing good milch qualities.

"I find it desirable to select brood sows from spring or early summer litters, inasmuch as they are generally better developed than those of winter litters. It will be obvious that the former have not only better weather for growth, but the advantage of field exercise and an abundance of green food, which all tend to promote milch qualities.

"In animals of both sexes, special attention should be paid to the formation of those portions of the body the joints from which are of the highest market value. This is a point not so uniformly observed by the judges in our show-yards as the importance of the subject indicates that it should be."

Mr. Spencer finds that the Small Whites will not stand forcing food when young; that the middle and large breeds are not so liable to cramp or rheumatism, will thrive upon richer food, and, now that they are so much improved in quality, become fit to kill at almost any age. He quotes an instance of one of his large-bred pigs which was recently killed at four months old weighing 6st. (of fourteen pounds), and which showed plenty of lean and great tenderness of flesh, being full of quality as well as pedigree. To obtain the most satisfactory results, he considers that the best plan is to keep the pigs fatting from the moment they are born, and to sell them at about nine months old.

AMERICAN OPINION.—The white pigs generally known as Yorkshires do not appear to be very largely bred in the West Central States of America; and Mr. Coburn, of Kansas, the popular author of "Swine Husbandry," says that he has never met them in the West, nor has he ever conversed with anyone having any positive practical knowledge of them. In his work, however, he quotes a Report presented to the Convention at Indianapolis by Professor Jones and Messrs. Kennedy and Barker, of Indiana. These gentlemen, after referring to the fact that the White Yorkshires are traced in almost every modern breed, state that they believe them to be the only pure and distinct breed of pigs, with the exception of the Black, that are now bred on the American continent.

The Report, after referring to the Chester White, the Thin-Rind, the Berkshire, and the Poland-China, continues: "All these breeds seem to have borrowed some of their good qualities from these original white hogs, and all are made up from crosses of the white and black hog; hence the character of the English, or white, hog crops out occasionally in almost every breed known in this country or in England. Accordingly, we believe it may be said that they are the purest

breed of hogs, and the best, in this country or in England,
from which to make crosses in forming a new or reliable.
breed.

"The Yorkshires are the most valuable swine to breed from,
or to cross with, that we have ever met with in this country,
and for these reasons:

"1. They are of a size, shape, and flesh that are desirable for
the family or the packer's use.

"2. They have a hardy, vigorous constitution, and a good
coat of hair protecting the skin so well, either in extreme cold
or hot weather, that it rarely freezes or blisters.

"3. They are very quiet and good grazers; they feed well,
and fatten quickly at any age.

"4. They are very prolific and good mothers, and the young
never vary in colour, and so little in shape that their form
when matured may be determined in advance by an inspection
of the sire and dam. This we have learned by a practical
experience of many years in breeding, slaughtering, packing,
and consuming."

THE YORKSHIRE BREED.—According to Sidney, whose
work was written over forty years ago, the White
Yorkshire pigs are closely allied with the Cumberland;
while the Manchester, the Suffolk, Middlesex, Coleshill,
and Windsor pigs were all founded upon the Yorkshire-
Cumberland, some of them being merely pure Yorkshires which had been re-christened. That the Yorkshire and
Cumberland pigs were largely intermixed, there can be no
doubt; but we very much question whether the improved race
of to-day goes back beyond the herds of Tuley and Wainman,
to which we have referred. The pigs of other counties, more
especially those of Lancashire and Lincolnshire, have long been
famous for their size; but these, too, have been improved by
the happy Yorkshire cross, and the general type of the Large
White pig has become so universal that, as it were, by unwritten consent, the breeders have agreed to the term "Large
White" instead of the "Yorkshire" breed.

The Large White pig has been wonderfully patronised by the
artisan classes in both Lancashire and Yorkshire, and to this
day these persons breed and exhibit them at the multitude of
shows held in the small towns in both counties; and it is not

an uncommon occurrence for a mill hand to sell his youngsters at five guineas each, and to obtain from ten to twenty pounds for a sow. These people generally insist upon a great number of teats in a Large White sow, for they are not contented with litters of ten to twelve; and, what is more, they breed as early as possible, frequently sending the gilt for service at six months old. As the Large White pigs usually exhibit plenty of growth, there is little fear of loss, as would be the case in small breeds, when prices are low; indeed, however indifferent trade may be, young pigs from highly-bred Large White sows, especially when the sire is a pedigree boar, seldom lose money. Mr. Fisher has quoted cases of pigs which he has killed, one weighing 18½st. at seven months old, and another weighing 34st. at twelve months. Numerous instances could be cited of animals having scaled 80st. to 90st.; and we have seen sows shown by the Earl of Ellesmere, which, the late famous Worsley pigman, "Jack," has assured us, exceeded 90st. in weight. Sidney quoted a Large White prize sow that weighed 11½cwt., and measured, in length, 7ft. 2in. from the twist to the end of the nose; while the girth behind the shoulders was 7ft. 8in.

It may be mentioned that, even at the principal exhibitions of the present day, the Large White pig is occasionally seen with black or blue patches upon the skin, although the hairs upon these somewhat objectionable places are invariably white. The breeders do not consider that the spots show any impurity of blood, for they have been a characteristic of the race from the time of Tuley downwards, and undoubtedly point to a black cross, which must have been used in their manufacture.

OPINION OF COLONEL CURTIS.—In spite of the undoubted quality of the small races of pigs, with their fine bone and comparative smallness of offal, the large breeds are gradually beating them; and it is only a matter of time for the Large White to take its place as one of the leading breeds of America.

Colonel Curtis, who was a recognised authority upon stock in the United States, once stated that, like Professor Sanborn, he had changed his opinion respecting the comparative value of the small and the large breeds, and that, having bred almost all varieties, he could not consistently recommend the small races, which, he contended, return to the farmer less money than the larger ones. Pigs of the large breeds are generally twice the size of those of the small when born, and this gives them such a start that the small pigs can never overtake them. They begin life with larger stomachs, or, in other words, with a greater capacity for eating, digesting, and assimilating food. They are machines of greater power, and, as a matter of course, are capable of yielding greater results."

Colonel Curtis added: "Now, when a year old, it will take at least two of the small ones to make the weight of one of the larger, and I am not sure but that, as they average, it would take a part of a third. There are at least two lives to be kept going, two sets of legs, and other organs to be grown and supported; and it is manifest to me that this double set of machinery takes more fuel or food than a single one. This is not all the difference: the power of digestion and assimilation in the larger pig is so much greater, that it will grow on food which the smaller one would reject, or of which, at least, it would not eat enough to thrive. A strong appetite, and ability to consume food, go together, and one is the natural sequence of the other. This may be illustrated in the fact that I can winter a Duroc-Jersey, and keep it in good condition, on bright clover hay. This may be the case with other large breeds.

"The difference between a pig of a small breed and one of a larger breed is only in the quality and cost of the two carcases. Both of these points, in my judgment, should be reckoned in favour of the larger breeds, because they would not be so fat, and hence would be more palatable, and could have been produced from cheaper kind of food, and hence at less cost. This, however, is not a fair example for the large breed. The two pigs should be killed at the same age—say, nine months—when it will be found that the one of a large breed will weigh twice as much as, and bring double the price of, the other, having cost no more for care, and probably not more than 25 per cent. more for food."

Years ago, there were other large breeds in addition to the Yorkshire, all of which were recognised in their particular districts. These were the Lancashire, Cheshire, Lincolnshire, Westmoreland, Cumberland, and Norfolk; but, as we have remarked above, they are almost all, as it were, absorbed into the improved Large White, although, unquestionably, many still exist in a few localities in each county. The Lancashire was, perhaps, nearest of all to the Yorkshire in type, and weighed equally as well. The Cheshire pig, however, was smaller, and thicker through the jowl, corresponding more closely to the present type of pig which is largely bred in that county. The Lincolnshire, an enormous animal, was more remarkable for its quantity of bone and coarseness of flesh. It generally had plenty of hair, a long snout, and huge flopping ears. Its chief qualities were its hardihood and prolificacy. The Cumberland pigs were not so large as the Yorkshire nor so symmetrical; their backs were very much arched, their skin was largely spotted with black, their quality coarse, and, in almost every respect, they were identical with the large pigs of Westmoreland, which were equally ugly, provided with huge legs, bony flanks, and very little capacity for laying on flesh. The Norfolk breed, which is almost extinct, was not so large as the above, and was considered by some to be a cross of the Lincoln upon the local variety.

The Middle White Breed.

The *raison d'être* of the valuable Middle White breed has often been discussed; and the committees of some of our largest societies have more than once refused to give it classes in their prize lists, for no other apparent reason than that it was originally produced by crossing between the large and small varieties, and was, therefore, suspected of being of no breed at all. Its good qualities, however, are so great that no agricultural committee could possibly omit it without doing great injustice to the farming community; for there is a wide gap between the small breed and the large, which would only be filled up by resorting to the Berkshire, or to a cross upon mongrel pigs.

At the Keighley Show, in 1852, Joseph Tuley, then at the zenith of his career, exhibited some extraordinary animals in the large breed class, amongst them being his famous sows, "Matchless," "Jenny Lind," "Sontag," and others. It so

happened that the judges could not agree, as some of the
animals were not considered sufficiently large for the class;
and as the merits of these were so extraordinary—entirely
forbidding recourse to disqualification—a committee was sum-
moned, when, upon the judges declaring that, if removed
from the large class, the pigs would not be eligible for the
small, it was decided to provide a third class, and to call it
"the middle breed." In this way, classes for the Middle White
pig were established, and for them, as for the pig itself, we
are indebted to Tuley, the Keighley weaver.

In establishing the middle breed, this pig-breeding genius
took a second cross with a boar of the Small White breed and
sows of the "Matchless" strain of the large breed, with a
result that the progeny were as heavy as the large variety,
although in form, and in lightness of offal and head, they
much resembled the finest specimens of the small breed. Those
points Tuley established at the time, seeing clearly that they
were exactly what a useful—and especially a poor man's—pig
should possess. Indeed, throughout this part of Yorkshire, at
the present day, this is the identical type of animal kept by
the industrial classes, who are so largely engaged in the
production of pigs and poultry.

Fisher, than whom no better judge of a pig ever existed,
says that the Middle York pig should have a coat of bright
white or pale straw-coloured hair, without any admixture
of black, even upon the few small blue spots upon the skin,
which are admissible, as they are regularly found in all herds
of white pigs. The sows are admirable breeders, and very
careful mothers, generally farrowing with their first litter at
a year old, and rearing them uncommonly well. In size they
equal the finest Berkshire, but have deeper sides, and are
considered to be quicker feeders, making greater weights at
the same age.

At one time, the Whites almost invariably carried off the
principal prizes at the fat stock shows; but they have since
been beaten by Berkshires, which fact may create in some
minds a belief that the blacks are better pigs. The fact is,
however—and it is one to be deplored—that at many exhibitions the white pigs have been very poorly shown as
regards numbers, in some cases only two or three entries

having been made in each variety, and these of comparatively inferior quality.

At the Chicago Fat Stock Exhibition in 1884, the prize for the best carcase of pork was taken by a Middle White pig, which dressed 90 per cent. of its live weight, whereas the Berkshires and Poland-Chinas only reached 84 to 86 per cent., all, however, being exceptionally high averages.

Fig. 18.—Front View of the Head of a Typical Middle White Sow.

The form of the head of the Middle White is still a moot point with some persons, and, therefore, we give some illustrations, which were drawn from life for this work by the late Mr. Harrison Weir, and which show the type modern breeders prefer to produce. It will be observed that the snout is neither the short, chubby snout of the small, nor the elongated nose of the large breed; but, as it were, something between the two, which happily coincides with the general characteristics of the Middle White variety. We have

observed, however, upon reference to portraits of some of Mr.
Wainman's prize pigs, that the snout, as bred by him, although
shorter than that of the large breed of to-day, was comparatively long and slightly "dished" (somewhat pug-like).
The short, broad face is very much admired by large numbers
of breeders, many of whom are attracted to the middle and
small varieties by this point alone; but, as Mr. Howard has
remarked, the long snout is an indication of prolificacy and

Fig. 19.--Profile of the Head of a Typical Middle White Sow.

vigour, neither of which qualities have we found, in our own
experience, to be too intimately connected with this somewhat
attractive type of head.

The sow whose portraits (shown in Figs. 18 and 19) provide
us with the type we have chosen, was drawn when in very
thin condition, and when suckling a large litter of pigs; and
it will be noticed that she is somewhat lacking in fulness
of face. Had she been in show condition, she would have
exhibited more cheek and fat between the eyes, which would

have made the face appear a little shorter, and the head thicker.

At the time when Sidney wrote (1860), the Middle York was about the same size as the Berkshire, but he remarks that they had smaller heads, and were much lighter in bone; they were better breeders than the Small Whites, but not so good as the Large Whites, occupying, even in this respect, a position between the two breeds. At that period, there had not been much demand for these pigs beyond their own locality, and they had not then been given separate classes at many of the leading agricultural shows.

Our own opinion of the Middle White pig is derived from an experience of many years in breeding and feeding the variety, and is therefore based upon results, and not upon mere fancy or predilection. We have found it in every sense a most admirable pig, whether for breeding or feeding, and one which cannot fail to return a profit to the farmer or the cottager, if it is kept under anything approaching fair conditions.

Among the disadvantages of pig-keeping is the trouble that the animals occasion by their unruly disposition, and the difficulty in keeping them within bounds. The Middle White, however, is one of the gentlest races in existence. It may be turned into a pasture without fear of its breaking bounds if there are any fences worthy of the name, although it will be found that the majority of pigs, kept without regard to race, are such determined foragers that ordinary hedges and fences fail to restrain their determination to roam. Few persons who have not grazed pigs can appreciate the importance of this quality; for the Middle White pig is second to none in its value as a grazer, and in its capacity to maintain high condition upon grass alone during the few best months of the year. Unless a pig has a quiet and contented disposition, the pig-keeper is often unable to turn it into his fields; in which case, there is an extra cost for food or labour, while the pig itself does not receive that benefit, and maintain that vigour, which are almost alone acquired by regular exercise and by herbivorous feeding. So gentle are well-bred Middle White sows, that we have never hesitated to allow two to farrow in the same sty, and to rear their young in common; and we have frequently owned both gilts and sows that would permit us to handle their mouths and to ascertain the state of their teeth.

These pigs generally bring litters of from ten to thirteen. We have seldom found them either to exceed or to fail to reach these figures, and we may quote an instance of a litter of pigs from one sow, seven of which were gilts. These, being unusually fine animals, were kept until they had produced their first litters, and, excepting in one instance, where there were only two little pigs, not one of the gilts produced less than five; in fact, the number of pigs farrowed by the remaining animals were six, seven, eight, ten, and ten. The gilts come into service exceptionally early, but we have found that they do not make such fine animals as those which are not served until they nearly approach twelve months old.

With regard to their feeding properties, we may cite another experiment, made between three highly-bred young pigs of the Middle White breed and three common pigs of the district, which were long-snouted, larger in the bone, and deficient in hair, although exceedingly vigorous and hardy. These animals were all of one age, and a good ration was provided daily for each. By degrees, the common pigs lost condition, and became noisy and extremely troublesome, continually asking for more food, as pigs only can. The pure-bred animals, on the contrary, maintained their condition and increased in weight; they were always contented, and so quiet in the sty, even amid the noise made by their neighbours, that their existence might scarcely have been suspected. As the feeding proceeded a little longer, the common pigs fell away so rapidly that we were obliged to revert to the old feeding, and to give them double the quantity of food that was consumed by the Middle Whites. Were it necessary, we could state numerous instances of this nature, where common pigs have been fed upon identically the same amount of food that was given to well-bred animals, and where they have entirely lost condition, sometimes being brought round only by gentle nursing and great perseverance, and at others slipping their pigs and eating them.

There is great similarity between the Middle and the Small White in two respects: both are small-boned, and carry a

MIDDLE WHITE SOWS.

Bred by Mr. Alfred Brown. Winners of First Prize and Breed Cup, Smithfield; and First and Reserve to Champion, Birmingham and Norwich, 1902

The exhibition Middle White should have a fine skin quite free from wrinkles, a moderately short head, a full jowl, a wide and deep chest, straight and well-set legs level with the outside of the body, a long, level, and wide back, a full but not flabby belly, and broad, full, and deep hams.

minimum amount of offal. A butcher who has no knowledge of breed would, perhaps, be disinclined to pay a higher price for a pure-bred hog of this race than for a common pig of similar weight; but we are of opinion that there is at least 10 per cent. of difference between the amount of actual meat value in a fat Middle White pig and that in a fat pig of a nondescript type.

We remember sending, on one occasion, several fat Middle Whites to an ordinary market auction. They yielded no higher prices than other pigs; indeed, in one or two cases, they returned less, for buyers are often prejudiced against pigs which carry fancy points, and of the properties of which they are ignorant. One of the buyers, however, upon this occasion —a butcher—was so delighted with his purchases, upon cutting them up, that he made a special journey for the purpose of purchasing more. He stated that the offal was unusually small, and that the meat was very plentiful in comparison with the bone.

The Middle White is admirably adapted for crossing upon the common pigs of the country, and we can point to at least a score of herds in our own district that have been materially improved by the introduction of blood from this pure race. The cross generally produces a pig that matures earlier, fattens more quickly, and requires a smaller quantity of food than the common pig, and, instead of unsightly, gaunt, flop-eared, large-boned animals, a much more symmetrical beast is produced, and one of which every man who has any eye for shape must be much more proud.

We have only had the advantage of trying a cross between the Middle White and one other breed, the Suffolk. For this purpose, three high-class animals—if one might judge by appearance—were purchased at the sale of a well-known prize-taker in Suffolk, some few years ago. Unfortunately, one died upon its way home, and, instead of being in farrow, as it was stated to be, it was in a high state of disease, brought about, doubtless, by excessive fattening. The others, apparently healthy, were afterwards proved to be greatly impaired in constitution, and turned out thoroughly bad breeders; and the crossbred pigs produced, like those of the Suffolk, were deficient in growth, in vigour, and in every good property that the pig should possess. Perhaps the experiment was unfortunate; but, as it was made with what was supposed to be the flower of the Suffolk race, we can only suppose that in-breeding or high feeding, or the two combined, had destroyed what had previously been the art and aim of the breeder to construct.

STANDARD OF EXCELLENCE.—The following is the standard of excellence of the Middle White breed formulated by the National Pig Breeders' Association and the British Berkshire Society:

Colour white, free from black hairs or blue spots on the skin; head moderately short, face dished, snout broad and turned up; jowl full, wide between ears; ears fairly large, carried erect and fringed with fine hair; neck medium length, proportionately full to the shoulders; chest wide and deep; shoulders level across the top, moderately wide, free from coarseness; legs straight and well set, level with the outside of body, with fine bone; pasterns short and springy; feet strong, even and wide; back long, level, and wide from neck to rump; loin broad; tail set high, moderately long, but not coarse, with tassel of fine hair; sides deep; ribs well sprung; belly full, but not flabby, with straight under line; flank thick and well let down; quarters long and wide; hams broad, full, and deep to hocks; coat long, fine, and silky; action firm and free; skin fine, and quite free from wrinkles. Objections: black hairs, black or blue spots, a coarse mane, inbent knees, hollowness at back of shoulders, wrinkled skin.

AMERICAN OPINION.—In the Report to the Convention, Messrs. Jones, Kennedy, and Barker state that the Middle-bred Yorkshire pig had, in their opinion, attained nearer to perfection than any other breed known to them. These pigs are not generally distributed through the West of America; but, when thoroughbred specimens have been introduced, they are held in great esteem, as well for exhibition as for family use. The Middle Whites are especial favourites with packers, who buy their stock on foot, for the reason that the breed yields larger proportionate net weights than any other hog grown for their use. These pigs are small in bone, but large in flesh of the best quality, evenly and proportionately spread over the whole frame,

One animal was weighed by the members of the Convention, as they thought her worthy of special note, possessing, as she did, a strong combination of good qualities typical of her ancestors. This pig weighed, in good flesh, but not really fat, 475lb., and measured 6ft. from the root of the tail to the top of the face between the ears, and exactly the same distance round the body. She stood only 6in. from the ground, and was 2ft. 6in. high, with a body nearly straight below and well arched above, showing great strength of back and loins. The surface of the body was smooth, and well covered with a short, smooth coat of white hair.

The gentlemen already referred to add: " We know of no breed of hogs in this country but what might, in some degree, be improved by occasionally crossing with the thoroughbred Yorkshire, which has been bred back pure since 1860. We have seen whole districts in which the swine were nearly all lop-eared, rough-skinned, black, sandy, and spotted white or blue, where, in a few years, by introducing a few of these pure-blooded white hogs, the general stock was made white, given erect ears, and their skin made smooth. Such a result cannot be obtained by Cheshire Whites alone, but it can be accomplished by the thoroughbred Yorkshire. These pigs are so thoroughbred and positive that they carry their own colour when crossed with almost any other breed, even if it is entirely black. Hence it is difficult to find a breed of swine, in this age of their improvement, in which the White Yorkshire does not crop out in some particular."

The writers add that the pure White Yorkshire and the Black Essex, or Neapolitan, may be bred together in such a way as to duplicate the colour of any other breed of hogs to be found among us. Upon these premises, the committee which drew up the report claim that the White Yorkshire, as now established in America and England, is the most thoroughbred hog known.

The Small White Breed.

The Small White may not inaptly be termed the "fancy" breed of the pig genus. It has in the past, in more ways than

one, been brought to a state of perfection that no other race
of pigs has attained; and, lest we may be misunderstood,
we will explain what we mean by so emphatic a statement.
In the first place, the word "perfection" is used in a
qualified sense, inasmuch as, properly speaking, the perfect
pig would be the one that proved the most profitable and the
best adapted to the requirements of the breeder. This fact
no one will dispute. When, however, we speak of the perfection of the Small White, we mean that in form of body,
in head, and in hair, it is unequalled, and has for some
years been regarded as the most astonishing result of the
breeder's art, receiving more admiration from the public at
large, who have nothing but appearance and symmetry to
guide them in forming a judgment, than all the other races
put together. In other words, the Small White, both from
its excessive depth and breadth, in comparison to its size,
and from the extraordinary formation of its head, is as great
a triumph of the art of the breeder as the bulldog and the
pug, both of which it, in a measure, resembles in the face.
During the last two decades, however, the breed has greatly
deteriorated in character and declined in numbers. It has,
as a matter of fact, been "found out."

The Small White will never maintain the highest position
in the herd of the farm or in the sty of the poor man.
It is pre-eminently a gentleman's pig, and it will only
occupy this position so long as the striking—nay, the curious
—points that are its characteristics are well maintained.
The reason for this is not far to seek: while wealthy people
readily pay large prices for animals that differ in a striking
degree from those of their neighbours, they will only be
burdened with common pigs when these yield an acceptable
profit.

The Small White can be fattened upon a smaller quantity
of food, and what is more, it will maintain a state of excessive fatness upon less food, than any other pig. It has
often been a cause of astonishment to visitors at the Royal
Agricultural Show, that these pigs could be brought to such
a state of obesity, and that they should yet be able to eat
sufficient food to maintain that condition. The fact is, however, that when a Small White pig is fit for exhibition, it

gets very little but grass, and frequently nothing more, if the grazing round the homestead is good and the weather suitable. We have repeatedly seen fat pigs of this breed which in summer were entirely grass fed, and which during other portions of the year received less than half the quantity of food given to common pigs of small size.

Between the years 1855 and 1860, the late Prince Consort exhibited a remarkable race of Small White pigs, which took almost all the leading prizes during that period. They were known as the Windsor breed, and were, naturally, very fashionable among amateur breeders who had been touched with the pig fancy. The formation of this race was not identical with that of the breed that we now term the Small White, for it was, without doubt, built up by selections cleverly made from the herds of some of the principal breeders of the time, the pigs used not being exactly similar in type to those which they produced. Considerable difficulty, however, was found in maintaining the breed, and in the course of a few years it became a thing of the past. The Small White pigs since exhibited by Queen Victoria, and which were brought to a standard of considerable excellence by Mr. Tait, one of the shrewdest judges and breeders of stock in our country, were identical with those which are recognised as the standard varieties. The head of a typical Small White is shown in Fig. 20.

STANDARD OF EXCELLENCE.—The following is the standard of excellence of the Small White breed formulated by the National Pig Breeders' Association and the British Berkshire Society.

Colour pure white; head very short and dished, snout broad and turned up; jowl very full, and broad between ears; ears small, short, and erect; neck short and thick; chest full and broad; shoulders full and wide; legs short, set well outside the body, fine bone; pasterns short and springy; feet small; back broad, level, and straight; loin wide; tail high set, small and fine, with tassel of fine hair; sides deep; ribs well sprung; belly deep, and near the ground; flank thick, and well let down; quarters wide and full; hams deep, wide, full, and well rounded; coat fine and silky; action firm and free; skin fine, quite free from wrinkles; general appearance of animal—small.

K

thick, and compact when compared with other breeds. Objections: black hairs, black or blue spots, coarse hair, inbent
knees, hollowness at back of shoulders, wrinkled skin.

With regard to disqualifications for faults, we think that
age should be decided by the teeth, and that pigs which, upon
examination, are found to be over age, should be disqualified.
Excessive fatness in breeding animals might very properly be

Fig. 20.—Head of a Typical Small White Pig.

noticed by judges—by what amounts to disqualification—
simply passing them by.

EARLY BREEDERS.—Fisher, than whom, as we have shown,
there was never a better authority, said that he could not
trace the progenitors of the Small White pig to an earlier
date than 1818, when it was bred by Mr. Charles Mason and
Mr. Robert Colling, both of Durham; but how they became
possessed of it, it is, at this distant date, impossible to say.
Fisher's own acquaintance with the breed dated from 1824,

when he found it in the possession of the Earl of Carlisle and Major Bower. Both of the latter obtained breeding stock from the well-known Mr. Wiley, of Brandesby, whose stock came from Messrs. Mason and Colling. The breed was then known as the Chinese, and, although it was in many ways vastly different from the original Chinese pig, there is no doubt that race had something to do with its manufacture.

The Small White seems to have been adopted by many of the earlier and most celebrated of the breeders of Shorthorn cattle, who were at considerable pains to perpetuate and disseminate it through the country. Fisher stated that, although the race greatly improved the common breeds of the country, yet it was not very highly esteemed by the farmers, as it required special treatment, and was too short in the leg for the straw-yard, where, from its want of activity, it was often in considerable danger of being injured by horses and cattle. When boars, however, were kept in something like store condition, and had plenty of exercise, they largely improved the bacon made from common farm pigs, by reducing the offal and increasing the thickness of the flitches.

THE SOLWAY AND SUFFOLK BREEDS.—There appears to be no doubt that the Small White pig, long known as the Small York, which we have to-day, was greatly improved by a Durham breed, then known as the Solway; this latter was a local variety descended from another branch of the Mason-Colling pigs, which also provided at least one of the progenitors of the Windsor breed. The result of the use of the Solway was that the Small White was improved in size as well as in constitution. The flesh contained more lean in proportion to the fat, and the hair was finer, more profuse, and more vigorous in its growth.

The variety we have referred to, known as the Solway, took its name from Solway Hall, in Cumberland, and, strange to say, it had an influence not only upon the Small Whites, but upon the larger breeds of white pigs exhibited in later years by Mr. Wainman, of Carhead. One of the best crosses ever made at Carhead was from pigs purchased from Mr. Unthank, of Netherscale: he obtained a number of animals, the produce of a cross between a Solway and a pig of some other variety, which grew to a much greater size than the Solway

K 2

itself. This was one of those fortunate but inexplicable incidents that had so much influence upon the northern races of pigs, and through them upon the white pigs of the country. Fisher states that the Solways in his hands at Carhead grew on with their years, and gives an instance of a boar that took several prizes at the age of one year as the best of the *Small* breed, when he was never beaten. At two years he was, on account of his increased size, advanced to the *Middle* breed classes, and won the first prize at the Royal Agricultural Society's Show in 1864. This, however, is not all, for in the following year, and again at the Royal Show, he won the first prize as the best boar of the *Large* breed.

A few years ago, there was a variety of small white pigs known as the Suffolk, but it has now been absorbed into what we may call the English Small White breed. It had evidently descended, at least upon one side, from the white breeds to which we have referred; but, although neat in form and in type of head, and unusually small in bone, it was deficient in hair and exceedingly delicate in constitution. This variety gained a considerable number of prizes in a few years, and was frequently used for crossing with the Small Yorks (as they were then called).

SIDNEY'S OPINION.—Sidney, who, for the purposes of his work, maintained a large correspondence with the leading pig-breeders throughout the country, thought it was quite safe to assume that the best Small White pigs of modern times had been bred from the Yorks and Cumberlands. He found that both varieties had continually been intermixed with great advantage. The Small White of his day was a pig with a short head, small, erect ears, broad back, deep chest, short legs, and fine bone. It was easily converted into bacon, and was found useful for the manufacture of small pork, whereas the young pigs were admirably adapted for roasting; and, what is equally true of the breed of the present day, he found that three or four could be well fed and kept in good condition upon a quantity of food that would barely keep a single specimen of the Large York. One of Sidney's correspondents, Mr. Mangles, who had a considerable reputation in his day, said that, by judiciously crossing the Cumberland and the

York, he obtained a breed that combined size and aptitude to fatten with early maturity. The Yorkshire variety furnished quality and symmetry, and the Cumberland size.

LORD DUCIE'S BREED.—Lord Moreton, who is heir to the Earldom of Ducie, only continues the good work of the improvement of British pigs which has been so conspicuously connected with that title. Lord Ducie's breed of Small White pigs has been known for nearly two generations, and the noble earl, in founding and maintaining his herd, drew largely and judiciously from most of the leading strains of the earlier days, and, in its turn, his own strain has greatly contributed to the improvement and permanence of the breed. Some of the best of the Cumberland variety that were purchased are stated to have been medium in size, with short noses; fine, well-made ears, carried a little forward, and were medium in size; necks full, handsomely formed, and free from creases; short legs; straight, well-developed backs and ribs; capital hams; and plenty of fine, vigorous hair.

AMERICAN OPINION.—Mr. Coburn, in his "Swine Husbandry,"· does not deal with the Small White breed, as we understand it in England, but he has devoted a chapter and an illustration to the Suffolk pig, assuming it to be the recognised small variety. The authority he largely quotes, the Hon. John Wentworth, an Illinois breeder, states that the Suffolks are *the most popular breed in England,* and that they are *the English nobleman's hog;* but both statements are absolutely contrary to facts. We shall deal with this question in our remarks upon Suffolk pigs, now known in England as Small Blacks, although, as we have remarked, there were a few well-selected herds of Small White pigs of a similar type to the blacks, which were entirely distinct in type from the original white Suffolk. These, however, had no peculiar right to the title of a Suffolk breed, or to be exhibited as representatives of the English Small White pig. Suffice it to say, that the American National Convention of Swine Breeders, in reporting upon Suffolk swine, found it convenient to quote Sidney, who stated that the improved Suffolk, like the Windsor and other ephemeral varieties, were founded upon the Yorkshire-Cumberland stock; and that some of them were merely pure Yorkshire pigs transplanted and re-christened. He added,

too, that not one of the local varieties which were so frequently quoted, if we except the Suffolk, was worthy of
cultivation, and that the Suffolk pig itself was only another
name for a Small Yorkshire.

The following are the characteristics of the Small White
breed as recognised in America: Head small, very short;
cheeks prominent and full, face dished, snout small and very
short, jowl fine; ears short, small, thin, upright, soft and
silky; neck very short and thick, the head appearing almost
as if set on front of shoulders; no arching of crest; chest wide
and deep; elbows standing out; brisket wide but not deep;
shoulders thick, rather upright, rounding outwards from top
to elbow; crop wide and full; sides and flank long; ribs well
arched out from back; good length between shoulders and
hams; flank well filled out, and coming well down at ham;
back broad, level, and straight from crest to tail, no falling
off or down at tail; hams wide and full, well rounded out,
twist very wide and full all the way down; legs small and very
short, standing wide apart, in sows just keeping belly from the
ground; bone fine; feet small, hoofs rather spreading; tail
small, long and tapering; skin thin, of a pinkish shade, free
from colour; hair fine and silky, not too thick; colour of hair
pale yellowish white, perfectly free from any spots or other
colour; size small to medium.

BREEDING.—In breeding Small Whites, the aim should be to
maintain the characteristic points of the variety, the chief
of which are: form (much in little), fineness of bone, quantity
and quality of hair, shortness of snout, and aptitude to
fatten. The form of the animal is such that the most
esteemed parts are well developed, but, if either is sacrificed,
the progeny will suffer in appearance, and the breeder in
results. Vigour of constitution should be maintained by
judicious crossing, and, if the Small White pig is sustained
in all its purity—the fancy, as well as the useful points—the
breeder's art will sometimes be severely tested. The greatest
danger is in using stock pigs without pedigree, however perfect
they may be in appearance, and no breeder should venture
to cross with any other variety, unless with the full knowledge
that his chances of success in producing show pigs are
hopeless.

BERKSHIRE SOW.

Bred by Mr. Alfred Brown. Sire, son of "Sir Alfred"; dam, daughter of "Halle Starlight." Winner of First Prize, Breed Cup, and Reserve to Championship, Smithfield; First and Championship, Birmingham, and First Norwich; all in 1902.

No other race can produce such a proportion of lean meat as the Berkshire, and as a bacon pig it has no equal, the bone being fine and compact, and the offal very light. It should have an abundance of fine and compact hair; a short, fine, and well-dished face, broad between the eyes; thick, round, and deep hams; and a small tail.

CHAPTER XI.

THE BERKSHIRE BREED.

THIS variety of British pigs, which is one of the most valuable of known races, is as popular in the South-west of England, and covers as large an area of country, as the white pig—which is equally popular, if not more extensively bred—in many of the Northern Counties. As its name suggests, the Berkshire is largely bred in the county of that name, but it is quite as extensively produced, and as much admired, in the neighbouring counties of Wilts, Gloucester, and Buckingham, and is largely found in Somerset and Devonshire in the south, and in Warwickshire and several other of the Midland Counties. It has, too, been gradually gaining ground throughout England, Scotland, and Ireland; and, as the exhibition system grows, it finds admirers in almost every district, whether other varieties have previously been favourites or not. In America, the Berkshire pig is much more extensively bred than with us, and there is in that country not only a very much larger number of breeders of pigs of an exhibition type, but also a Berkshire Pig Association, supported by a large body of members, although English breeders, to whom the Americans originally came for the foundations of their herds, long lacked the spirit and energy necessary to establish a similar organisation.

The Original Berkshire.

There is no doubt that the original Berkshire was a larger animal than it is at present, and that its colour and marking

were in no sense defined. Although principally black, it was
a black-and-white pig, which frequently threw young ones
marked with red or sandy-coloured spots. It was much coarser
than it is at present, but was undoubtedly improved, between
the years 1820 and 1830, by Lord Barrington and other
breeders, whose method of crossing has, so far as we can learn,
never been handed down. Sidney states that almost all the
best herds of Berkshires of his day were derived from Lord
Barrington's pigs, which were of medium size, and certainly
smaller than the old race of the district. In those days, the
majority of well-bred Berkshires were not only marked with
white, as they are at present, but also had a small quantity
of white hair behind each shoulder; and Mr. Sadler, of
Cricklade, who was a most successful breeder, exhibited 300
Berkshires upon his farm to an Agricultural Association,
every one of which was marked in this way.

It has frequently been stated that the Berkshire was made
by a cross with the Neapolitan, but we are inclined to doubt
the statement that the true Berkshire, the animal so famous
for the large proportion and streakiness of its lean, owed
such a remarkable and valuable quality to a race which
imparted fat so generally to British pigs.

An idea of the weight of Berkshires, thirty years ago, may
be obtained from the fact that some of Mr. Sadler's prize
pigs, under seven months old, weighed 240lb. each, although
they were turned into an orchard daily while fattening. The
same gentleman exhibited, and won the first prize at
Smithfield, with a fat pig, which weighed 856lb, the length of
her body being 6ft. 4in., and her girth 7ft. 6in. At that time
the ordinary weight of a well-bred Berkshire bacon hog, when
ready for the butcher, was about 50 stones of 8lb., but for
the curing of the best hams these were generally considered a
little too large.

Sidney states that Fisher Hobbs, to whom the present high
state of perfection of British pigs is largely due, used the
Berkshire as a cross upon his Essex pigs, and that it
materially assisted in imparting size and constitution; and
so strong was the influence of the Berkshire, that at least
twenty-eight years after the cross was made, some of the
young Essex reverted to their alien ancestor, and were, in fact,

exact types of the true Berkshire pig. Sidney also adds that, at the time he wrote, it was generally agreed that the Berkshire was well adapted for hams and bacon, but not for the production of small fresh pork. This evidence does not coincide with that of the breeders of to-day, who, probably on account of a more recent cross with the Black Dorset or the Suffolk, have imparted a qualification that is valuable in the production of pork.

The Modern Berkshire.

Strangely enough, the Berkshire pig has not attracted the admiration of continental breeders. In France, the only British race which is used, and which the breeders permit themselves to class with their own Craonnaise, is the Large White; and in Germany, where there is no native breed of eminence, the Berkshire is entirely ignored in favour of the White races. The English breeders of Berkshires have themselves principally to thank for this want of appreciation, for it is a noticeable fact that, during the last forty years, the Berkshire has been losing size; and, if we except two or three herds that have been kept up to the old standard, we cannot but remark that the lengthy, broad, well-haired pig of forty years ago has been replaced by an animal of fashionable shape, but which, from crossing and careless breeding, is imperfect in the colour of its skin, too short in the snout, short in the body, and altogether smaller in type than it ought to be. Unfortunately, the majority of the judges at our leading pig exhibitions are practically unacquainted with the real type of some of the varieties of pigs, and, knowing this fact, the breeders have produced for their edification an animal that has much to commend it, being beautifully formed and attractive in many of its points; but it is not a Berkshire pig. In the production of animals of this kind, the Dorset, the Essex, and the Suffolk have been used in the past, and, this being the case, it is scarcely necessary to suggest that the improved (?) Berkshire has, in many instances, degenerated into an animal as remarkable for its fat as the old Berkshire was for its proportion of lean. The crosses, too, in addition to the production of this undesirable point, have decreased the quantity and quality of the hair, and they have prevented the development of the pig into that size and length for which its flitches were at one time so remarkable.

Another cause of the alteration of the Berkshire type has undoubtedly been the action of some of the principal agricultural societies, who have provided prizes for a certain number of young pigs under a given age, instead of giving more classes and better prizes for breeding pigs to be shown in breeding condition. This system is carried out to the present day, and is patent to every stock-breeder. It has but one result, and that everyone deplores. In place of awarding the prizes to pigs that the stock-owner would select for the improvement of his herd, many judges show their preference for animals which, by their excessive development of fat, display, as it were, the greatest indication of early maturity. It will be admitted that, where two pens are placed side by side, the pigs in the one, having been forced from birth, and fatted until each animal has become a perfect model of what a butcher's pig should be, will, in the hands of a weak judge, be almost certain to obtain the prize over the other, containing a number of animals in little more than store condition, which are consequently smaller, less tractable, and far less handsome.

There is, therefore, but one course to adopt : rules should be drawn up with sufficient clearness to enable either the judges or the veterinary inspector to disqualify every animal that is not shown in actual breeding condition. In such a case, exhibitors will find that their best course is the one we have indicated, viz., to exhibit animals that are, in every sense of the word, such as they themselves would prefer to purchase if they were making a selection for the improvement of their own herds. At the London Dairy Show in 1884, a pig department was included in the scope of the exhibition, at the instance of the author of this work, and it was a special feature of the schedule that every animal should be exhibited in breeding condition. The consequence was a much smaller display of fat pigs, with a much better and more satisfactory result than is customary. We have referred to this point because there is no doubt of the fact that Berkshire breeders, in many instances, have used crosses of the races we have named above, in order to develop early maturity, which the judges have so distinctly admired.

Among the chief points of objection to the Berkshire have been its dark skin, which, it has been stated, butchers who purchase porkers dislike—and its comparative slowness as a feeder. There is certainly something in the latter objection, even with the true Berkshire; and breeders of the variety should hesitate before attempting to perfect the breed in this particular. Its principal feature is its value as a bacon pig, and the nearer it approaches perfection in this point, the more widely will it differ from the white pigs. It is generally found that a fast-feeding Berkshire has alien blood in its veins, and that, when killed, it exhibits far too large a proportion of fat. So long as the Berkshire pig is blessed with a hardy constitution, and is of fairly large size, with plenty of length and deep flanks, very little can be said, providing that it is as prolific as it ought to be; but if a breeder finds that the pigs in his herd exhibit delicacy, shortness of body, absence of hair, and excessive fat, and that the sows bring small litters, it is quite time that he disposed of them, and made some effort to obtain a fresh start with Berkshires of the true type.

The late Mr. John Fisher once stated that the best cross with the Berkshire he ever saw was between a Berkshire sow and a boar of the hairy Cumberland breed, which was called the " Solway." The progeny, which were both blue and blue-and-white, were as good as the finest type of Berkshire pigs, if we except size and colour; but, more than this, they were better in form than the Berkshire is admitted to be. No Berkshire breeder, however, would permit of a cross of this kind, which, from its colour alone, would quite shock his taste, for he is so much attached to the white points of his breed, the white mark on the face, on the tip of the tail, and on the four feet, that he is generally inclined to discard animals failing in any one of these particulars. There may or may not be a higher reason for insisting upon these points than, *primâ facie*, appears to the outside public, but we have no hesitation in believing that they have considerable influence in restraining the progress of the breed, and, as it were, in preventing it from attaining greater perfection. We have repeatedly seen animals of the finest type passed by judges because of their deficiency in the white points; and, although it will be admitted that pigs that have been crossed are thus rendered imperfect in colour, yet we are convinced that a practical judge can as easily determine the purity of the Berkshire race, whether the white marking happens to be present or not.

Instructions to Judges of Berkshire Pigs.

The following instructions to judges were suggested by the Committee on markings of the British Berkshire Society:

"We recommend that a perfectly black face, or a black foot or black tail, should disqualify a pig in the show yard. White or sandy spots on the top or sides of the animal, or a decidedly white ear should be a disqualification. Any description of colouring, staining, or clipping should also be a disqualification. White on the ear or under the throat, or on the underlines of the body, should be considered objectionable. A rose back should be an objection. Either too much or too little white in the place of the proper recognised markings should be an objection, also to be noted in the competition."

Opinions of Breeders.

We have been favoured with some valuable remarks upon the Berkshire pig by several of our best breeders, to whom our thanks, as well as those of the public, are largely due.

Mr. Arthur S. Gibson, of Bulwell, Nottingham, who is a good authority on Berkshires, says: "The first point I claim for Berkshires is great hardihood of constitution. I believe no breed can equal them in this desirable quality. With the exception of the Large White breed, I believe they are as prolific as any variety of pig. My *average* of reared pigs is a fraction over eight—seven to eight for a gilt, and nine to ten for a sow, is quite as many as the mother can properly rear. We hear of over a dozen pigs being reared, but at the age of nine weeks they are not worth so much as nine which have been well reared. When it can be obtained, there is nothing to

equal milk, given in some proportion with other food, for the
pigs, from the time they are weaned. I cannot obtain milk,
but without it I find fourths flour, mixed with maize meal, and
scalded, the best food for young pigs; indeed, I like all food
scalded. A few roots, such as potatoes, mangolds, and swedes,
boiled and mixed with the meal, will be found economical, and
are much relished. Pigs of all ages like a run in the yard
or field every day, and it is confinement that ruins so many
good ones. My pigs, old and young, as well as those I am
exhibiting in breeding classes, have their *entire liberty* through
the whole of the day during six months of the year.

"There cannot be two opinions as to the quality of Berkshire bacon, for it is not equalled by that of any other breed.
No other race can show such a proportion of lean flesh. The
pigs may not exhibit such fast growth when young, but they
are certainly thrifty, and have great aptitude for fattening on
coarse food, and, when they have attained maturity, can be
kept in a thriving condition with the roughest fare. I find
a well-fed Berkshire will give from 86 to 90 per cent. of
carcase to live weight. At the age of sixteen months, I have
had them weigh 33st. (imperial), and, without exception, they
always weigh more than the butcher's estimate. For crossing
purposes, I believe no blood is so much resorted to as the
Berkshire. In this neighbourhood, some few years ago, if I
except one small lot, there was not one herd of Berkshires
except my own, although they are now becoming general and
spreading northwards, and are especially popular in Scotland.
I have not been able to supply the demand for boars for
Scotland during the past summer, these being required mostly
for crossing purposes. The old slouch-eared pig is the most
common in the neighbourhood, especially among the colliers,
but some of this type, mated with a Berkshire boar, produce
a really good pig, and mostly white.

"I consider that a good specimen of a well-bred Berkshire
boar has the grandest possible carriage, and more nearly
approaches perfection than any pig I know. One great
difficulty in the present day is to breed Berkshires with true
markings and skin, and the right class of hair. This no
doubt arises from alien blood, and especially from the Suffolk.
This cross is plainly seen in many of the pigs exhibited with

the short noses and almost hairless skins. A Berkshire should
have abundance of fine hair all over him, and a snout which
is neither too short nor too much turned up; the ears should
be medium in size, well carried, and wide between. The
Berkshires will bear the necessary forcing for exhibition
without ruining their fecundity, as it does in other breeds.
A good young sow, with which I was second at the Royal of
1884 in a class of thirty, against much older pigs, was shown
almost continuously, from the time of service until she farrowed, the day after she came home from the Leicestershire
Show, when she had eleven healthy pigs, all of which did well.
This was no exceptional case."

Mr. William Ashcroft, of Hayes, Beckenham, formerly a
practical breeder of Berkshires of a very good type, favours
us with the following remarks: "I must preface what little
I can say with the remark that I have never bred anything
but Berkshires, having had a partiality for them because I
have found them hardy, easily fed, and well adapted for all-
round purposes, as they will either make porkers of 8st. to
10st., larger pork of 10st. to 15st., or bacon pigs of from
15st. to 40st.

"I have sold Berkshire pigs at all weights to butchers, who
were always willing to pay me the top price according to
weight, and who give me to understand that they prefer them
to white or any other black pigs for quality of meat and
percentage of lean. Whenever I have sent store pigs to
private individuals, who keep two or three pigs for making
bacon for their own table, and who may have previously kept
white pigs, I have always heard that the change gave great
satisfaction. I have given precedence to quality of flesh,
because I think that is one of the main features of the
Berkshire breed, and one which ought never to be lost sight
of. I deprecate most strongly absence of hair in a Berkshire pig, or that short, square, neat, and, if I may be allowed
the expression, 'punchy' appearance, which makes one strongly
suspect crossing with smaller black breeds. Among other
failings brought about by crossing with alien breeds are loss
of lean in the flesh, delicacy of constitution, and a tendency
to breed a deficient number of pigs; and in-breeding is equally
to be deprecated on this account.

"Berkshire sows are often blamed for failing to bring a sufficient number of pigs at a litter. I think, however, that, although they may not be quite so prolific as white pigs, the tendency to produce medium litters, of, say, six to eight, may be corrected by breeding less closely. If unrelated stock pigs are used, the litters will come pretty regularly twice a year, and average about eight to eleven. Close breeding also encourages a tendency to lung disease and a difficulty in getting sows to stand to the boar; and the sows become careless and bad mothers. It is well, therefore, to be careful, in breeding Berkshires, to change the sires frequently, and to select them from strains that have generally brought a numerous progeny, and such as exhibit the characteristic white markings of the breed, the strong, open, and lively countenance, not too long and very little dished. The breeder should strive for length and depth, and correct by selection, as gradually as he can, the rather drooping quarters which are a characteristic fault of the Berkshire pig. The Berkshires are capital pigs for a straw-yard, and, if kept in a growing condition until they are five or six months old, they will then make the best of bacon hogs at about eight months. Most of my pigs have been sold as porkers at fifteen to seventeen weeks old.

"As to management in feeding, farrowing, &c., what is necessary for one breed is, of course, equally necessary for another.

"What I have desired to draw your attention to is the point that the hardy and well-fleshed character of a Berkshire should not be lost by crossing or in-and-in breeding, and I think that breeders for exhibition purposes have not been sufficiently careful in this direction. More attention should also be devoted to increasing the prolific character of the breed."

The late Mr. Joseph Smith, of Henley-in-Arden, one of the oldest and most experienced breeders of the Berkshire, wrote us as follows: "I have bred and exhibited the Berkshire with great success for the last thirty years, and I have much pleasure in giving you a short account of the result of my experience. The true Berkshire should be black, with four white feet, a white blaze in the face, and a white tip to the tail. The head should be short and nicely dished, and the nose wide. The coat should be of soft (not, as is frequently seen, of coarse and hard) hair, and the skin should be soft, clean, and free from scurf. In defining the form of the Berkshire, I would apply the four 'L's'—long, low, lusty, and level—and, with these qualifications, Berkshire pigs will always thrive and satisfy their owner. They are good mothers, bringing from nine to twelve at each farrow, and making the finest pork at from three to four months of age. If, however, they are intended for bacon, and are well fed, they will, at twelve months old, weigh from eighteen to twenty score pounds, and show more fine-fleshed lean in proportion to their fat than any other variety of pig, their bacon being of the primest quality. In breeding, I have always a good demand for young ones; I put my gilts to the boar at the age of eight months, and take two farrows a year from the sows, which I keep as long as they prove prolific. In introducing fresh blood, great care is necessary. I always avoid extreme crossing, using none but the oldest and best strains. The herd book will be of great service in this respect, especially to young breeders in selecting their breeding stock.''

Mr. W. H. Wykes, so long the manager of the Countess of Camperdown's Berkshires, states that he finds the cheapest and quickest method of fattening the Berkshire pig is to give barley meal in the proportion of two parts mixed with one part each of bean meal and wheat meal. This mixture should be scalded, then cooled to a lukewarm temperature, and given three times a day. When the pigs are first put up to fatten, the food should be given thin, and gradually thickened until it is almost of the consistency of paste. When they have reached this stage, the pigs should have a little drink once a day: water may be given, but, if it is convenient, milk is preferable. For a pig at the age of nine months, one bushel of meal is sufficient to commence with; but this should be increased to two bushels, or even to two and a half for a pig of large size, as a liberal diet pays much better than a scanty one in fattening pigs. A most important point to consider is that of warmth, which is so essential that a large percentage of the food is otherwise wasted in maintaining the animal heat of the system. It is also advisable to provide a few coal cinders, which should be occasionally thrown into the sty; these will be eaten by the pigs, and will materially assist digestion.

Mr. Wykes also states that he has killed Berkshires under six months old, which weighed ten score pounds, and three fat pigs exhibited by the Countess of Camperdown, at Bingley Hall, Birmingham, in November, 1884—winners of the first prize and two cups—weighed over twenty score pounds each, their age being ten months and three weeks. Mr. Wykes finds that the best crosses are made by the Berkshire upon the Black Suffolk and the Sandy Tamworth. In dealing with a scale of points, he is of opinion that it is necessary, in view of the fact that breeding classes are now recognised as essential, to formulate two, one for fat pigs, the other for pigs intended for breeding purposes. As there is much sound sense in the scale of points suggested by Mr. Wykes, we have pleasure in quoting it:

SUGGESTED SCALE OF POINTS FOR BERKSHIRES.

	FAT PIG FOR THE BUTCHER.	SHOW PIG FOR BREEDING PURPOSES.
Head	12	20
Neck	15	12
Shoulders	12	10
Back	18	14
Ribs	14	10
Hams..	21	16
Hair and Skin	8	12
Markings	0	6
	100	100

The late Mr. Richard Fowler, of Aylesbury, whose success in the show pen with one of the oldest types of the Berkshire gave weight to his opinion, said that he never began to fatten for exhibition until two months beforehand. He preferred bran, barley meal, and milk. If fattening was commenced too soon, the pigs not only went off their feet, but also off their

L

food. It was Mr. Fowler's opinion that the best way to
ascertain the carcase weight of a fat Berkshire was by mul-
tiplying the live weight by 5 and dividing by 8; but we cannot endorse this, and we should be sorry to think that the
dead weight of the Berkshire was no more than five-eighths
of its live weight. On the contrary, many of the pigs exhibited at Smithfield and elsewhere have scaled over 80 per
cent. in the carcase. Mr. Fowler was also of opinion that
the Berkshire would improve any other breed of pigs, and
more especially the Tamworths, when crossed upon them.
He considered judging by points to be a great mistake, and
that the real questions upon which to decide between pigs
were quality, shape, and breeding; but, inasmuch as every
judge would necessarily give more weight to one of these
properties than to another, this would practically be judging
by points.

Mr. Nathaniel Benjafield, of Motcombe, Shaftesbury, a well-
known and practical breeder and prize-taker, says that he
prefers Berkshires that are long and deep in body, and provided with long thick hair of fine quality. The head should
be well set on the body, and furnished with good-sized ears,
which should be well fringed at the edges, and hang nicely
forward, but not droop, as is sometimes seen in Berkshires
of otherwise good type. The snout should be slightly dished,
"and," continues Mr. Benjafield, "I do not object to a little
white hair in one of the ears; at the same time, the skin
should be of a good dark colour, excepting only the blaze in
the face, a white tip of the tail, and the four white feet, the
marking not being carried too high up. I should, however,
never object to a really good animal if one of the feet had a
trifle too much, or even too little, white for the orthodox
taste. I would naturally prefer that the ears were entirely
free from white, but I have known some exceptionally grand
pigs which have been so marked that I would not, on that
account, throw them out of the prize list.

"I prefer sties with open yards for Berkshire pigs, and, if
possible, they should face the south. I think there is no
better plan for improving grass land than to use movable
sties in summer. If they are provided with four wheels, with
a bed for the pig in one end, which is covered and floored,

the other portion of the sty being simply enclosed, the droppings of the animals will fall immediately upon the ground, and wonderfully improve it. A house of this kind could be moved daily, or at the option of the owner, and, like the land itself, the pigs would benefit materially, for they are exceedingly fond of grass, and invariably thrive well upon it."

The Berkshire in America.

We have already mentioned that Berkshires are largely bred in America, and we might have added that there has been more written upon this breed by Americans—who have taken considerable trouble to investigate the history of the Berkshire pig—than by our own people. Two of the foremost authorities in the United States are the Hon. A. B. Allen and Hon. F. D. Coburn, the author of "Swine Husbandry," and now one of the most prominent men in Kansas, and the Berkshire owes much of its present position to their able and persistent advocacy of its merits. Mr. Coburn states that, now the Berkshire is known and appreciated in America, it stands second to no race of pig in the estimation of intelligent pork-producers in that country. While the Berkshires of to-day are probably much improved, as compared with those of sixty years ago, the spirit of improvement is still abroad, and the standard of perfection is high. According to this writer, the good qualities that serve to make the Berkshires American favourites are:

1. Great muscular power and vitality, which render them less liable to disease than many other breeds.

2. Activity, combined with strong digestive and assimilating powers; hence, they return a maximum amount of flesh and fat for the food consumed.

3. The sows are unequalled for prolificacy, and are careful nursers and good sucklers.

4. The pigs are strong, smart, and active at birth, and consequently less liable to mishaps.

5. They can be fattened for market at any time, and they may be fed to any reasonable weight desired.

6. Their flesh is the highest quality of pork.

7. Power of the boar, when used as a cross, to transmit the valuable qualities of the breed to his progeny.

8. Their unsurpassed uniformity in colour, marking, and quality.'

Although some of our own breeders have doubts as to the value of the Berkshire in crossing upon some of our native breeds, it appears that in America there is scarcely a medium or a large breed upon which they cannot be crossed with advantage, chiefly owing to their vigour and hardiness, and Mr. Coburn tells us that, crossed with the Poland-China, they make the best feeding hogs possible. There is, perhaps, great reason for this, as the Poland-China is a variety of comparatively recent manufacture, and, like the Dorset of England, would be certain to produce by this cross a pig of unusual excellence.

American Standard of Excellence.

In the year 1875 the American admirers of the Berkshire pig organised the American Berkshire Association, which is stated to have for its object "the collection, preservation, and dissemination of reliable information on the origin, breeding, and management of Berkshire swine, and the publication of a herd book or record of Berkshire pedigrees." In the early days of this Association a hundred-dollar prize was offered for the best original essay upon the origin and management of Berkshire pigs, and this was awarded to the Hon. A. B. Allen, above mentioned, who shortly afterwards prepared the Report upon the Berkshires, as adopted by the Swine Breeders' Convention. The following are the characteristics and markings of Berkshire pigs as agreed upon by the Convention:

Colour black, with white on the feet, face, tip of the tail, and an occasional splash of white on the arm. A small spot of white on some other part of the body does not impute impurity of blood, yet it is to be discouraged, to the end that uniformity of colour may be attained by breeders. White upon one ear, or a bronze or copper spot on some part of the body, suggests no impurity, but rather a re-appearance of original colours. Markings of white other than those named above are suspicious, and a pig so marked should be rejected.

Face short, fine, and well-dished, broad between the eyes; ears generally almost erect, but sometimes inclining forward with advancing age, small, thin, soft, and showing veins; jowl full; neck short and thick; shoulders short from neck to middling, deep from back down; back broad and straight, or a very little arched; ribs—long ribs well sprung, giving rotundity of body; short ribs of good length, giving breadth and levelness of loins; hips of good length from point of hip to rump; hams thick, round, and deep, holding their thickness well back and down to the hocks; tail fine and small, set on high up; legs short and fine, but straight and very strong, with hoofs erect, legs set wide apart; size medium; length medium, extremes are to be avoided; bone fine and compact; offal very light; hair fine and compact; skin pliable.

The scale of points, as arranged in 1877, is as follows:

Colour	4	Brought forward	47
Face and Snout	7	Sides	6
Eye	2	Flank	5
Ear	4	Loin	9
Jowl	4	Ham	10
Neck	4	Tail	2
Hair	3	Legs	5
Skin	4	Symmetry	6
Shoulder	7	Condition	5
Back	8	Style	5
Carried forward	47	Total	100

The Hon. A. B. Allen and the Siamese Cross.

In his essay Mr. Allen appears to have traced the history of the breed more closely than any British writer, and he is convinced, from investigations that he personally made in the county of Berkshire, that the fixed type of the breed is more than a century old, it having been kept and bred up to a point of considerable excellence before the year 1780. In 1807, Sir William Curtis exhibited a Berkshire pig weighing 904lb., and there are numerous instances on record of animals

of the race which, in the early days of the breed, reached
from 700lb. to 900lb. Mr. Allen is of opinion that the
original cross upon the old Berkshire, which contributed so
largely to the formation of the improved breed, was with the
Siamese, and he gives very cogent reasons in support of this
opinion, based upon considerable experience in breeding both
the Berkshire and the Siamese itself.

It will be remembered that the portrait of the Siamese
sow in the illustrated work of Professor Low is something
between a slate and a plum colour. The fore feet and a
portion of the fore legs are white, slightly shaded with slate.
The body is tolerably long, the loins are broad, the hams
are deep and thick, the face is decidedly dished, and the
ears are erect and closely resemble those of the Berkshire.
To compare with this, Mr. Allen describes the Siamese swine
which he himself owned and bred for a number of years.
In colour they varied between a black, a dark slate, and a
rich plum. Two or three of the feet were generally white,
but the white did not extend to any other part of the body.
The head was short and fine, the jowl thin, and the face
dished, and the ears were short, slender, and erect. The
shoulders and hams were large, round, and smooth; the back
was broad and slightly arched; the body was moderately long,
deep, well-ribbed up, and nearly as round as a barrel. The
legs were fine and short, the tail was slender and curled near
the rump, the hair soft, silky, and thin, the skin fine and of
a dark hue, and always free from bristles. The flesh was
firm, very tender, and carried less lean than the Berkshire.
The Siamese were extremely hardy, and never injured by unusual heat or cold. Their disposition was quiet and gentle,
and they would partially fatten upon good pasture or coarse
raw vegetables. These pigs could be fatted for the butcher
at any age, and matured at from twelve to fifteen months,
weighing, when fully ripe, from 250lb. to 300lb., sometimes
even reaching 400lb. Whether or not Mr. Allen is right in
his inference, matters very little at this late period; but it
is interesting for breeders to know how a popular and successful race has been formed, and our American friends have
recognised this fact with regard to one of our most important
British races more distinctly than we have ourselves.

It appears that the first Berkshires exported to America were purchased in 1823 by a Mr. John Brentnall, an English farmer who settled in New Jersey, and whose stock has done good service for the Berkshire breeders of the States. The next importation was not made until 1832, when another English farmer, settled in Albany, brought over Berkshires from the old country. Since then numerous breeders of eminence have visited England, and taken back with them specimens of the race that were as fine as any our country has produced; and thus it is that the American breeder has excelled us in the production of our principal race of black swine, and has been largely enabled to send us in return so much bacon and so many hams.

Mr. Allen is a strong advocate for the Berkshire, which he considers should be the favourite swine among American breeders, and he adds that his countrymen ought to take every possible pains in breeding, rearing, and fattening Berkshires in such a manner as to produce a superior quality of smoked meat, both for the home and the foreign market. "Improved methods of curing and packing," he continues, "should likewise be adopted, so as to enable us to get as high a price in the English market as the best Irish bacon commands."

Shape of the Snout.

There has been considerable argument, both in this country and in America, upon the shape of the snout of the Berkshire pig. It is advocated, and with much reason, that the pug-shaped or turned-up nose is characteristic of small races of pigs, which are more addicted to the production of fat than of lean, suitable for bacon and hams. It is useless to argue, in these days of economical feeding, that the snout of the pig should be adapted to his requirements in feeding upon the land, for the shape matters little where animals are continually shut up in sties or paved yards, whereas, if it is the practice of a breeder to turn them into his fields to grass, he is invariably careful to ring them, so that they are prevented from damaging the pasture and searching for roots. It seldom happens that a farmer finds it necessary to turn swine into his arable fields for any other purpose than that

of picking up the waste that lies upon the surface; and we
question very much whether the result, from a feeding point
of view, of allowing pigs to dig, even where they could do
no damage, would in any sense prove a benefit to the farmer.

The shape of the snout should undoubtedly be guided by
the purpose for which the pig is bred.   It is the aim of
breeders of Berkshires to produce long, deep flitches of bacon,
good hams, and meat well marbled or streaked with lean.
As these products are without doubt obtained from pigs of
the true Berkshire type—*i.e.*, from the comparatively long-
snouted animals—there should be no further argument upon
the subject.   Unfortunately, there are fanciers of pigs who
are not practical breeders, and who, admiring the pug nose
or short snout, which they conceive to be characteristic of
high breeding, would place these points upon pigs of every
race.

Colour and Marking.

With regard to the colour of the Berkshire there are scarcely
two opinions in England at the present moment, and, although
we have seen at the chief exhibitions exceedingly fine pigs
that have been absolutely black, the colour should more
closely approach that rich plum to which Mr. Allen has
referred.   This was the favourite colour forty years ago,
even when considerable latitude was permitted as to the
admixture of white or red hair, and it cannot be denied that
animals of this colour are not only superior in quality of
flesh, but they have a grander style, thinner skin, and finer
hair.

Another point of contention is the white marking upon the
feet, face, and tail; and it is urged, more especially by
American authorities, that the presence of these points is
indicative of purity of breed; and, *au contraire*, that their
absence generally points to a cross that has been made at no
very distant date.   There is certainly some foundation for this
argument, more particularly as we know that the black pigs
of the Eastern Counties have been crossed with the Berkshires; but, as we have previously remarked, we believe that
a thoroughly good judge should be able to tell, and can tell,

a true-bred Berkshire when he sees it. If we thought that the difficulty of procuring markings of so definite a character did not interfere with the more rapid improvement of the breed, there would be no need to say anything upon this question; but, whether it be in the breeding of cattle, or poultry, or of pigs, we are convinced that the narrowed sphere which, consequent upon the necessity for fancy points, is at the disposal of the breeder for the selection of stock for breeding purposes, cannot fail to influence the result. It frequently happens among swine that the best animals are too faulty, either for stock or exhibition, whereas as pigs of high type they are of surpassing excellence.

It may perhaps be well to notice a practice which happily is now becoming rare. There are exhibitors of Berkshires who so far lose sight of the honourable part they ought to play as to send their stock before the judges blackened and oiled. It may be that a white patch on the skin would otherwise have disqualified. Acting as judge at one of the greatest of our summer exhibitions, we declined to award a prize to the best boar so disfigured. Our colleague, however, thought otherwise. An umpire was consequently sent to decide, with the result that the animal gained the prize. The umpire, however, was himself an advocate of the detestable practice, which we believe is now forbidden by the Royal Agricultural Society.

Mr. Humfrey's Opinion on Modern Points.

Mr. Heber Humfrey, who has long enjoyed a reputation as a breeder of Berkshires, writing to us upon the question of a standard of points, says that the standard published in America, in 1877, was considered a good one, but that English breeders have since been forced by the consumer to change their opinions upon one or two points. Mr. Humfrey finds that whilst the high-class consumer purchases the same kind of breakfast bacon as he used to do—that in which the fat is mixed with a good proportion of lean—the middle and lower-class consumer has, during the last few years, required the same article, so that really the fat bacon has little or no sale at all. This is not only the opinion of Berkshire breeders, but it is general throughout the country; and the agricultural labourer himself, once content with bacon that was absolutely all fat, has now become extremely fastidious in his tastes, and has produced a marked effect upon the sale of pork of all kinds. Mr. Humfrey attributes this, in some measure, to the introduction of tinned meats, which consist almost entirely of lean, and thus it is that this gentleman suggests certain alterations in the type of the Berkshires. He says: "We should like to see them broad across the snout and between the ears, as well as between the eyes. The jowl is fairly valued at four points, but its description as 'full and heavy' is incorrect according to our present requirements. I think 'full' would meet the case, for the old description very often means a corresponding deficiency in the flank; and as the one part is worth fourpence-halfpenny a pound, and the other ninepence, I cannot think that judges should expect us to cultivate an extra heavy jowl. Again, with regard to the back and side, I find that the trade says, 'Leave out the word "short" as a point of merit in the back.' I think it is right to recommend a medium length in the male animal and much greater length in the female; this point, of course, being coupled with a well-sprung rib and a good loin. We require a long, deep side, but it must not be a flat one; and many of the specimens exhibited in the show-yards during the last few years prove that such a qualification is attainable without loss of quality."

Mr. Humfrey sums up the chief points thus: Body long and low, having short, strong legs; head short, with the bridge of the nose nearly at right angles with the forehead; tail twisted; whole body perfectly black except the bridge of the nose, the tip of the tail, and the feet, which are white.

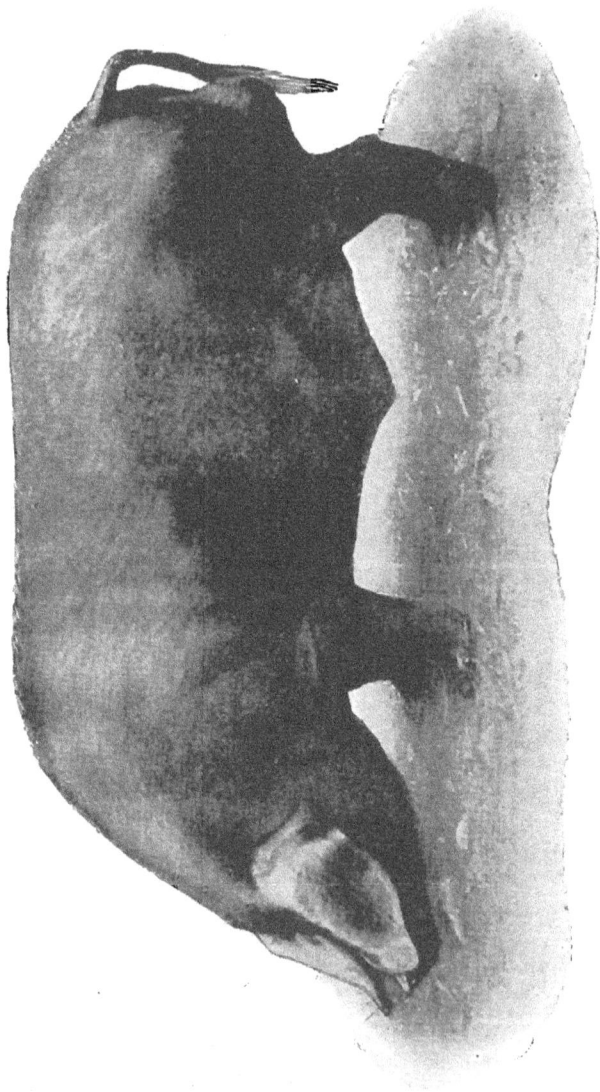

LARGE BLACK BOAR.

Mr. J. Robinson's "Borstal Conqueror" (Herd Book No. 1177), bred by Mr. John Warne.

Gentle in disposition, a good feeder and grazer, the Large Black is a variety of great utility, providing a goodly proportion of lean meat, while the least valuable parts are relatively light. It should have a fine and soft skin, clothed with a moderate quantity of straight, silky hair.

CHAPTER XII.

THE REMAINING BREEDS.

The Large Black Breed.

BREEDERS of the Large Black variety believe that it is one which has some claim to antiquity, inasmuch as it has been bred for generations in different parts of the country by different families; but while this may be the case, it cannot with much reason be insisted that these various herds are in any way allied, or that they in any sense of the word represent one distinct variety. There have undoubtedly been many breeders of Black swine, some of which were of large size and great length, with lop ears, adapted for the production of the large flitches of bacon of old times, and others of small size with short, erect ears, and great aptitude to fatten early. The latter were, and still are, quite opposite in type to the former. They produce more fat in proportion to lean, but are more useful for the production of pork than of bacon. It has been shown by the members of the Large Black Pig Society that in the past great weights have been achieved, and that at the present long, deep-sided carcases, weighing from 160lb. to 190lb. per side dead weight, are now comparatively common. At Smithfield Show, Dec., 1905, the champion pen of two pigs weighed 8¼cwt. at 10¾ months—equal to 360lb. carcase weight. The proportion of lean to fat is large, or, as admirers of the breed believe, larger than in the case of any other variety, while the worst parts of the carcase —the shoulder, the jowl, and the offal—are relatively light.

The Large Blacks are regarded as being of*gentle disposition, good pasturers, as well as good feeders, and a variety the utility of which cannot possibly be disputed. These, however, are early days, and although there are many specimens which are meritorious in size and form, the variety must be regarded as one which is only in the process of manufacture.

The society which promotes the breeding of the Large Black pig was formed in 1898, after a meeting held in the Royal Show ground at Birmingham. At the Smithfield Club meeting of December, in the same year, the Articles of Association, Bye-laws, and Scale of Points were formulated, while the Certificate of Incorporation dates from early in the following year.

The Large Black Pig Society (the Secretary of which is

Fig. 21.—Trademark adopted by the Large Black Pig Society.

Mr. W. J. Wickison, 12, Hanover Square, London, W.) has adopted a shield as a trademark, of which Fig. 21 is a copy.

The chief counties in which the Large Black pig is bred are Cornwall and Devon in the West, where size has been more or less continuously maintained, and Essex and Suffolk in the East, once famous for the Small and Medium-size Blacks, which were of entirely different type. Herds are, however, being formed in all parts of the country, and an export trade has already been established.

SCALE OF POINTS. — The following is the scale of points: Head medium length, and wide between the ears, 5; ears long, thin, and inclined well over the face, 6; jowl medium size, 3; neck fairly long, and muscular, 3; chest wide and deep, 3; shoulders oblique, with narrow plate, 6; back long

and level (rising a little to centre of back not objected to),
12; sides very deep, 10; ribs well-sprung, 5; loin broad, 5;
quarters long, wide, and not drooping, 8; hams large, and
well-filled to hocks, 10; tail set high, and not coarse, 3;
legs short and straight, 5; belly and flank thick and well-
filled, 8; skin fine and soft, 4; coat moderate quantity of
. straight, silky hair, 4. Total, 100. Objections: Head, narrow
forehead or "dished nose"; ears thick, coarse, or pricked; coat
coarse or curly, bristly mane. Disqualification: Any other
colour than black.

HERD-BOOK ENTRIES.—The Regulations for the Registration
of boars and sows provide that no boar or sow is eligible
for registration if known to have a cross of any other breed
within four generations, or if showing any white spot or
mark. The name and registered number of the sire and dam
must be stated, and a distinctive ear-mark must be given to
each animal entered for registration. In the case of animals
entered by persons other than the breeder, the correctness
of the pedigree must (if required) be certified by the signature
of the breeder. Applications for registration are received
subject to inspection of the animals entered, if deemed
necessary. All entries must be made on the Society's printed
form, and must be accompanied by the necessary fees
as follow: Members.—Registration fees: Boars, 5s. per head;
sows, 2s. 6d. per head. Non-Members.—Boars, 10s. per head.
Boars and sows must be entered for registration at time of
service, or at not exceeding twelve months from date of birth.
Double fees will be charged for registration of animals exceeding the age of twelve months at time of entry.

The Small Black Breed.

Whether the Small Black pig known as the Black Suffolk is an
improved Essex, or whether it is an amalgam of the various
black breeds, in a proportion only known to those who have
produced it, it is now impossible to say; but our own impression is, that it has been produced by one or two crosses
upon the original Essex pig, as manufactured by Lord
Western, and subsequently improved by Mr. Fisher Hobbes.

Undoubtedly the first breeder of eminence to take this particular type in hand was Mr. Crisp, of Butler Abbey. We believe that he was also the first to call them Suffolks, and, not content with a black breed which he named after his county, also exhibited white pigs, which, it is now admitted, were improperly called Suffolks.

The original pig of the county of Essex was as distinct in type from that subsequently known by that name as it is possible for two animals to be. The old race was gaunt, bony, coarse, and flat-sided, without any of the qualities so conspicuous in the improved race as first produced by Lord Western. We believe it was in the year 1830 that his lordship purchased, in Italy, a boar and a sow of the Black Neapolitan breed, which, after failing to perpetuate the pure race, he crossed upon some well-selected pigs common to the county of Essex. The result was highly satisfactory, and the new variety was commonly successful at agricultural shows. The parti-coloured markings of the original Essex were obliterated, the Essex Neapolitans became entirely black, they improved in form, quality, and aptitude to fatten, and were for a time the most popular pigs in England; but by degrees their constitution failed, they became less prolific, and lost some of the most important features originally obtained by Lord Western.

It was then that Mr. Fisher Hobbes came upon the scene, and, taking up the Essex pig, still further improved it, and raised it, if possible, to a higher degree of popularity. Fisher Hobbes was a tenant of Lord Western's, and consequently enjoyed unusual facilities for crossing his stock. He established his pigs upon a firm base, and called them the Improved Essex, under which title they were long known in the show-pen—at all events, from the year 1840, when Mr. Hobbes obtained the chief prize.

Although the Improved Essex were superior to the original Essex Neapolitans of Lord Western, both in constitution and in size, they were not what we should call hardy pigs, certainly not so robust as the Berkshires; and it is questionable whether they ever reached such a pitch of excellence as the last-named have attained. Like the Small Whites, they possessed a special aptitude for fattening, and this propensity would appear to be indelibly connected with a somewhat delicate constitution. The Essex, like the more modern Suffolk, were appreciated for their early maturing property, and for the fact that they imparted this valuable quality to every race upon which they were crossed; and there is no doubt that they were largely used, not only upon the Dorsets, the Sussex, and other local varieties, but upon the famous Berkshire itself. The great improvement of which they were the subject points once more to the fact that it is to the introduction of foreign blood that our breeders owe so much; and it also suggests that it is to the keenness of British stock-breeders, and their appreciation of quality, wherever they find it, that the agricultural world is indebted for the finest animals in existence.

The popularity of the Small Black breed is by no means so great as it was; indeed, since the First Edition of this work was published it has almost entirely dropped out of public notice, and it is now no more seen at the great shows of the year. At the same time, the variety has so many good qualities that it should be more widely known. It is particularly adapted for those who prefer a small variety of pig, and we question very much whether it is not more suitable than the Small White, as it attains a greater weight, and, in carefully-bred strains, proves to be stronger in constitution. Its chief fault, however, is that it is too fat, and, bearing in mind the present taste of the public for lean bacon, we doubt whether the breeders of both varieties would not be well advised to attempt the introduction of a larger proportion of lean.

The Small Black is a symmetrical, strongly-haired pig, with fine bones and short legs. It has considerable width between the ears, which are short and carried slightly forward. The jowl is broad, and the chap unusually well developed, a point which is much appreciated by the connoisseurs of this delicate part. The hams are thick and good, the sides deep, but wanting in length. The snout is decidedly short and slightly dished, but the face has not the same character as that of the Small White pig. Great mistakes have been made by breeders in the past in fattening their stock pigs too heavily, and the result of our own experience of the variety —certainly not a fortunate one—we attribute to the loss of stamina, owing to excessive fatness in the original stock.

Breeding pigs require a certain amount of condition, without
which they will not produce vigorous young; but the state of
obesity in which Suffolks were at one time shown in breeding
classes at exhibitions, and even maintained when they were
breeding, was sufficient to undermine their own constitution
and to destroy that of their progeny. In view, therefore, of
the fact that the Small Black is so easily fattened, the
breeder should exercise great care and judgment in the
management of his stock pigs.

AN AMERICAN ERROR.—American breeders appear to have
determined that we have in England a leading breed of swine
known as the Suffolk, which is a *white* pig, but this is an
entire fallacy. The only legitimate breed known in connection with the county is the one we have referred to, now
distinctly known as the Small Black breed. The fact that
one or two Suffolk breeders at the outside have at one time
kept a herd of exceptionally good Small White pigs, which
they have chosen to call Suffolk, is unworthy of notice, inasmuch as they were identical with the Small White pigs of
England, hitherto generally known as the Small Yorks. We
regret that American breeders should have made the mistake,
and that they should have gone so far as to prepare standards
and illustrations of a race unrecognised in the county from
which it is supposed to hail. Mr. Coburn, in his very
practical work, "Swine Husbandry," quotes Sidney, who says
that the Suffolk is only another name for the Small Yorkshire
pig; and yet the former name is retained and adopted, while
the White York, or what we may call the genuine breed,
is to a large extent ignored.

The Small Black breed of pigs is now particularly identi-
fied with the county of Suffolk. Messrs. Crisp and Stearn,
and, later on, Mr. G. M. Sexton devoted much time and attention to the development and improvement of the breed, and
certainly paved the way for such success as was more recently
achieved. The Small Blacks very much resemble the Small
Whites in size and symmetry, although in colour they naturally
present a perfect contrast; and they certainly have more
growth than the latter. It is very seldom, however, that the
hair is seen so curly or so woolly as in some of the highly-
bred specimens of the white breed.

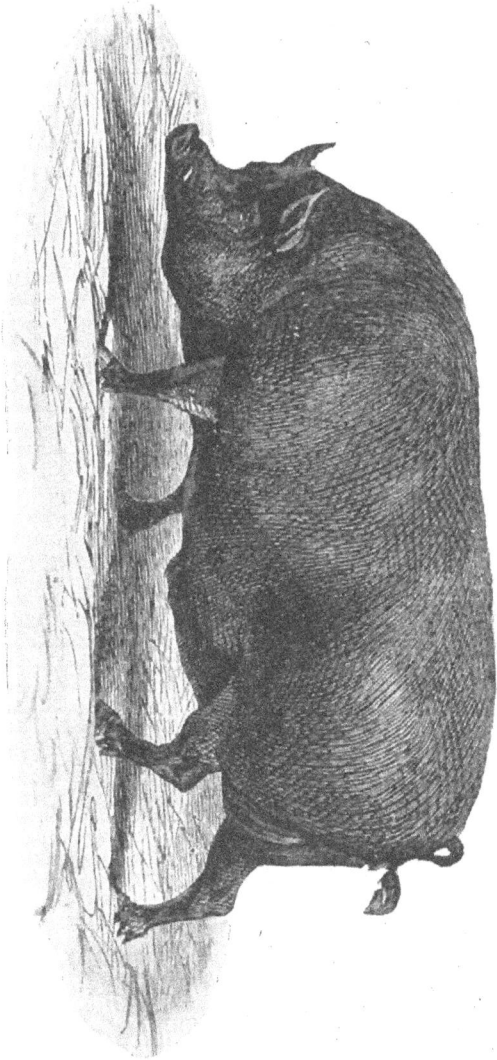

BLACK ESSEX BOAR.

The Essex Pig, formerly popular, has been merged in the black breeds of the country. As a recognised breed it no longer exists. It played an important part in the manufacture of some of our native breeds. The original Essex was gaunt, bony, and flat-sided, but it was greatly improved by crossing with the Neapolitan.

A Breeder's Experience.—Mr. J. A. Smith, of Akenham, Ipswich, who has exhibited the Suffolks with considerable success, says: "The colour of black pigs is so much better adapted to our climate, that it seems to me inexplicable that the pigs bearing it have not beaten the white porcine stock quite out of the field. White pigs cannot be turned out to seek their living in the meadows or clover leys in the summer, but the Small Black breed is well adapted for this purpose. Provided that the weather is not extreme, the Suffolks can be turned out to graze upon good pasture during six months of the year, requiring but little supplementary food. They are naturally *contented*, and will sleep and grow fat when large and ordinary breeds of pigs (nearer to the original type of the wild boar) are tearing up the pastures or breaking through the fences.

"The grazier's problem to produce a maximum amount of flesh with a minimum consumption of food is most satisfactorily solved by the Small Black breed of swine. It is marvellous how rapidly they can be forced into early maturity. Another point in their favour is their extreme lightness of offal. Many of my pigs have lost less than one-fourth in weight after they have been dressed for the shambles. The bones are as small as they can be, consistent with locomotion. Thus, the head and feet, which are always sold at a reduced price, represent but a small proportion of the carcase weight. The flesh is short, sweet, and juicy, and Suffolk hams are known and esteemed throughout England.

"One objection urged against the breed is their tendency to produce an undue proportion of fat, consumers complaining that the bacon is not sufficiently streaky. This seems to me to be the only point in which the Suffolk Small Black pigs suffer by comparison. Pork lard, however, unlike mutton fat, is generally a marketable commodity. When properly managed, our pigs are very prolific, fifteen or sixteen being frequently found in a litter, although ten or eleven is considered a fair number; and we find no difficulty in arranging for two litters a year. Our boars are especially esteemed for crossing with the ordinary farmyard pigs, the produce of the latter having a greater tendency to fatten in consequence."

Mr. J. A. Smith says that the Dorset and Sussex differ

M

materially from both the Essex and the Suffolk pigs, showing greater resemblance to the Neapolitan, their ears pointing
forward, and distinctly differing from the Small Black breed
with which Messrs. Crisp, Stearn, and Sexton obtained such
great success. The cardinal points of the Small Black pig are
quality of flesh, early maturity, and lightness of offal. A
reddish tinge upon the skin should be strongly deprecated,
as it denotes a cross, although perhaps a very distant one.
The hair is another important index to the breed; if it is
found in abundance, it generally indicates a cross with Berkshires of the purest blood, having plenty of fine hair,
entirely free from that woolly appearance which is so often
met with in white pigs.

"I like the head short," adds Mr. Smith, "but not too
insignificant; the nose short and slightly turned up, and the
ears short and erect; the back wide and straight; the tail
set on high; the carcase compact; the legs symmetrical, deep,
and square; and the shoulders wide and full. There should
be just sufficient bone to support the carcase. In pigs intended
for breeding purposes, certain other points should be considered: the teats of the sow should be ample and regular;
and, in the male animal especially, I think the organs of
generation should be well developed, although, as a rule, these
points are scarcely considered by the judges at the principal
shows. As an example of the importance attached to this
matter by some judges, I may remark that none of the prize-
winning sows at the Suffolk show of 1884 could claim the award
as breeding sows.

"In dealing with points, I conclude that they should be
placed in the following order of merit: Back, loin, hams,
shoulders, depth of carcase, head, legs, quality of hair and
flesh, colour, proportion of offal, early maturity. In breeding
pigs, however, the breeding capabilities should stand first and
foremost, and I consider that in every case activity should
be highly valued."

The Black Dorset Breed.

A race of pigs of an exceedingly useful type has been
known for generations in Dorsetshire as well as in the neigh-

bouring counties; but, if we except the highly-improved pigs which were so long bred by Mr. Coate, and which will be referred to later on, they are not such as the naturalist or the skilled breeder would describe as a distinct race. Although Mr. Coate, of Hammoon, near Blandford, to whom Dorset breeders owed a great deal, made use of the local breed for the importation of stamina into his herd, there can be no doubt that this debt has been more than repaid by the aid of the blood of his own herd, which has done so much to improve the pigs of the county, and, we might also say, to convert them into a recognised race.

It is generally believed that, prior to the introduction of Mr. Coate's strain of blood, the Dorset pigs owed most of their quality to the Berkshires of a neighbouring county. There is, however, little or no evidence upon this point; and we can only deal with the Dorsets of to-day, which, in every sense of the word, are, when carefully bred, pigs of high character, and exactly of the type that we might expect to find from the manner in which they have been produced. They show a fineness of quality, and a smoothness and soft-ness in the skin, which are not found in the Berkshires, which are seldom found in the Small Black or Suffolk, and which distinctly point to the Neapolitan or the Chinese cross. The skin is neither a black nor a dark purple, but has a bluish tint, which does not manifest itself in any of the other varieties. The hair, too, is neither vigorous nor strong, and it is chiefly in these points that the Dorset requires improve-ment. The hair and skin indicate a want of vigour and con-stitution—qualifications of the utmost importance in the pig of the farmyard.

Although we have in past years seen some perfect specimens exhibited by Mr. Coate, the ordinary Dorset pig is narrow between the ears — which are rather small and delicate, and carried forward—and the snout is medium in length, rather pointed at the tip, and quite straight. As in almost all pigs of this fine quality of flesh, the bone is exceedingly small, but the legs are often long and the sides short.

The Dorset is easily fatted, and matures early, although the bacon is too destitute of lean for the present taste, and fat pigs are consequently less suited to the market than the more

M 2

lengthy and meaty Berkshires. The sows are usually very prolific, extremely gentle mothers, and excellent nurses of their young. They cross well with the Berkshires, gaining in constitution as well as in the lean properties of their meat.

The name of Coate is so intimately connected with the Dorset variety, that our description would have been incomplete without reference to the extraordinary pigs which Mr. Fred. Coate, of Sturminster Newton, has so frequently exhibited at Smithfield, and without some remarks from his own pen.

Mr. Fred Coate says : " The black Improved Dorset originated as follows. A friend of my father's brought home two sows. from Turkey, and gave him one of them. They were very hairy, and of a wild type. My father at once put a pure Chinese boar to his sow, and was fortunate in obtaining from the cross two or three gilts, which he again crossed with the Neapolitan boar; and since then he has kept his pigs as near to the type thus manufactured as he possibly could. For the sake of retaining vigour and preventing decadence of the strain by in-breeding, he has occasionally used a sow carefully selected from some good herd of the Dorset blacks, and so our variety has been maintained up to the present time.

"My father has exhibited pigs in London every year since 1850, and has never failed to obtain a prize, except upon one occasion. At the last Smithfield Club show [these remarks were written in 1884] he was second for the champion cup, which was won by myself, but he has gained ten champion prizes, besides cups and medals in great number. Dorsets of this breed weigh at six months old about eight score, at twelve months sixteen to seventeen score, and at eighteen months twenty-four to twenty-five score."

The great success of Mr. Coate with his Black Dorsets is almost unique, for as long ago as the Baker Street exhibitions of the Smitheld Club he was exhibiting with a success as marked as that which he still later enjoyed. In the year 1859, the *Illustrated London News* gave a portrait of Mr. Coate's winning pigs, remarking that it was almost " a case of *veni, vidi, vici* "; for—although opposed by the lamented Prince Consort; Mr. Druce, with his Improved Oxfords; Mr. Morland, with his Chiltons; Mr. Bone, with his Berkshires;

Mr. J. K. Tombs, with his Whites; and Mr. Cattle, with his Lincolns—his pigs, whether exhibited as Improved Dorsets or Improved Hampshires, and whether shown by himself or by Mr. Baker, swept all before them, one of the pens taking the gold medal of the society.

Sussex, like Devonshire, has long been noted for a local race of black pigs, which are evidently of a similar type to what the Dorsets were before they were improved by the introduction of Mr. Coate's strain. In neither county are they cultivated with care, nor is there any attempt made to breed them to a type. They have therefore no pretensions to be classed as pure races, and may be dismissed upon the same grounds as those on which we are unable to deal with the so-called Lincolnshire, Hampshire, and Cheshire breeds, all of which have been enthusiastically admired in the past, just when some clever breeder had taken the pains to improve them and to exhibit them at the local fat stock shows.

The Tamworth Breed.

The Tamworth has, from time to time, been described in various ways, and perhaps more in accordance with the taste and fancy of the writers than with the intrinsic merit of the breed. One of the most skilful breeders of the past said that it was a large, coarse, leggy pig, with a straight and thinly-set coat, and a dark chestnut-coloured skin, more or less spotted with black, and that its head was long, tapering, and wedge-shaped, its tail never curled, and it had two great faults. The first was, that no fence which had been invented could prevent it from getting at and digging up his potatoes; and the other, that it was too prolific by half, bringing 50 per cent. too many young ones. In support of the latter statement, the gentleman quoted that of a friend, who declared that his stock increased too fast, and that the longer he fed them, the further they ran into debt. Such, then, is the description of the breed by a most competent authority, in spite of the fact that it is claimed to be of great excellence.

Sidney, writing in 1860, speaks of the Tamworth as the Staffordshire breed, which was rapidly growing out of favour with the farmers on account of its want of aptitude to fatten. Then, as now, a cross between the Tamworth and the Berkshire was considered most valuable; and a practical breeder of that day spoke of the cross as producing the most profitable bacon pigs in the kingdom, the Berkshire blood contributing a tendency to feed, and securing the early maturity in which the Tamworth is deficient. The Tamworth was described by Sidney as a red or red-and-black pig; but the evidence of breeders who can carry their memories back many years earlier, contradicts this statement. Sidney, however, does not appear to have investigated the matter very closely, for his remarks upon the breed are meagre in the extreme.

STANDARD OF EXCELLENCE.—The following is the standard of excellence of the Tamworth breed formulated by the National Pig Breeders' Association and the British Berkshire Society:

Colour golden red hair on a flesh-coloured skin, free from black; head fairly long, snout moderately long and quite straight; face slightly dished, wide between ears; ears rather large, with fine fringe, carried rigid and inclined slightly forward; neck fairly long and muscular, especially in boar; chest wide and deep; shoulders fine, slanting, and well set; legs strong and shapely, with plenty of bone and set well outside body; pasterns strong and sloping; feet strong and of fair size; back long and straight; loin strong and broad; tail set on high and well tasselled; sides long and deep; ribs well sprung and extending well up to flank; belly deep, with straight under line; flank full and well let down; quarters long, wide, and straight from hip to tail; hams broad and full, well let down to hocks; coat abundant, long, straight, and fine; action firm and free. Objections: black hair, very light or ginger hair, curly coat, coarse mane, black spots on skin, slouch or drooping ears, short or upturned snout, heavy shoulders, wrinkled skin, inbent knees, hollowness at back of shoulders.

OPINIONS OF BREEDERS.—Tamworths have been more generally exhibited at the Birmingham Cattle Show than at any other meeting in the United Kingdom; at least, such is our belief, after having visited the winter show for the past thirty years; and many years ago, when speaking upon the subject with

Mr. J. B. Lythall, the then secretary, this gentleman remarked that Mr. John Lowe, who was so long and honourably connected with the Society as its honorary treasurer, could, perhaps, give earlier information than anyone now living. Mr. Lowe, who was then approaching eighty years of age, subsequently told us that he could remember the breed for nearly fifty years, his father having kept them long before that period. He considered the Tamworths to be an ancient and perfectly pure race, entirely distinct from any other variety of pigs known in England. In Mr. Lowe's earliest recollection, they were identical in colour with the Tamworths of to-day— a pure, unspotted, clear sandy—and he had frequently known them to be fatted to forty score, after having reared five or six litters of pigs.

Mr. Lowe, who has long since been dead, was of opinion that the Tamworth had been much improved, for, seventy years ago, it was often required to get most of its own living, it was a good grazer, and especially vigorous in hunting for acorns. It was then, to use Mr. Lowe's words, "fit to jump a five-barred gate," and its proportion of fat to lean was very much less than it is to-day, although modern Tamworths are particularly valuable on account of the large quantity of lean they carry. In the early days the Tamworths were always considered the most useful of farmers' pigs; but Mr. Lowe admitted that they required a considerable time to fatten, which, however, the modern breed does not. When crossed with the Berkshire they threw pigs of very much better form, but the cross was never advantageous from an economical point of view; and, strange to say, the produce was generally divided into two colours, one of which followed the Berkshire, and the other the Tamworth—in other words, setting aside the fact of their improvement in form, the two varieties never seemed to blend. At the early period of which Mr. Lowe spoke, the Tamworth was considered by the farmers of Warwickshire and neighbourhood to grow into a sovereign quicker than any other pig. They were especially hardy and vigorous, they produced larger litters than the Yorks or the Berkshires, and reared their young with less trouble and more satisfactory results.

Mr. Lowe added an amusing anecdote, which, from the

practical point it raises, is worth repeating. He said that, many years ago, some distinguished foreigners visited the Birmingham show, where the chief prize for Tamworths was taken by the Mayor of Birmingham for that year, who was an enthusiastic breeder of the variety. The visitors were taken round the exhibition, and, when shown the pigs, expressed their disapproval, both of the Improved Whites and of the Berkshires. When, however, they reached the Tamworths, and examined the Mayor's prize pen, they manifested both surprise and admiration, declaring that those were the only valuable pigs they had seen in the exhibition. Mr. Lowe asked the reason of this preference, stating that in England the Yorkshires and Berkshires were the leading breeds. The answer given by the visitors was, however, that, in their country, the farmers generally lived a number of miles from a market, usually eight or ten, and that they invariably preferred pigs of the Tamworth type, inasmuch as they were found to be "the only animals which could do the distance." There is certainly something in this point, but modern breeders find that it is the quiet, contented animal, whether it be a pig or a bullock, that lays on flesh the most rapidly; and that, on the contrary, those animals which are never contented, and, from their roaming and energetic nature, are always searching for food, and expending their vitality, are the swine which consume so much food to waste, and are invariably long in fattening.

Mr. F. C. Fidgeon, of Tamworth, a very old breeder of the variety, and one who has a great knowledge of its early history, remarked at the time of the first appearance of this work that he had known the Tamworths for nearly sixty years, and that he had never noticed so great a change in them as during the previous seven or eight years. A generation ago they had been a lean pig, of a very dark red colour, which never varied in shade, and had been provided with a very long snout. Later, however, the colour altered in shade, having become a sandy or lighter red, a point which is apparently growing in favour; "although," said Mr. Fidgeon, "I question very much whether this colour is as pure as the original dark red, for I believe it has been obtained by a not very distant cross with a white pig."

Mr. Fidgeon added: "As a breeder, the Tamworth is very good indeed. My sows generally bring from ten to fifteen young pigs at a litter; they suckle their young ones well, and, where a breeder desires to sell his litters at a pound a head when taken from the sow, I believe there is no other class of pig that will realise the price so quickly. The Tamworth is an exceedingly fleshy pig when fed for bacon. One of the principal reasons why I keep the breed is because the bacon is so fine in the grain and so well interlarded with lean. The gilts will breed very early, but it is my practice never to use them for stock until they are twelve months old; and, as a consequence, I generally have a very good sow with a large litter of pigs. To the latter I give bran or sharps mixed with skim milk. With regard to the shape of the snout, I certainly should prefer to see it shorter; but that seems quite foreign to the breed, as they are particularly long in this point. Perhaps I should also say that the Tamworth pig requires age to feed well, as, if it is put up to fatten too soon, it grows instead of feeding. But there is not much difficulty in feeding it up to twenty or twenty-five score when it has reached the age of twelve months. I speak of ordinary feeding, and not of feeding for the purpose of exhibition."

Mr. G. M. Allender, whose support of the Tamworths at the Royal Agricultural Society's exhibition was most spirited, and who commenced to breed pigs nearly forty years ago, found that the bacon-curers complained of the modern Berkshires as being too fat, all black, and deficient in the sides. For bacon, he regarded a long, deep animal as the most profitable; and, bearing this fact in mind, he said, before his death:

"I determined to try the Tamworths, and I most certainly do not regret this decision. They are prolific and good mothers, and, as an example, I may quote my three young sows, which were first, second, and third at the Royal Shrewsbury meeting. These farrowed in August with their first litters, although only nine months old, and had twenty-seven pigs between them, two of the three having, at the time I write, just farrowed with their second litters of eleven each. They would also appear to carry this prolificacy with them when used for crossing, for I have two half-breds, a cross with the white breed, which, at twenty-two months old, have given me fifty-five pigs. I find that the Tamworths feed early, and I have hams now hanging which exceed 20lb. each in weight, smoked and dried, although they were cut from pigs which were only twenty-seven weeks old. A Sheffield butcher, who bought my second prize Smithfield pen of fat Tamworths, together with several other prize pens, said that the Tamworths were his best bargain, and that their flesh contained more lean than any of the other prize pigs which he killed."

Messrs. Mitchell Brothers, of Birmingham, who won the first prize with Tamworths at the Birmingham Fat Stock Show, in 1884, informed us that the pigs were farrowed on the 13th June, and were from a gilt twelve months old at the time of farrowing. They were fed upon wheat and barley meal, scalded and mixed with skim milk, and when exhibited they were only five and a half months old, but weighed above sixty score, or twelve score per pig—a weight which in itself is sufficient to stamp the Improved Tamworth as an extremely useful and precocious variety. Messrs. Mitchell believe that the best of the five pigs weighed at least twelve score dead weight. These gentlemen have bred and fed a large number of pigs of the Berkshire, White, and Tamworth varieties, and their reason for selecting the last-named is that they find the pigs will stand more forcing than any other breed, that they feed and grow at the same time, and that when killed they weigh extremely well, and have a larger proportion of lean meat. During two years, Messrs. Mitchell bred and fed three hundred pigs, all of which were by Tamworth boars from Berkshire and Middle White sows. These were all fed with warm food, and sold to butchers, and averaged from ten to twelve score, although under nine months old.

The Duroc-Jersey Breed.

There are in America two sub-varieties of the red pig, one of which is known as the Jersey Red, and the other as the Duroc. It appears, from the Report of the Swine Breeders'

Convention, that the origin of the Jersey Red is unknown, but that they have been bred in the State of New Jersey for upwards of fifty years, and are considered by many farmers to be a most valuable breed. They are large, reaching from 500lb. to 600lb. in weight, and they are now extensively bred in the middle and southern portions of the State. With regard to colour, it would appear that sufficient care is not taken to breed them uniformly, for, while they are of a dark red in some districts, they are a lighter sandy colour in others, and are sometimes found marked with white. The Convention did not believe that they descended from the Tamworths, as there is no record of an importation of this British race; but they are supposed to be descended from the old race of Berkshires, although they are much coarser than the improved pig of this variety.

The points of the Jersey Red—to which we refer in succession to the Tamworth, as we have no doubt whatever that it has descended in some way from that breed—are a red colour, large lop ears, a snout of moderate length, a small head in proportion to the size and length of the body, heavy tail, coarse hair, great length, and thin legs, which are unusually long.

It is certainly surprising to find the above description proceeding from an American association, for we should prefer not to recognise as a distinct variety a pig which embodied characteristics so antagonistic to those of the improved and useful races.

A RED PIG CLUB.—The Duroc Reds, which found an able champion in the well-known Colonel Curtis, are chiefly bred in Saratoga County, New York, where they have been known for more than forty years. In 1883, an organisation known as the Duroc or Jersey Red Swine Club was formed with a view to improving the red pigs and establishing a registry of pedigrees. The standard drawn up by the club was as follows: "The true Duroc, or Jersey Red, should be long, quite deep-bodied, not round, but broad in the back, and holding the width well out to the hips and hams. The head should be small compared with the body, and the cheek broad and full, with considerable breadth between the eyes. The neck should be short and thick, and the face slightly curved, with the nose rather longer than in the English breeds; the ears rather large, and topped over the eyes, and not erect. Bone not fine, nor yet coarse, but medium. The legs medium in size and length, but set well under the body and well apart, and not cut up·high in the flank or above the knee. The hams should be broad and full, well down to the hocks. There should be a good coat of hair, of medium fineness, inclining to bristles at the top of the shoulders; the tail should be hairy, and not small; the hair usually straight, but, in some cases, a little wavy. The colour should be red, varying from dark, glossy cherry-red, and even brownish hair, to light yellowish-red, with occasionally a small fleck of black on the belly and legs. The darker shades of red are preferred by most breeders, and this type of colour is the most desirable. In disposition, they are remarkably mild and gentle. When full grown, they should dress from 400lb. to 500lb., and pigs at nine months old should dress from 250lb. to 300lb.''

Since the step above indicated was taken, the majority of the breeders connected with the association were convinced that the Durocs had prior claims to the title of a distinct breed to the Jerseys; and, recognising that both came from the original stock which Colonel Curtis claimed to be a Red Berkshire, the two breeds were united, and are now known under the name of the Duroc-Jersey. The name of the Jersey Red arose some fifteen years ago with the late Mr. J. B. Lyman, at one time Agricultural Editor of the *New York Tribune*. There is no doubt that the adoption of a standard by the club referred to has led to an improvement in the breed by the inevitable crossing between the hitherto distinct varieties, and that uniformity of colour and other points will ultimately be established.

An Illinois Breeder on Red Pigs.—Mr. Thomas Bennett, of Rossville, Illinois, a large breeder, who retains, on the average, about one hundred breeding sows, says that he prefers to breed so that his young pigs come in at one year old; but he admits that if the gilts do not farrow before they are eighteen months old they make finer sows. He feeds his red sows upon grass, with grain or maize, in summer; and in winter he gives roots instead of grass. The sows are usually half-fat at the time of farrowing. The boars are generally allowed a grass run for exercise and the maintenance of condition. After farrowing, the sows are fed upon bran or wheat offal. If grass is plentiful at the time of parturition, the sows are allowed to graze every day, and, when the young pigs commence to eat, they are fed with oats daily, together with a little milk or meal slop; this is gradually increased until they are weaned, when they receive steamed oats, rye, and maize, in equal parts, mixed with one gallon of oil-cake. This mixture is placed in water for steaming.

Mr. Bennett recommends mangels very strongly for growing pigs or for fattening. In the latter case, he prefers that the ration should consist of two-thirds maize. Illinois breeders usually feed upon maize, oats, and rye, for very little barley or pulse is grown. Almost all the food for the young Duroc-Jerseys is steamed or otherwise cooked, and this is considered most beneficial, although for older pigs the practice is not so generally adopted.

In answer to our question, "What breed makes the best weight at six months, and what should the pigs weigh?" Mr. Bennett remarks, "We have in America a few breeds that I think are almost equal in this respect, viz.: (1) Poland Chinas, (2) Small Yorkshires, and (3) Duroc-Jerseys. Any one of these breeds will gain a pound a day in weight from the date of birth, if they are properly managed. Extra feeding and good management are necessary to make a pig at six months old weigh 200lb., but there are some breeders who can reach this weight. The Tamworth, if kept until it is twelve months old, will, on the average, gain more than any other breed."

American breeders of red pigs will notice that Messrs. Mitchell's pen of five prize pigs, which we have previously mentioned, far exceeded the weight quoted by Mr. Bennett, although they were under six months. Referring, however, to the White York breed, Mr. Bennett adds: "The reason why these are not kept is because they are not adapted to the Western country. They will suit the kid-glove farmer, but, as a rule, we have no use for them. The farmers in the United States want a hog that will make his own bed, grind his own corn, and take care of himself. Some farmers call the White hogs 'Band-box' hogs [we are not told the reason

why]. It is the custom here to feed cattle and hogs together,
The former first consume the maize, and after it has passed
through their system the latter feed and fatten upon it.
Sometimes the feeding yard is dry and comfortable; sometimes
it is six inches deep in mud, or 20deg. below zero. In every
case, the animals are without any more shelter than that
furnished by the heavens above them. Thus, you see it takes
a hog with a good constitution, and a good coat of hair
upon it for protection, to exist under our system. The best
cattle-hog we have to-day is the Duroc-Jersey."

The Poland-China Breed.

The origin of the Poland-China pig, like that of many other
races of modern live stock, is shrouded in obscurity; and,
although we have read numerous accounts in American
journals, purporting to be faithful descriptions of its manufacture, we have as often seen contradictory statements
made by men of position and authority. How, when, and
where the Poland - China was manufactured will probably
never be known; but it is sufficient to state that it is a pig
of high merit, of a fixed type, and of unusual character,
and that for these reasons, and these alone, we have
thought it necessary to give it a place in this work. Nor
have we done so without some little examination of the animal
itself: for, although it is a purely American pig, it is now
largely bred in Germany; and we have there had the advantage of seeing imported stock of the highest class, which fully
confirmed us in expressed opinions that it was a really useful
animal.

In his work upon " Swine Husbandry," Mr. Coburn, of
Kansas City, gives a short history of the Poland-China, in
the form of descriptions written by some of the oldest and
most reliable of American breeders; and from these it appears
that, at different times, the variety has been called by a very
large number of names, some of which were evidently taken
from the district in which it was produced, or from the
breeders themelves. Mr. Coburn thinks that the main crosses

used in its formation were local stock, known as Warren
County pigs, these having been crossed with the China, or
perhaps with the Bedford breed. The latter is evidently a
variety of the English Large White, but which the writer in
question describes as coarse, large, and slow to grow and
fatten; the Chinas, however, are of exactly opposite qualities.
Mr. Coburn concludes that the cross was made chiefly in
Butler and Warren counties, in Ohio.

The evidence of Mr. Cephas Holloway (who settled in the
former county in the year 1807), read before the National
Convention of Swine Breeders in 1872, is worthy of notice.
This gentleman said that the first introduction of China pigs
into his part of Ohio was in 1816; a friend of his having
visited Philadelphia in that year, where he saw some " Big
Chinas," which he purchased and brought home. The boar
and two sows were entirely white, the third sow having some
sandy and black spots upon her skin. These animals were
stated to be either imported or bred from China stock. They
were used for crossing upon the best pigs then existing, and
the produce became known as Warren County pigs. Two
other varieties of swine were introduced into Warren County
in 1835, these being the Berkshires, which were black, with
white marks upon the feet, tail, and face, and with sharp-
pointed, upright ears; and a pig termed the Irish Grazier,
which was white, with occasional sandy spots. Both the Berkshire and the Irish varieties were crossed upon the Warren
County pig, and from the mixture, in the course of time, was
evolved what is now known as the Poland-China. China pigs
were subsequently introduced into the district; but these,
although equally good in fattening qualities, were neither so
prolific nor so large, and generally had small, erect ears.

There have been many claimants to the manufacture of the
Poland-China; but Mr. Coburn remarks that "no one man
can say he had more to do in the formation of this breed
than another: it was the result of the labours of many : it
grew out of the introduction of the China hogs by the
Shakers, of Union Village, the crossing with the Russia and
the Byfield, and subsequent crossing with the Berkshires and
the Irish Graziers." Since 1841 or 1842, these breeds have not
existed in the counties in question; they have not, therefore,

been crossed upon the modern pig for upwards of sixty years. In the Report upon the race which was adopted by the Swine Breeders' Convention, we find that, among other crosses, the Berkshire blood was liberally introduced before 1840; and that for fifty years no new blood had been introduced into the breed, nor had any efforts been made to obtain a new supply of the blood of either breed previously used. The Report concludes with this remark: "Thus we have a breed thoroughly established, of fixed character, of fine style, and unquestioned good qualities, which can be relied upon for the production of a progeny of like qualities and character."

OPINIONS UPON THE BREED.—Among the claimants to the manufacture of the breed, one declares that the Berkshires were never used at all—a point which we believe is not accepted in America by those who are best qualified to form an opinion. Mr. Coburn says: "It is of much greater importance to millions of people to understand the fact that there has been produced a race of swine which many intelligent men consider to be the best pork-making machines known—nearer, indeed, to what the farmers of the Great West require than any other single breed in existence. Their size, colour, hardiness, docility, and good feeding qualities make them favourites, when purely bred; and where fineness of contour, quick maturity, and a little less size are demanded, it is found that, if sows are put to Berkshires, they produce some of the best possible of feeding hogs."

In a well-known book, "The Western Farmer," the author says that, in the breeding of swine, the Berkshire breed is probably destined to supersede all others in the northern localities of the United States; but for the intermediate portion, or between 38deg. and 42deg. of north latitude, the Poland-China is destined to hold the favourable opinion of the Western farmer. Again, in "The Hog," Dr. Chase remarks that Poland-Chinas are not suitable for a small farm where a few pigs are annually fattened; but, for the farmer raising and fattening a large number every year, this breed has no superior.

Mr. Conover, of Butler County, U.S.A., a prominent breeder of the variety for many years, gives an example of how the Poland-China weighs. A sow, which was born on the 10th of

June, farrowed on the 18th of April following, with eleven
pigs. In the succeeding October, the sow and young weighed
2,735lb., all being in good show condition. The sow was
fattened, and her net weight proved to be 535lb. The young
sows were kept for breeding, while the five young hogs
averaged 282lb. at eight months and twenty days old.

Mr. Levi Arnold, a famous Michigan breeder, reports that
he had a sow weighing 762lb., the width of her back being
28in. Mr. Arnold says : " We feel proud of the Poland-China,
as it is emphatically an American-bred hog. It is of fine
bone and large size, combining, more eminently than any
other, the excellencies of both large and small breeds. It is
docile, an excellent feeder, breeder, and suckler, capable of
fattening readily at any age and attaining great weight at
maturity; in fact, it will grow and lay on fat as long as
anyone will continue to feed it. Poland-Chinas usually dress
from 300lb. to 350lb. at from ten to twelve months old, and
500lb. to 600lb. at eighteen months old. In colour they
are dark, generally black, with small white irregularly-placed
marks upon the body; but there is usually some white
upon the face, or upon the legs and feet. The bodies are
long, the necks short, the jowls heavy, and the legs short
and well spread. The back is straight and broad, the sides are
deep, the shoulders heavy and square, hams broad and deep,
frequently overlying the hock joints, and the ears thin and
drooping at the points. There is no flabby, thin, belly meat,
but plenty of thickness in the front of the ham, and very
little offal when fat. Pigs of this breed are strong and hardy,
and will make more pounds of pork per bushel of corn than
any other pig known in America."

Mr. Ellsworth, another large breeder, residing in Illinois,
says : " I think the Poland-China makes the best weight at six
months, when it should weigh from 200lb. to 260lb., although
I have had larger. One of my pigs at ninety days weighed
100lb., and one at 165 days weighed 225lb.; both being in
nice thrifty condition, but not over fat. We do not like the
White Yorkshires in Illinois, because their skin is too thin,
and white pigs do not thrive so well as black ones. We find
a Poland-China a much more profitable hog, and the Berkshire to be preferred."

N

The Record Company's Scale of Points.—The American Poland-China Record Company published the following comments:

(1) Poland-Chinas are the best-known machines to manufacture corn into pork. They thrive better than any other breed upon a diet mainly composed of corn.

(2) They are docile, hardy, prolific, and almost invariably good mothers.

(3) In large pastures they are not inclined to waste flesh by over-exercise.

(4) The Improved Poland-China has as small offal in proportion to carcase as any hog in the hands of the general farmer.

The following are the characteristics of the improved Poland-China hog, based upon a standard of 100 points:

	POINTS.
Short nose (dished)	4
Small head	4
Width between eyes	4
Ear fine, thin, drooping	5
Neck short, thick, and well arched	4
Jowl large	3
Shoulders broad and deep	9
Girth around heart	9
Back straight	6
Depth of body	9
Ribs well arched	9
Loin wide and strong	7
Width of body at belly	5
Hams broad, full, deep	12
Coating, fine and thick	4
Legs short and of good strength	6
	100

Mr. A. C. Moore, of Canton, Illinois, to whom we are indebted for some interesting letters upon the Poland-China, claims to be, if not the originator, at least the first to breed this pig to a type. He says, that after eight years of labour he sold, in 1854, to one firm, thirty-two head of seventeen months old, and twenty-five head of nine months old pigs, the one lot averaging 469lb., and the other 207lb.; while, in the following year, he sold to another firm forty head, averaging 439lb. It was then that Mr. Moore concluded that he had *the* pig which was destined to take the leading place in the country, and which was uniform in almost every respect. In his scale of points, he gives the highest place to breadth and depth of shoulders, depth of sides, and hams, and the lowest to shortness of nose. He claims for the breed that it has no equal for early fattening qualities and continued growth; that it will thrive better upon grass and clover than any other breed; that it is a good breeder, a kind mother, and a good suckler; and that it is the best hog for the farmer and the packer, because of the proportion of its weight between the shoulders, its small head and feet, and the small percentage of offal it produces.

The Chester White Breed.

The leading native race of white American pigs originated in Chester County, Pennsylvania, and is believed to have descended and to have derived its principal properties from some Large White pigs which were taken from Bedfordshire, in England, to West Chester, in the year 1818. These pigs were largely used by the farmers of the State in the improvement of the common white pigs of the country, and, in course of time, by continual crossing and selection of animals which approached a high standard, the Chester White was produced. It is believed by American stock breeders to be one of the best races of pigs of the present day.

Some idea of the quality of this variety may be gained when we state that Berkshires and Suffolks have both been introduced into Chester County, and are said to have been thoroughly tried by the stock-breeding farmers; but they did not find such favour as, and are stated to have yielded less valuable results than, the breed to which we have specially referred.

It has been asserted that the pork-packers of America have objected to the Chester White as being too large for their trade; but the breeders claim that it can be fattened at any age with very little trouble, and made to produce any weight and size that the dealers may require. For example, pigs born in the spring are usually fit for the packer in the following autumn; but, by continuing to feed them well, they can be made to reach weights varying between 70 and 120 stones of 8lb., one animal that weighed 1300lb. alive having been exhibited at the Philadelphia Exhibition. Mr. Coburn says that the best Chester Whites stand in the same relative position to the Poland-Chinas as the Essex did to the Suffolks, if we except the colour—inasmuch as, if a Chester White were partially black, it might be mistaken for a Poland-China; and, on the other hand, a perfectly white Poland-China could hardly be distinguished from a Chester White.

The Chesters are classed with the large breeds of pigs; they produce their young true to colour under all circumstances of crossing; they are remarkably clean in their habits, unusually docile, and make an excellent foundation for crossing smaller breeds, imparting size when the sow is used, as well as early maturity—whereas the boar, in crossing upon common races, imprints fine quality upon his offspring. A capital cross is made between a Small White boar and a large Chester sow, the young pigs inheriting excellent points and the very highest of feeding qualities.

The points of the Chester Whites, as adopted by the National Swine Convention of America, are as follows: Head short, broad between the eyes; ears thin, projecting forward, and lapping at the point; neck short and thick; jowl large; body lengthy and deep, broad on back; hams full and deep; legs short, and well set under for bearing the weight; coating thinnish, white, straight, or if a little wavy not objectionable; small tail, and no bristles.

The Cheshire or Jefferson County Breed.

There is a breed of swine known in America as the Cheshire, or Jefferson County, breed, which originated in Jefferson County, New York, and is believed to have been derived from an importation of pigs from the county of Cheshire, in England. Mr. Sanders, editor of the *Breeders' Gazette,* who was well known to many English breeders, said that the Cheshire was simply an offspring of the York. He bred the former for several years, and declared that he produced all the types of the Yorkshire, from the large to the small, short-snouted variety. It bred true to colour, and he never remembered producing a pig with a black hair upon it, although his pigs were crossed with all breeds, including the Black Essex. He noticed, however, that blue patches frequently appeared upon the skin, even in the purest specimens; and, although the peculiarity would sometimes disappear for one or two generations, it would ultimately crop up stronger than ever.

The type that Mr. Sanders succeeded in fixing upon the Cheshires was almost identical, in size and quality, with the most approved medium Berkshire; and so marked was the resemblance in all but colour, that they were frequently, but facetiously, styled White Berkshires. They carried plenty of hair, and a delicate pink skin, and the meat was excellent.

The Victoria Breed.

Colonel F. D. Curtis, of Saratoga County, New York, is the originator of another breed, known in America as the Victoria. This was produced from the native Byfield hogs, subsequently crossed by the York and the Suffolk varieties (by the Suffolk, we presume the English Small White is intended). The Victorias resemble the Middle White breed in many respects, and are descended from a famous sow named Queen Victoria, which doubtless gave them their name. A special committee of the Swine Convention reported highly in their favour, and asked for them a hearty welcome and a fair trial at the hands of American breeders.

The points of the Victoria are as follow: Colour white, with a good coat of fine soft hair; head thin, fine, and closely set on the shoulders; face slightly dished; snout short; ears erect, small, and very light or thin; shoulders bulging and deep; legs short and fine; back broad, straight, and level, and the body long; hams round and swelling, and high at the base of the tail, with plaits or folds between the thighs; tail fine and free from wrinkles or rolls; feathers or rosettes on the back are common; skin thin, soft, and elastic; flesh fine-grained and firm, with small bone and thick side pork.

These pigs easily keep in condition, and can be made ready for slaughter at any age.

The Pigs of France.

During a number of visits to French exhibitions and stock-breeding farms, we have noticed that although, in some instances, admirable specimens of native races are to be found, yet the very best pigs produced and exhibited in France are either of a pure English race or of a cross between the British and native races. The leading French breed is, un-doubtedly, the Craonnaise, which, in these later days, has been bred to a high standard of excellence—at least, for France—and with continued improvement we have no doubt that it will at least equal some of our English breeds. In the North of France we have noticed that the Augeronne has been very carefully cultivated, and that although the pigs of this race are well exemplified in the accompanying Plate, yet a few of the best breeders have distinctly improved upon it, and produced an animal which, though still too high upon the leg, and long-snouted, is well-haired, thick through the heart, broad in the loin, and carries exceedingly good hams. We may specially men-tion the Augeronne pigs bred by M. l'Abbé of Ommeel, Orne. Some of the porcine stock shown by this gentleman were much shorter in the head than the ordinary Augeronne; their skin was pink, and the hair unusally fine; but the ears were coarse, and flapped over the sides of the face, being a distinct contradiction to the other properties of the animal. Side by side, however, with the Berkshires of M. Desvignes of Bazouges, Sarthe; of Mr. Souchard of Verron; of M. Dumou-tier, of Claville, Eure; and the White English pigs of M. Quetel, of Saint-Come-du-Mont, Manche, they were much behind, and only served to show how well-bred pigs of the French races fail to compare with those of a British race, even when produced by a Continental breeder.

. THE PIGS OF NORMANDY.—According to M. Gustave Heuzé, author of "Le Porc," French pigs are divided into seven groups, those in the first four being white, and having pendent ears.

The first group consists of the common pig, under which heading the local swine, known by the following names, are classed: Bretonne, Augoumoise, Nivernaise, Poitevine, Bourbonnaise, Berrichonne, and Vendéenne. The common pig of France needs no description.

Under group two, the Normandy breed, are included the following sub-varieties: Cotentine, Mancelle, Cauchoise, Nonant, Alençonnaise, and Perche. The pigs of Normandy are an improvement upon the common pigs above referred to. They are, however, very far from being perfect, being furnished with coarse, hard hair, a thick skin, deficient hams, thin collars, and flat sides; although the flesh is exceedingly good, and leaner than is generally found in the pigs of England. They are appreciated by the peasantry for one quality especially, viz., they can be easily driven to distant markets without appreciable loss of any kind; and the peasants claim that this could not be done with the improved races of Great Britain. These pigs frequently attain considerable weight, and are very prolific.

In the valley of the Auge, the farmers have made great progress in the improvement of their pigs, and have produced a variety which, although essentially Norman, has been classed by French writers under a separate heading. This is the Augeronne, to which we have referred. In this race, the head is shorter than in the common breed of Normandy, and the snout approaches more closely to that of the Middle White breed of England; but the ears are large, and fall over the side of the face. There is a deficiency in the colour, the bone is too coarse, and the legs are too long; but the hair, which is white, is silky, short, and abundant, and the back is broad and tolerably straight. The Augeronne fattens with considerable aptitude, and pigs of a very fair size are fatted at the age of from fourteen to sixteen months, the live weight at this period frequently reaching 650lb. Pigs of this race, moreover, are largely kept in the dairying districts of France, where they are fed for the Paris markets upon buttermilk and skim-milk.

THE CRAONNAISE.—The next French group is that comprising the pigs of the Craonnaise race, the stronghold of which is the Department of Mayenne. This famous French breed is also known by several synonymous names, among them being those of Charollaise, Poitou, and Angevine; and also by the

names of Bourbonnaise, Mancelle, Nivernaise, Vendéenne, &c.,
which appear to be applied to it as well as to the races previously named.

The Craonnaise is a large upstanding pig, the head not
being carried so close to the ground as in other French races.
The ears are smaller, but pendent; the body is unusually long,
straight, and well formed, the shoulders being wide apart,
the collar thick, and the loin broad. The legs, too, are short
and muscular, and the skin fine and well covered with hair
of nice quality. One failing of the Craonnaise is, that it is
not so prolific as some of the less approved races; but, according to Heuzé and Léouzon, young pigs weigh, at fifteen
to eighteen months, 400lb. to 550lb., and the *charcutier* of
Paris esteems its flesh as superior to that of any other French
race of pig. A *charcutier* (practically a vendor of cooked as
well as raw flesh) is a French shopkeeper who has no equivalent in this country. He prepares and sells pig-meat in a
variety of tasty forms—a system which English pork-butchers
might adopt with the greatest advantage.

The Craonnaise race is common in Anjou, and in portions
of Du Mans, Bretagne, Poitou, and Angoumois; and there is
probably no variety that is more easily improved by a cross
with the best breeds of this country. Unfortunately, all sorts
of pigs are termed Craonnaise, just as, in England, white
pigs are often called Yorkshires whatever their quality; and
in this way, as, indeed, in England also, improvement is
retarded in those distant counties, where the highest quality of
porcine stock is seldom seen upon the farm.

THE LORRAINE.—The fourth group of French pigs comprises
those of the Lorraine race, the sub-varieties of which
are designated by such names as Vosgienne, Alsacienne,
Ardennaise, Flamand, Picarde, Artésienne, &c. The Lorraine
pig is largely found in the Departments of the Moselle and
the Meurthe, as also in Alsace and the Ardennes. It is of
medium size, a greyish-white colour, and often marked upon
the body with black spots or patches, while the ears are
large, and the head is long. It is larger than, but by no
means so well formed as, the pigs of Normandy. The body
is also longer and narrower, the sides being flat and the
bones coarse. Heuzé says that the Lorraine is slow in growth,

being often badly fed during the first months of its life; but the meat is excellent in quality, and the fat much esteemed.

The remaining three French groups comprise pigs which are black and white, or pied, in colour, and have ears neither erect nor actually falling over, but what may be described as semi-erect.

THE PÉRIGORDINE.—The fifth group is that comprising pigs of the Périgordine race (as shown in the accompanying Plate), also known as the Périgord, Limousine, Gascogne, Bayonne, Landaise, Navarine, &c., and is extensively bred in many departments of France, including that of Haute-Vienne. These pigs are large, and have strong muscular limbs, which are much shorter than they used to be. In colour they were formerly a grey-black, but, by crossing with the Bourbonnais, a pied, or black-and-white, pig has been produced, and this colour the breeders prefer; indeed, they go so far as to insist as much as they can upon there being a saddle, or black band, across the middle of the body. The Périgordine is fairly precocious in growth, gentle in disposition, and entirely rustic; but it is certainly a coarse pig, although highly esteemed in the districts in which it is bred.

THE BRESSANE.—The sixth group of French pigs is that known as the Bressane, and the varieties under this heading are known as the Bresse, the Bugey, the black Bourbonnaise, and the pied Charollaise. The Bressane pigs are found in Franche-Comté, Bourbonnaise, Dauphiné, Bresse, Beaujolais, &c. In colour they are almost black, with a white saddle or band across the middle of the body. The sows are most prolific and capital mothers; but there is no special qualification in the variety that distinguishes it from the races previously described.

THE CORSE.—The seventh and last group, that of the Corse, appears to be the only black pig claiming to be a distinct variety that is bred in France. In the Corse, which is practically the race of the Pyrenees, the ears are erect, the skin is brownish-black in colour, and covered with fine black hair, of which, however, there is an insufficient quantity. The head is large and long, and the body fairly well formed; but the hams are deficient in thickness. The flesh, nevertheless, is excellent, and the skin fine, and young pigs of the breed are fatted with ease. It is claimed by some authorities in France that the flesh of this black pig is superior to that of any other race.

M. Heuzé states, in his work, that, although French pigs are too coarse in bone, they have the advantage of supplying a meat of the highest quality, and of being more prolific than pigs of the English races; moreover, he declares that, when crossed upon superior breeds, they lose their principal defects, and acquire qualities that place them side by side with the most useful and economical pigs known in Europe.

OPINION OF A FRENCH JUDGE.—We have been favoured with some remarks by M. Demole, of Geneva, one of the most popular and able of French judges. M. Demole states—and he has conveyed similar remarks to the French Inspector-General—that the number of exhibits in the principal French shows alone proves that the system of crossing with the English races is now general, and that it is an evident sign of the utility and profit derived from it.

With regard to the English-French cross, M. Demole says: "I look at the matter in this light. The forest feeding of pigs has considerably diminished in France, while in many districts it has altogether disappeared; and the peasant has now no further need of long-legged animals capable of making long journeys in search of food. The short-legged type has, therefore, been accepted by our people, and, wherever the English pig has been introduced, it has quickly taken its place in the piggery, more especially in the sty of the poor man. It has merited and justified this preference by its form, its precocity, and its facility for fattening. The above facts are admitted upon all hands, and I need not, consequently, attempt to prove them.

"With us, the English pig is enthroned; but we find that it has two faults, viz., the fat is less firm than that of pigs of French races, and the flesh is not sufficiently marbled with lean. Evidently, therefore, in these respects the French pig is preferable to the English; and thus it is that we elect to make a cross, in order to obtain the most desirable form, precocity, fattening properties, quality of meat, and firmness of fat. Our breeders have had the good sense to abandon the small races, and to cross the large breeds of France, such as those of Normandy and Craonnaise, with the Berkshire of England."

At the Paris Exhibition, in 1878 (which we visited, as also in 1880, 1890, and 1901), the first prize in the first class was awarded to a Normandy pig ten and a half months old, and weighing 800lb.; whereas the first prize in the third class, and the prize of honour, were awarded to a Normandy-Yorkshire, eleven months old, weighing 795lb.—three pigs of the Normandy race, in the fourth class, weighing together 265 stones. "Thus, for the fourth time," says M. Demole, "the prize of honour was taken by pigs which were crossed with an English race."

In his Report to the Minister of Agriculture, following upon the Exhibition at Paris, in 1878, Mr. Demole spoke of the pigs of the English races, both large and small, as being much shorter on the leg than French pigs, and furnished with wider loins and more massive fronts. He also remarked that the English jurors, from the commencement to the finish, always preached *vieux cheval, vieux porc, et jeune taureau* (an old horse, an old pig, and a young bull) as an axiom of reproduction, and that, consequently, although the average age of the pigs in the British classes was fifteen months and two days, the average age of the prize animals in the same classes was twenty-three months and two days.

If we except a race of pigs in the South of Italy, which was described in the introductory chapter, under the name of the Neapolitan, there are no swine in Europe, beyond those already described, which merit recognition. Even in Italy, however, we have found English pigs held in the highest esteem, and on one occasion the fact of being an English breeder provided a short cut to the heart of a large Italian farmer. The pigs of Germany are divided into two classes— those of the North, which are at the present time largely composed of English blood, and those of the South, which resemble, in almost every particular, the Lorraine group (described under the heading "The Pigs of France"). We have had the advantage of closely examining swine in various parts of Switzerland, Holland, Belgium, Denmark, Sweden and Norway, and in no case have we found anything in the shape of an improved race, or which deserves mention on account of its possession of any particular quality. Setting aside the pigs of Great Britain, we do not believe that there are any in Europe—France excepted—that are worthy of the title of a pure race; whereas the majority of those with any pretensions to quality have been improved by repeated resort to British blood. Until a few years ago, in the countries that we have named, the type of native pigs was so primitive, that, considering the comparative shortness of the distance from England, we could only conclude either that many breeders were ignorant of the existence of pigs of superior quality to their own, or that they had simply no interest in the improvement of the animals they possessed. During the last two decades, however, matters have changed, and in almost every country our native breeds have been used with great effect—especially in Denmark, which has long been sending us good bacon as the result of her enterprise.

CHAPTER XIII.

PIGGERIES AND STIES.

FEW details connected with the management of pigs are of more importance than those relating to their housing, inasmuch as no other part of their management, not even feeding itself, has so marked and so direct an effect upon their health and general well-being. It has been the custom of pig-keepers of all ranks for generations to consider the housing of the pig as a matter of very slight importance; and even at the present day, when there is so much enlightenment in pig matters, and when the pigs of this country have obtained a reputation greater than that enjoyed by any other nation, we are far behind less astute stock-breeding people in our ability to design and provide piggeries. We are, indeed, equally content—and this applies almost as much to breeders of eminence as to the labourer himself—to keep swine of the highest possible excellence in such buildings and under such conditions as are a positive disgrace to the name of an English breeder. If, as is universally admitted, warmth and the exclusion of wet, as well from above as below, are the equivalents of so much food, it would seem that, even upon economical grounds alone, it should be the policy of a keeper of live stock to provide these qualifications at least; but a pig, of all domestic animals, is supposed to be not only the scavenger of the farm, but proof against the changes of temperature and the violence of the weather.

A cursory examination of the question with which we are about to deal will immediately show that, in building a piggery of any kind, it is imperative to provide a warm, wind-tight, and rain-proof apartment, which can be freely exposed to the cool air of summer, and which, by a simple arrangement, can be as easily made warmer in winter. Next to this, the chief features are a perfect system of drainage, and an adjoining court or yard into which the pigs can be turned whenever it is found necessary. Descriptions of several admirably built piggeries follow; but a few words upon the general arrangement of an ordinary sty, which will apply to either a large or a small building, will be found of service to the reader.

A pig-house, of whatever kind, should preferably be built of brick or concrete blocks. The roof may be either of tiles— and good ones only should be used—or of corrugated iron. Wood is perhaps the commonest material used in building piggeries in this country, but it has many drawbacks. It is not substantial; it is not proof against the teeth of the animals themselves, and it is not sufficiently weather-proof unless it has a double lining; and even when this is provided —whether the space between the walls is filled with any heat-retaining material or not—it furnishes a refuge for vermin, which are invariably quick in establishing themselves near a well-fed pig. If the building is made of substantial material, with the open portion facing the south, it will be at the same time dry and absolutely free from vermin.

There are various opinions as to the best manner of making the floor of the piggery. Concrete is, undoubtedly, a first-class material, and can scarcely be improved upon when it is finely grooved, for, in such a case, it is not only impervious to wet, but it conducts the drainage, and affords a firm foot-hold for the pig, which is the chief point of objection to the ordinary smoothly laid concrete. Many of the chief breeders prefer Staffordshire bricks, grouted in cement. These are of the greatest value, and we know of nothing superior to those made especially for the purpose. Stone floors, which are common in some parts of the country where stone is cheap, are disagreeable in many ways, but particularly so from their being continually wet, slippery, and unsightly. Wood is bad in any form, except when it is laid down upon the system so strongly recommended by the late Mr. Mechi — in spars. There are, however, objections to this plan, for, although it

means a saving of litter, a comparatively clean sty, and a
great gain in the form of manure, the apartment can never
be sweet, owing to the absorbent nature of the wood and the
open drain beneath; it is exposed to vermin, and provides a
disagreeable medium for the entrance of cold and draught.
If spars are adopted in a piggery, they should be at least 4in.
in breadth, with apertures of not more than an inch between
each. This point is very necessary, in order to prevent the
greatest inconvenience to the pigs themselves. The wood
should be tarred before using, and, if possible, it should be
seasoned.

The floor of a pigsty should always slope to one side, so
that the drainage may be carried off as completely as possible.
In some instances a drain is run down the centre, but this is
not absolutely necessary. The bed of the pig should be upon
the upper and higher side, although, without some inducement,
a sow generally prefers to make it upon the lower side, and
preferably upon the top of the drain itself. To prevent this,
and in order that she may be warm and dry, a wooden bench
should be provided in one corner. This should be fixed so
that the sow cannot move it. In some cases, however, she
declines to recognise it, and when this happens, the framework of the bench alone should be used to enclose a sufficient
space for the animal to lie in, when the litter provided for
her will generally be kept together, and be an inducement to
her to use it as her bed. Breeding-sties should also be provided with a rail erected to a height of 1ft., and standing 9in.
from the side of the wall. The purpose of this is to keep
the little pigs from being crushed, which frequently happens
when they get between the sow and the wall as she lies down.

In every well-built farmery there is some system for carrying
off and preserving liquid manure. With this system the
drains of the piggery should be connected, not by a common
earthenware drain pipe, but by well-made glazed pipes, laid
in cement, and provided with a trap, which will prevent the
ingress of the actual soil of the pig. In double piggeries—
that is to say, those which have apartments on each side of a
central passage—two systems of drainage are adopted. In one
case, a conduit runs the length of the building, under the
passage itself, the drainage from each sty being carried into

it, whereas a simple arrangement prevents anything but liquid passing from it. As, however, it is found in practice that solid manure does find its way into any drain, this conduit is furnished with a movable cover, in order that a man may descend and periodically remove with a spade the solid content as it collects. Under the other system, the drainage from the covered apartment is carried into the outside court or yard, and thence, with the rain which descends into it, into the ordinary drainage system of the farmyard.

It is important that, however the drain in the sleeping apartment may be made, water should not be able to soak through the floor, for nothing is more necessary than that this should be as dry as possible. To this end, it is advisable, before making a floor, to dig out the earth to a depth of two feet, and to fill in with broken brick, cinders, stones, or any other material that will act as a drainer and maintain the floor above in the driest possible condition.

It is very uncommon to see piggeries provided with doors through which egress is obtained into the court beyond. A door is as necessary for protection against the rains of summer as against the cold of winter, and it is quite easy to provide one that can be opened by the pig itself. A plain wooden door, slightly smaller than the aperture in the wall, may be provided, and hung upon two bolts or axles, fastened almost in the centre of its outside edges. If sockets are fixed into the wall, it will swing and maintain its position if the bolts have been fixed in the right place. When the pig wishes to pass through, it simp'y pushes its snout against the bottom of the door, which opens, and swings back into its place when the pig has passed. This simple plan should do away with all objections made against ordinary doors, and the trouble of opening and shutting them.

It has been truly said, with regard to pigs, that, "where the sun does not come, the doctor does," and we therefore think it requisite to provide a window on the sunny side for the entrance of light, which, with air, is so essential to animal life.

Whether bedding is or is not at all times used in a pig sty, is a matter of choice. In the summer, where a bench is provided, we think it is not necessary. In winter, however, straw is preferable, from the warmth it contributes. Wheat straw, as clean and bright as possible, should be used. There can be no more objectionable litter for pigs than barley straw, which promotes the breeding of vermin and an unhealthiness of the skin. Oat straw is better, but by no means equal to wheat. Some persons use sawdust, but, although we have tried it with success, we do not believe its use is advisable, inasmuch as, at some time or other, it gets into the food of the pigs and is consumed. Pigs eat a quantity of good straw, and benefit by it. They also thrive better if coals or cinders are occasionally given them.

The majority of model piggeries are so built that the space between the pig and the roof is entirely unoccupied; but among some of the improved buildings upon the Continent we have noticed that, at the height of about six feet, a rack is placed, and upon this, during the winter, large quantities of straw are stored, the chief object being the provision of additional warmth for the pigs beneath, although some considerable value is attached to the storage itself.

One great objection to large piggeries is the inducement which they offer for overcrowding. In no case should pigs be packed in large numbers within a comparatively small space. Indeed, though there be few pigs in a large building, cleanliness and constant attention are required to prevent the outbreak of disease at some time or other.

As a general rule, the yard or court of a piggery, if paved at all, is laid down with stones or flags, both of which are most objectionable. If bricks are not provided, as above described, concrete should be made, and almost any handy workman, under the direction of a master who understands what he wants, can lay this down. One part of the best Portland cement, with two parts of washed sand, and five or even six parts of stones, will make an admirable floor, but the stones or any other sharp, hard material that may be used should be absolutely clean, and free from any dirt whatever, or the cement will not bite.

Troughs.

One of the most important features in the piggery is the trough, which is made in a variety of forms. In Cheshire we have noticed that stone troughs are in general use, these being

fixed in the outside court within the wall, and so built that the
food can be poured in from the outside without coming into
contact with the pigs; but this is an exceedingly primitive
system, and one that cannot be recommended.   Wooden

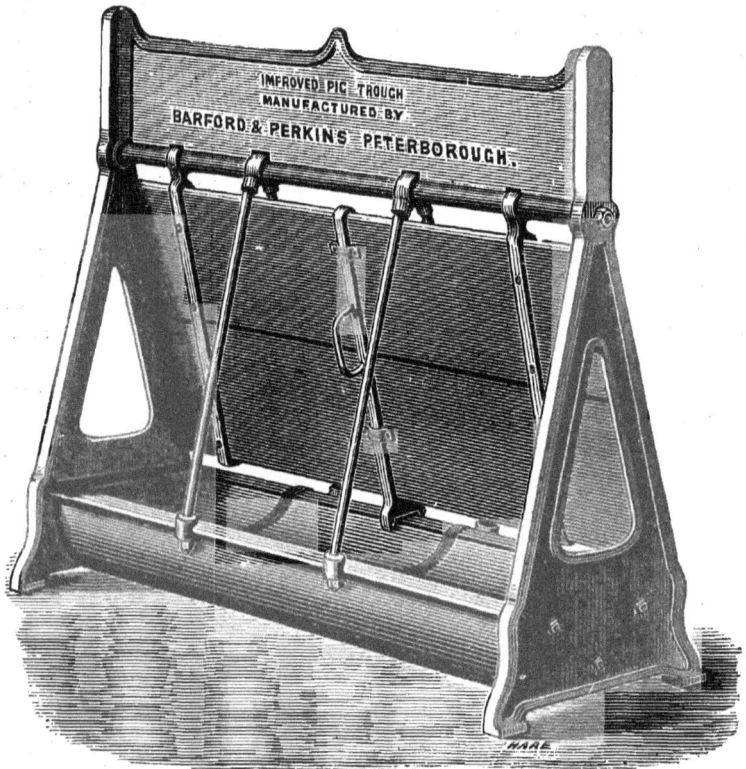

Fig. 22.—Iron Trough with Swinging Door.

troughs are still worse, for the bottom generally forms an acute
angle, and becomes a medium for the collection of dirt.   They
also contribute very largely to the waste of food.   If a wooden
trough is made, it should be of oak, and the edges should be
covered with iron, otherwise they will not last a year, as they
will be gradually bitten away by the pigs.   Movable iron

troughs are useful, but are much too easily turned over, and liable to breakage. A broken iron trough, however, can be utilised admirably by being placed in the corner of a covered piggery, and fixed in brickwork with cement.

Where a court is attached to a pigsty, the trough should be built within the wall, and so arranged that the attendant can fill it from without and, at the same time, shut off the pigs. There are numerous troughs suitable to this method,

Fig. 23.—French Trough with Swinging Doors.
One side is shown Open to the Pigs and the other Closed.

some of which are provided with hanging doors, made to bolt at the back or front of the trough. One of the very best we know is that shown at Fig. 22, and made by Messrs. Barford and Perkins, of Peterborough. The idea can, if necessary, be adapted to an ordinary trough built within the wall, a wooden shelter being hung above it, and provided with a bolt, which shuts off or exposes the trough of food as may be desired.

Another admirable plan is shown in the illustrations of the Beau Cèdre Piggery, at Lausanne (Figs. 42 and 43, pp. 212 and 213). This trough is semi-circular, and furnished with a shutter, which, when closed, is also semi-circular. The front

q 2

half of this cover only is drawn back, when the front section
is exposed to the pigs, the hind section then being covered,
and *vice versâ*. The whole trough can be covered, or either
section, at will. Upon the whole, this is the best trough that
we have ever seen, but its greater expense is such as to prevent
the general public from adopting it in preference to the
common open troughs, although the latter are infinitely more
wasteful and less useful.

Fig. 24.—Section of a Continental Trough with Swinging Shutter.

Fig. 23 shows a square trough (*i, g*), as used in France, which
is fitted with a plain wooden swinging door. In the engraving
the door at *p* is shown closed, with the trough open to the pigs,
and that at p^1 closed against the pigs; *q* is a strong wooden
latch, which fastens upon the two sides of the crossrail *r*. The
door swings upon hinges fixed to the rail *i*, and at each end
of the trough is an angular piece of wood (*y*), designed to
prevent waste.

Another useful trough (*a, b, c, d*), also built in the wall, as
used upon the Continent, is that shown at Fig. 24. It is made
of concrete laid upon stones (*f*). The shutter (*i*) used in this
case is of such a shape that it can be made to act as the door
described in the trough above, or swung (*j*), to allow of pigs
feeding from both sides at the same time.

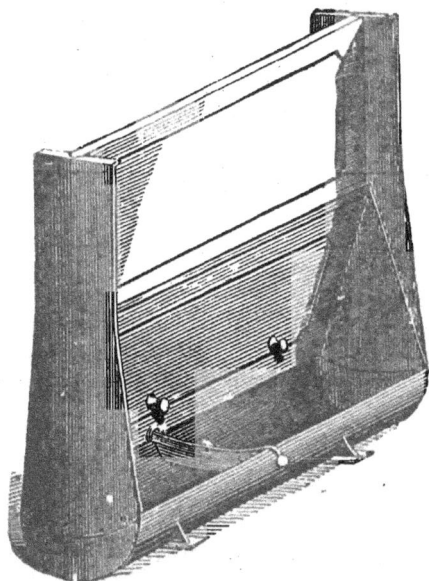

Fig. 25.—The Crosskill Iron Trough with Swinging Shutter.

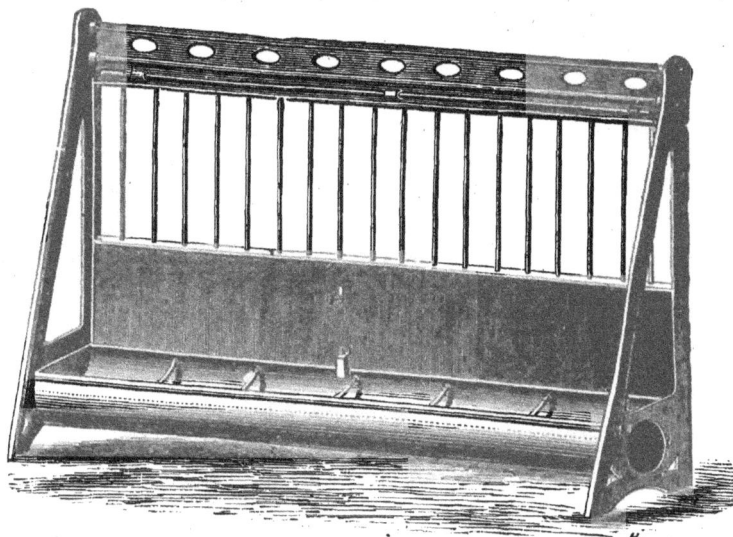

Fig. 26.—Trough with Swinging Shutter, adapted for a number of Pigs.

The Crosskill trough, shown at Fig. 25, made by Messrs. W. Crosskill and Sons, of Beverley, is also made of iron, with a partition in the centre. It is provided with a swinging

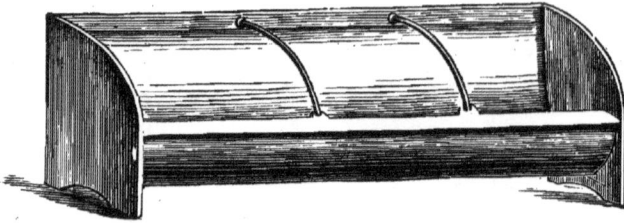

Fig. 27.—Movable High-backed Iron Trough for Three Pigs.

Fig. 28.—Movable High-backed Iron Trough for Four Pigs.

Fig. 29.—Common Iron Trough for Four Pigs.

door, which shuts off or opens the trough in a similar manner to that of Messrs. Barford and Perkins, although the door is fixed in a slightly different manner.

Fig. 26 shows a large trough for building within the wall of a sty, and adapted to a number of pigs; it is made by the St. Pancras Ironworks, 171, Pancras Road, N.W. In this case the swing door is not solid, but partly of open ironwork; and the ends are also open.

Fig. 30.—Glazed-ware Trough for Three Pigs.

One of the best small movable troughs is that with a straight high back, shown at Figs. 27 and 28. This is heavier, less liable to be overturned, and not so wasteful as the smaller open troughs, one of which is represented at Fig. 29—this being the trough commonly used throughout the country.

An exceedingly useful, cheap, clean, and ingenious pig trough is that made by Messrs. Oates and Green, of Halifax, and shown at Fig. 30. This is almost as indestructible as iron, being made of the hardest ware, glazed.

Fig. 31.—Circular Iron Trough for Small Pigs.

Fig. 31 shows a circular iron trough, with partitions, which is largely used in some parts of the country for litters of small pigs. This is not an expensive utensil, and has many advantages.

Necessary Points in a Piggery.

In his work upon "The Pig," M. Gustave Heuzé says that, as a rule—and this applies to England as well as to France—piggeries are damp, low, badly ventilated and lighted, and wretchedly kept. They are, however, well arranged when—(1) the soil is dry, and they are paved or cemented; (2) the floor is sufficiently sloped to carry the urine into the drain; (3) the sties are large enough to enable the animals to move about freely; (4) they communicate with an outside court which is provided for the pigs; (5) they have openings, or windows, to permit of the free circulation of air during summer; (6) the roof, the walls, and the doors are sufficiently substantial to protect the animals against the rains and cold of winter; and (7) the interior passage enables the attendant to feed the animals with a minimum amount of trouble, and to see them without the necessity of entering each sty.

This writer recommends that buildings devoted to pigs should be divided into three classes: the first for pigs intended for breeding and rearing; the second for the reception of store pigs; and the third for fatting pigs. There is considerable reason in this arrangement, but every pig-keeper should provide one unusually substantial sty for the boar, and, where boars are bred for sale, several compartments of this kind will be found necessary. We confess, however, that we should prefer to keep such an animal, as far as possible, from the sows, for many reasons. Practical pig-keepers will remember that boars are frequently troublesome when sows are in season, if they are kept in close proximity to the sties of the latter, and that, not infrequently, they jump completely out of their pens, where this is possible, and sometimes do a considerable amount of damage, in addition to giving some trouble to the attendant, and perhaps other people besides. Some breeders are content to allow their boars to have full liberty in the farmyard, and to run among the sows and gilts at all seasons; but this plan is distinctly wrong.

SIZE OF THE STY.—The apartments for breeding sows may be smaller than those for boars and for feeding pigs, which are generally kept together in numbers; and it may be taken for granted that the same sized sty that serves for fatting three or four store pigs of from nine to twelve months old will serve admirably for litters of young pigs, which, for example, may number from nine to twelve head. It will generally be found that the following dimensions are sufficiently large and suitable for particular cases:

A breeding sow—Large breed	7ft. by 7ft.
,, Middle White or Berkshire	6ft. by 7ft.
,, Small White or Small Black	6ft. by 6ft.
A boar .. Large breed	7ft by 8ft.
,, .. Middle White or Berkshire	7ft. by 7ft.
,, .. Small White or Small Black	6½ft. by 6½ft.

For litters of pigs, the apartments should be at least as large as follows:

For Large breeds	8ft. by 8ft.
For Middle breeds	8ft. by 7ft.
For Small breeds	7ft. by 7ft.

The sties should, in all cases, be in communication with outside courts or yards.

For fatting pigs, courts are not necessary; but the size of the sties should be, for two or three large hogs, 8ft. by 8ft., and for two small hogs, 7ft. by 7ft.

The size of the outside court is not a matter of such great importance. It is usually made as wide as the inside apartment and twice as long, but it may be extended, where found convenient, even to three times the length of the sleeping apartment. Doors should always be fixed in a piggery, to lead from the passage to the inner apartment; and they should invariably be strong, well made, and furnished with efficient fastenings. It is advantageous if they are made to open outwards, as, by this means, they act as a bar to the passage of an animal beyond the apartment into which it is intended to be driven, and the pigman is thus able to turn his animals into their proper sties with a very small amount of trouble.

In the majority of cases, the dividing wall of a piggery, in the passage, is no more than 4ft. or 5ft. high; but it may, if necessary, be carried to the roof, provided an opening is made, in the form of a window, to enable the attendant or any other person to see the pigs within. The floor of the passage should preferably be made of concrete, or any other hard, impervious substance that will not retain water or filth, and that will also

be a help to the attendant in wheeling the food to the various
troughs. If possible, water should be laid on; and, if placed in
some convenient spot, the passage, as well as the sties, can be
frequently cleaned out by the aid of a hose and pipe, or by
throwing water down in the ordinary way.

Feeding Apartment and Appliances.

In a large piggery, whether single or double, the feeding
apartment should be in the centre. This should be a dry,
airy, and cleanly-made room, provided with bins for the
storage of food, whether meal, roots, or liquid, and an

Fig. 32.—Steam Apparatus for Cooking Food.

apparatus for cooking when necessary. If milk, whey, or other
liquid food is used in large quantities, a concrete tank should
be erected in this apartment, and connected with the various
troughs by means of piping, so that, at a given moment, the
troughs can be filled without the necessity of the attendant
visiting each in turn.

In a very large piggery, a small tramway would be found of the greatest possible convenience: for if a truck, however small, can be run from the feeding apartment down the whole length of the building, it will save a large amount of labour; and the pigs will not only be fed much more rapidly, but they will, during the cold weather, receive their food in a hotter and fresher state than if it is brought to them direct from the feeding-copper in ordinary metal pails.

Fig. 33.—Messrs. Hill and Smith's Food Carriage.

Fig. 32 shows a capital apparatus for cooking food, which is made by Messrs. Barford and Perkins, and which, in a large pig-keeping establishment, is almost indispensable to the attainment of perfect success. It steams roots, meal, corn, and foods of all kinds, and boils milk or water. The makers claim that it will steam the food for fifty head of cattle, and as many pigs, at a cost of one shilling a day for fuel; and potatoes at a farthing per hundredweight. If cooked food is superior to uncooked food—as we believe it is—then there can be no doubt about the advantage of using an apparatus of this kind.

In almost every piggery a large quantity of wash is used, and, whether the food is cooked or not, this should be mixed in the vat with the ordinary food until the mass is slightly sour. We know of no utensil for the removal of this wash, or for the carriage of the food to the various sties in a large piggery, which, considering its cheapness, can be more strongly recommended than that shown in Fig. 33.

Another utensil commonly used in a piggery, more especially where the pigs are given a considerable quantity of cooked food, is a potato or root cleaner (Fig. 34). Some breeders and feeders use potatoes to a great extent for both store and fatting

Fig. 34.—Potato or Root Cleaner.

pigs; and we shall not be surprised if, in the future, they are much more largely used than they are at the present time, considering that the price of the potato is gradually and closely approximating to that at which it can be purchased with advantage for the feeding of stock.

To the majority of farmers, a pigsty is a matter of very little importance; and we have frequently recognised the fact that intelligent breeders of farm stock (like Lord Rosebery, who owns a fine herd) have been quite content to allow their breeding pigs to run at liberty in the straw-yard, among unbroken colts and cattle of all descriptions. Under these conditions, they

maintain that the pigs thrive and, in fact, do better than when
confined in sties; that they save the expense of the latter,
cost less for food, and lie better upon the heated straw
under the sheds than upon a bench or bed in a properly-made
pig apartment.

There may be—and we confess we believe there are—certain
advantages to be derived from this system; but they. are
distinctly outweighed by the disadvantages and . too frequent
losses. If this question were more carefully examined, in all
its details, by the most sanguine advocate of the straw-yard
system, we can scarcely believe that he would come to any
other conclusion than that at which we have arrived, viz., that,
to keep pigs in health and in good condition, and to manage
them in the most approved and practical manner, they should
be provided with such housing as that to which we have above
referred.

Lord Moreton's Piggery.

Lord Moreton's piggery, at Tortworth Court, Gloucestershire,
which is shown in Figs. 35 to 38, is one of the best that we have
seen in this country. The illustrations almost explain themselves; but it may be mentioned that the
building stands in a paddock of about two
acres, which is devoted to the pigs, the
animals being let out of their yards in turns
during suitable weather. Near at hand are
a rickyard and a second paddock, also about
two acres in extent, both of which are used
when it is necessary to give the pigs a
change of ground.

The piggery contains thirty sties, fifteen
on each side, with a passage running down
the middle of the building. The compartments on one side have boarded floors, while
those on the other side are of stone. The
roof is covered with double Roman tiles, one
glass tile being provided for every sty.
There are six cotes for breeding sows, all of
which are fitted with rails, to prevent the
sows from lying upon the young pigs. The

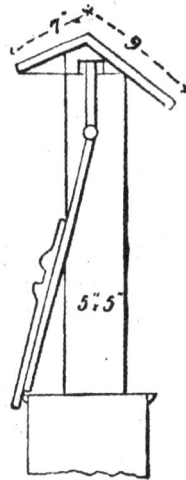

Fig. 35.—Section of
Ventilator in Lord
Moreton's Piggery.

Fig. 36.—Front and End Elevation of Lord Moreton's Piggery at Tortworth Court, Gloucestershire.

Height from Ridge to Floor, 14ft. 6in. Width, exclusive of Yards, 27ft.; inclusive of Yards, 49ft, 10in. End Door, 6ft. 7in. by 5'.

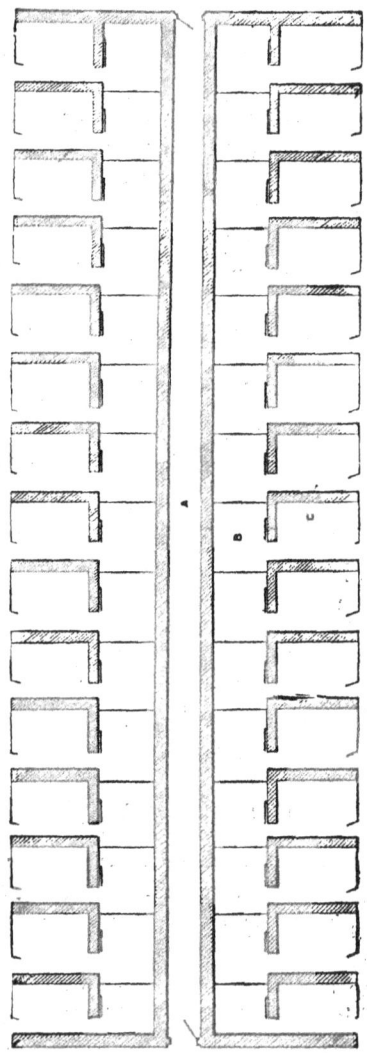

Fig. 37.—Ground-plan of Lord Moreton's Piggery at Tortworth Court, Gloucestershire.

A, Central Passage, 149ft. 3in. long by 5ft. wide; B, Sties; C, Outer Yards.

dimensions of various parts of the building are shown in Figs. 35 to 38. The troughs are 4ft. 6in. in length by 1ft. in breadth, and they are in every case fitted with swing lids 4ft. 6in. wide and 2ft. in depth. Sliding doors are provided between the sleeping apartments and the outside courts, which are shut in cold weather.

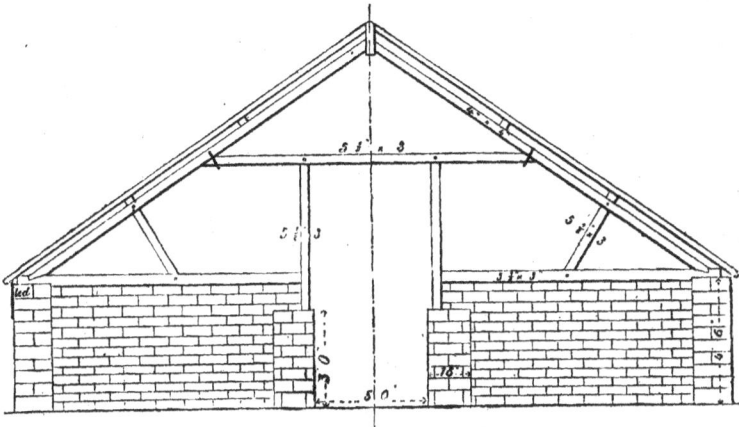

Fig. 38.—Cross Section of Lord Moreton's Piggery at Tortworth Court, Gloucestershire.

This building is worthy of the inspection of any person intending to build a model piggery; and Lord Moreton has not only done great service by his encouragement of our best races of pigs, but by the erection of a building at once so practical and so unique, at Tortworth Court, Gloucestershire.

Lord Ellesmere's Piggery.

For many years, some of the finest pigs of the white breeds in this country have been sent out by the Earl of Ellesmere, and we had hoped to present the plan of the piggeries at Worsley, where these grand animals are kept; but, although the buildings are practical in every sense of the word, they are somewhat disconnected, and, therefore, not sufficiently adapted for illustration. The principal piggery, however, very

much resembles that of Lord Moreton, and it is therefore unnecessary to give a full description of it.

In this building there are thirty-eight apartments, the average
size of which is 14ft. by 10ft. Each sty is provided with a
window, fitted with a sliding door, looking into the central
passage, and the breeding apartments are supplied with crush-
rails, 3in. by 3in., and standing 10in. off the ground. In many
instances loose wooden benches are used, and the iron swing
troughs made by Messrs. Crosskill and Son (Fig. 25) are
generally adopted. The floors of the sties are bricked, and
slope towards the outside, to carry off the drainage; and doors
are provided to admit the pigs into the open air.

Mr. James Howard's Piggeries.

The piggeries at Clapham Park, near Bedford, built by the
late Mr. James Howard, M.P., once a famous exhibitor of the
White breeds, form an almost exact square, as will be seen by
reference to Fig. 39.

The breeding-sties have a crush-rail 1ft. high, and 9in. from
the wall. There are no troughs, except in the case of boars and
invalid pigs, Mr. Howard having preferred the system of letting
out each pen of pigs in turn to feed separately in the yards D
and G. The farrowing sties for sows are shown at A; they are
fitted with rails, 1ft. high, and 9in. from the walls, to prevent
the sows from lying upon their young. The sties for young pigs
are shown at B; and the range of houses and yards for boars at
C. A yard and shed (D, D) are provided for in-pig sows. The
range of buildings at E is for pigs or, if necessary, for calves,
those at F being used for farrowing sows or for sows with their
litters. The yards shown at G, which are for the feeding and
exercise of the pigs, are well paved with brick; I,I, show steaming apparatus and food bins; J is an egg-shaped boiler, and K
a copper furnace close at hand; and L is the passage behind the
farrowing sties. There are no feeding-troughs in these piggeries,
except for boars and invalid pigs, the other animals being fed
in the yards, which are furnished with water troughs (M).
Buildings Nos. 1 and 2 are inclosed; but Nos. 3 and 4 are open
in the front.

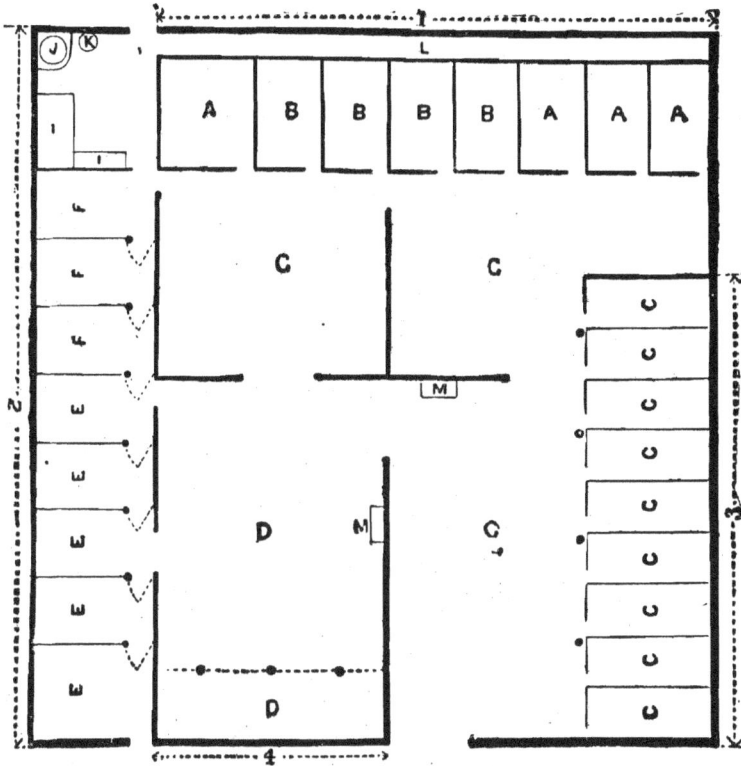

Fig. 39.—Ground-plan of the late Mr. James Howard's Piggeries at Clapham Park, near Bedford.

A,A. Farrowing Sties; B,B, Sties for Young Pigs; C,C, Houses and Yards for Boars; D,D, Yard and Shed for In-pig Sows; E,E, Sheds for Pigs (or Calves); F,F, Sheds for Farrowing Sows or Litters; G,G, Yards; I,I, Steaming Apparatus and Food-bins; J, Boiler; K, Furnace; L, Passage; M,M, Water Troughs. 1 and 2, Enclosed Buildings; 3 and 4, Buildings open in front.

A Swiss Piggery.

The piggery which was built by M. Gustave Auberjonois, at Beau Cèdre, near Lausanne, in Switzerland, and which is fully illustrated in Figs. 40 to 43, is the best of its kind

P

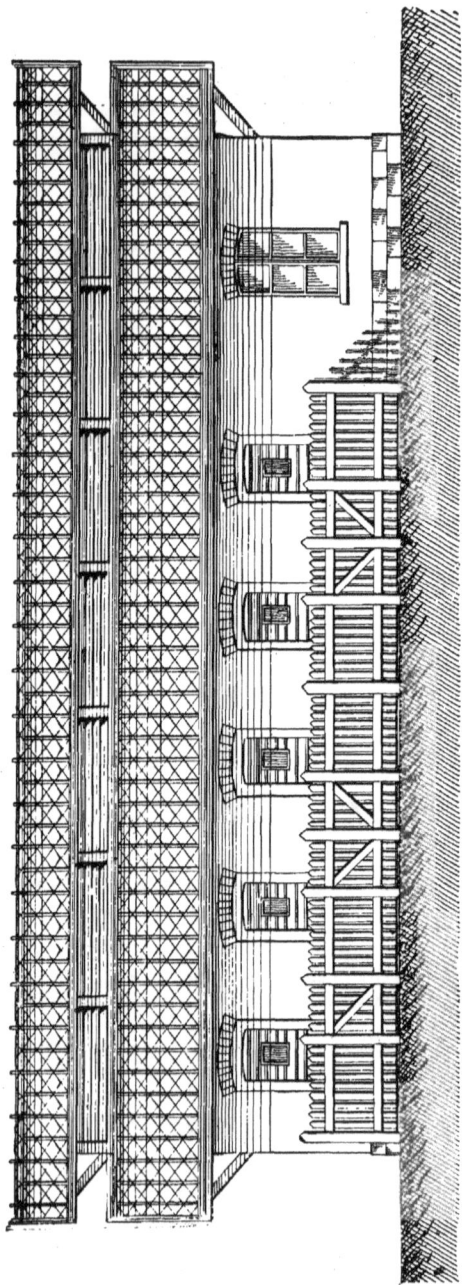

Fig. 40.—Front Elevation of the Beau Cèdre Piggery (near Lausanne, Switzerland), built by M. Gustave Auberjonois.

Substantially constructed of Bricks and Ornamental Tiles, this is one of the best as well as one of the handsomest of Piggeries.

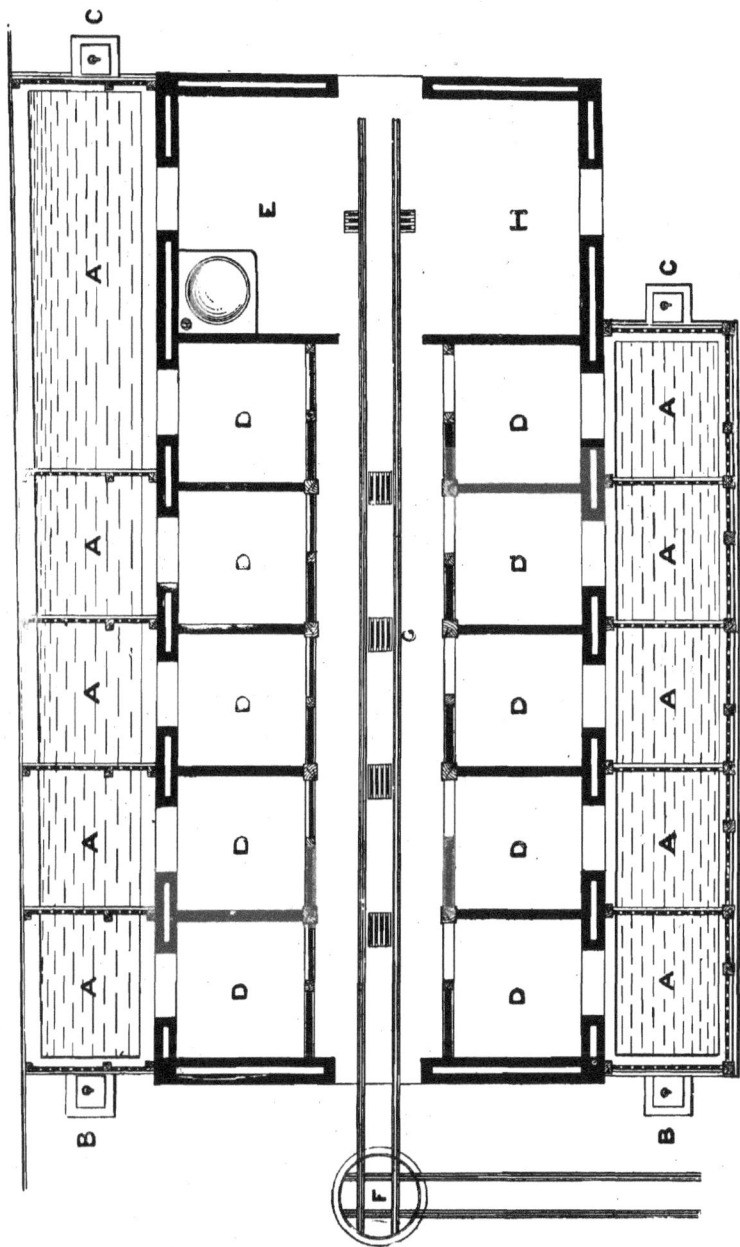

Fig. 41.—Ground-plan of the Beau Cèdre Piggery, near Lausanne, Switzerland.

A.A, Courts with Water; B,B, Water Inlets; C,C, Outlets and Catchpits; D,D, Sties; E, Food and Cooking Room; F. Turntable; G, Central Passage and Tramway; H, Food Store.

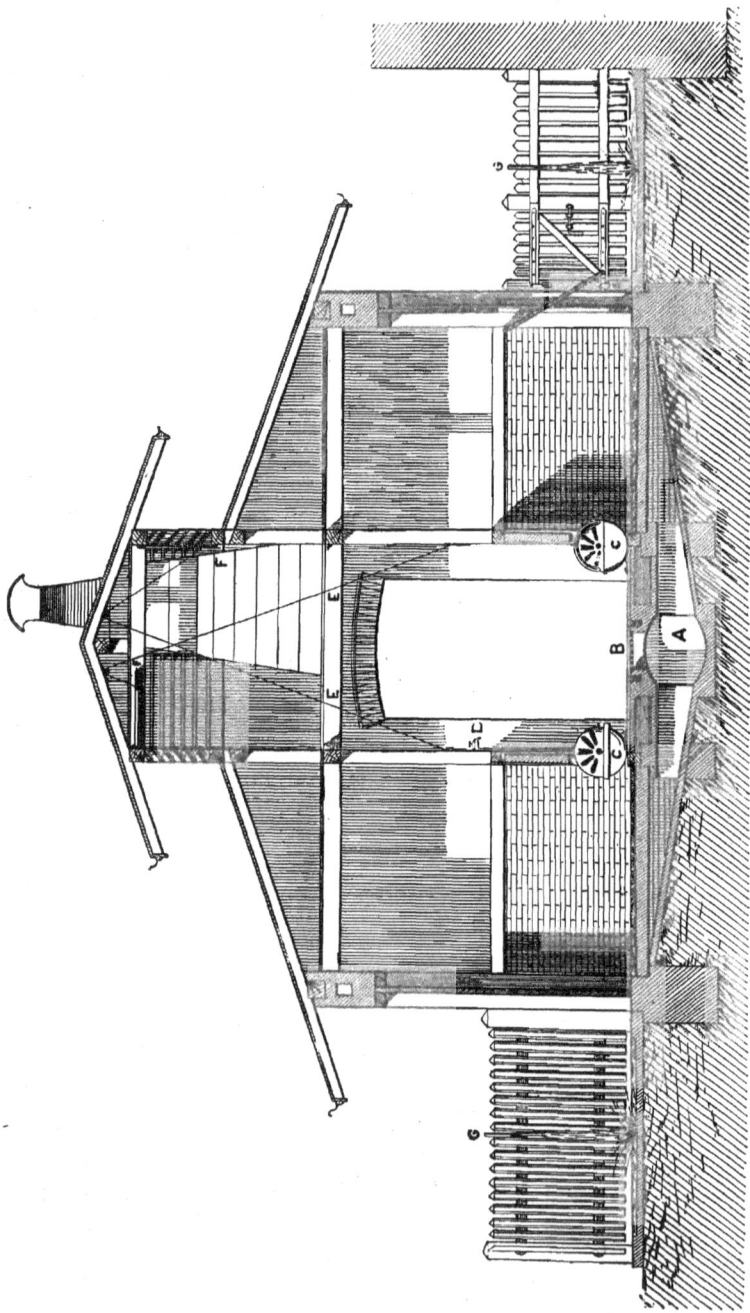

Fig. 42.—Cross-section of the Beau Cèdre Piggery, near Lausanne, Switzerland.

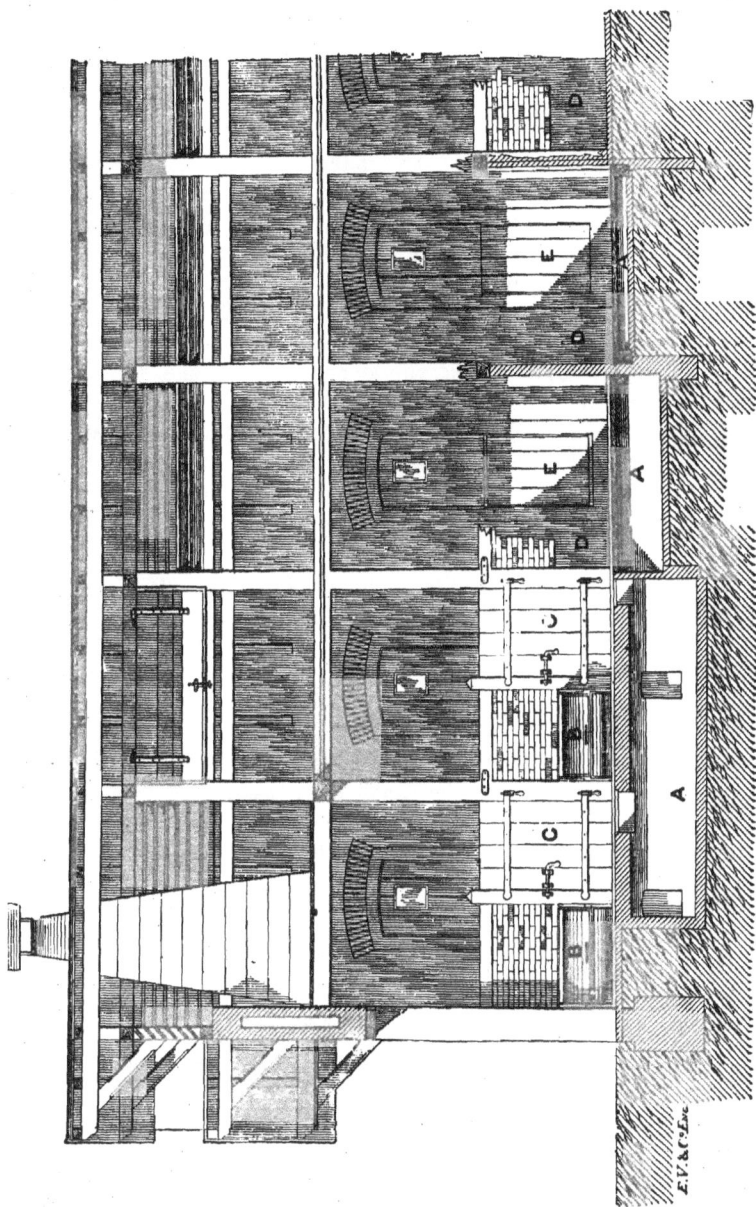

Fig. 45.—Longitudinal Section of the Beau Cèdre Piggery, near Lausanne, Switzerland.

A.A. Drains; B,B. Food-troughs; C,C, Doors from Central Passage; D,D, Interior of Sties; E,E, Swing Doors to Courts.

that we have had the opportunity of seeing. It is substantially built of brick and ornamental tiles; but, handsome as it is, the structure is designed more for use than for ornament. It cost about £350.

This piggery is divided into two parts, one being the food and cooking room, with stoves attached, and the other the piggery proper, with two ranges of sties, the inner portions of which are under cover. Down the centre of the building is a passage, paved with cement, the sties being on either side. The inside or sleeping apartments are about 8ft. square, and the floors are composed of thick planks, with a small space between each, through which the excrement of the animals passes. Below these floors are others, which are solid, and upon which the manure falls. The lower floors slope towards the passage, permitting the liquid manure to find its way into the drains running directly under the passage floor. These drains are sufficiently large to enable a man to stand in them up to his knees, and are covered, at intervals, with iron gratings fixed in the floor of the passage, serving to carry off the water when the place is cleaned. They are also connected with the larger drainage system of the farm, and, by means of water, which is liberally used, the whole of the manure of the piggery is carried away into a huge tank built for the purpose.

Rails are laid down in the passage for a tramway, which is used to bring the food, or for any other necessary purpose. The troughs have been already described (pp. 195, 196). An ingenious door communicates between the inside and the outside court. This door hangs upon bolts, which permit it to swing, so that the pigs can pass in and out of the sty of their own accord, the door always remaining shut behind them; they can also be kept inside or outside at will. In winter, the doors are always kept shut, partitions are placed in front of them, and the space between is filled with sawdust, the severity of the weather necessitating the keeping of the pigs as warm as possible. Movable ceilings are provided for each sty, straw being packed upon them in the winter, to increase the warmth; but they are removed in summer, when as much air is allowed to enter the building as can be obtained. The ventilation is materially assisted by a very ingenious contrivance in the roof, ventilators (shown in Fig. 42) being opened and shut at will.

The sties are entered from the passage, a door being provided for each, while, outside the building, each yard, which is about 14ft. in length, is covered with concrete, the two sides sloping towards the centre, where a drain is laid to carry away the excrement and the water which, except in midwinter, is constantly provided. This provision of water is a peculiarity of M. Auberjonois' system. Every yard is daily, during the warm weather, filled with water to a depth of three or four inches, and, as taps are provided for each, the operation takes only about five minutes. When necessary, the yards can be emptied in an equally short space of time. The pigs enjoy the water exceedingly, and are believed to derive considerable benefit from it.

We noticed that, in every part of the piggery where wood was used, it was invariably covered with metal. Near at hand, too, was a pig-weighing machine, into which an animal could be driven at any moment. The illustrations (Figs. 40 to 43) give a good idea of the arrangements of the entire building.

French Piggeries.

Inspector-General Heuzé describes an admirable double piggery which was built at the Agricultural School of Grignon, in France, and in which every sty communicates with an outside court. In this building, a passage, 7½ft. wide, divides the two ranges of sties, which are of different sizes, those at the four ends being some 9ft. square, with outside courts 11ft. square. Next in order to these corner sties are eight smaller ones, a pair adjoining each; these pairs are exactly the same as the single sties which we have described, merely being divided for the use of breeding sows. Across the centre of the building is a narrower passage running through from right to left, and crossing the principal passage referred to above. In this passage the food and cooking apartments are situated, whereas at the four corners are four sties midway in size between the small and the large sties mentioned above.

Thus, in all, there are sixteen different apartments, with courts attached, four being large, four of medium size, and eight small ones; the breeding sties are placed between the others, and, consequently, in the warmest parts of the building. The total length of the piggery is about 18½yds., and its width, including the passage, about 6yds.

There are three doors to each sty, one to enable the attendant to enter from the inside passage, a second for the egress of the pigs into the outside courts, and a third to admit the attendant, or the pigs themselves, from the paddock, in which the building is placed, into the courts. A small gutter, covered with a board, which is pierced with a number of holes, runs from back to front of each sty, and is connected with drains which run from one end of the passage to the other, and which communicate with the main drainage system of the farm. The troughs are placed so that they can be commanded by the attendant from the central passage. There is every convenience for cleanliness, light, and warmth, and the piggery is, in almost every particular, a model that might be followed with considerable advantage.

Another perfect French piggery is that built by **M. Hallo**, Director of the Colonie de St. Maurice, which we had the pleasure of seeing almost immediately after it was finished. This piggery is erected against the south side of a huge wall which forms the back of a very large shed. It was entirely built by the boys employed at the Colonie, under the direction, of course, of skilled teachers, whose duty it is to teach them various trades in addition to the agricultural work of the farm.

This piggery is in the form of a lean-to, the passage being at the back of the sties and next to the wall, with doors at either end, and a grand entrance in the centre, as shown in Figs. 44 to 46. The inside sties or cotes are under the roof, which is covered with handsome, well-made, red tiles. Each sty is 6ft. 9in. long by 6ft. 6in. in width; this is also the width of the outside court, and every court is 13ft. deep. A door leads from each sty into the court, while at the end of the court is a sliding gate, permitting the egress of the pigs into the pasture outside. Over the door of each sty is a small window for the admission of light. There is also a door

into every apartment, leading from the passage; while over-
head are rafters, erected for the purpose of storing straw in
winter, the object being principally to secure the warmth of
the animals.

The floors of the passages and sties are all of concrete,
whereas the courts without are laid with hard bricks placed
upon their edges. There is an open gutter laid in this brickwork, running down the centre of the court, and carried

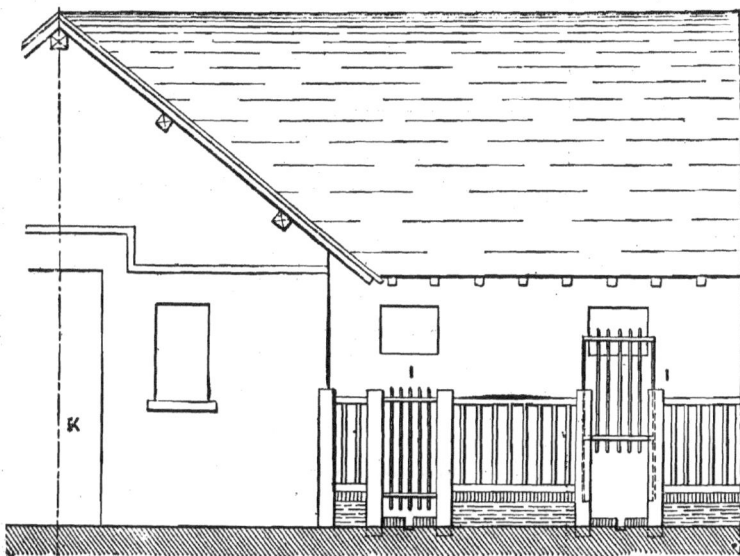

Fig. 44.—Front Elevation of the Piggeries at the Colonie Agricole de
St. Maurice, Loire-et-Cher, France.

I,I, Gates leading from Outside Courts into Park; K, Central Entrance.

from the two further corners. This gutter runs into a drain
below the gate, and into the ordinary drainage system of the
farm. The pig-troughs are similar to those used in the piggery
at Beau Cèdre, as previously described. The buildings are all
of brickwork, and are most substantially built, the divisions
between each court, as well as the ends, being of oak, neatly
and strongly made, and well tarred throughout. The gate
itself, which is made to slide up and down at will, is of iron.

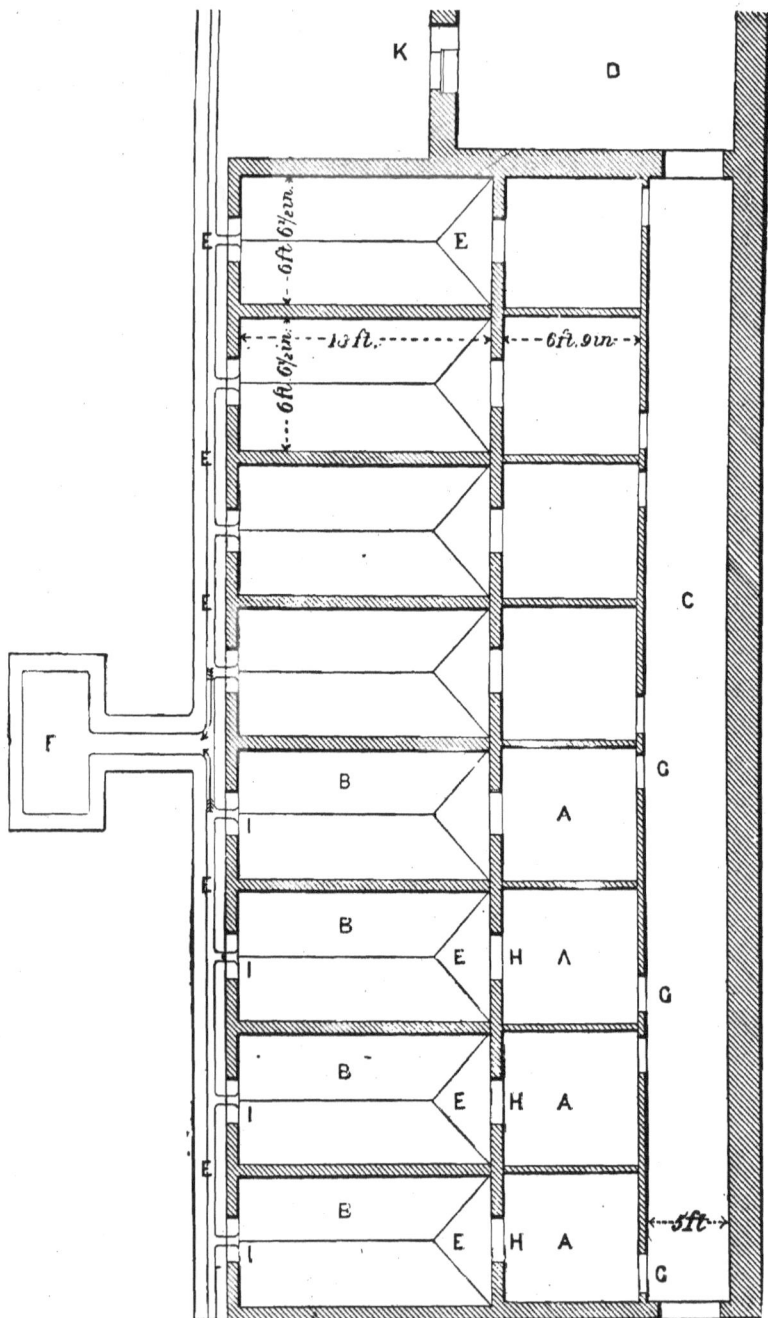

Fig. 45.—Ground-plan of the Piggery at the Colonie Agricole de St. Maurice, Loire-et-Cher, France.

A,A, Sties; B,B, Outside Courts; C, Passage; D, Cooking-house; E,E, Drains; F, Liquid Manure Tank; G,G, Doors in Passage; H,H, Doors into Courts; I,I, Gates into Park; K, Central Entrance leading into Cooking-house.

We are indebted to M. Louis Léouzon for the following description of his piggery, at La Poule, Loriol, Drôme, France, which is shown in Figs. 47 and 48. He says: "In the construction of buildings designed for the use of cattle, it is necessary to consider the rules of hygiene, the well-being of the animals, and the facility of service. My piggery incontestably fills these three conditions; and, in order to

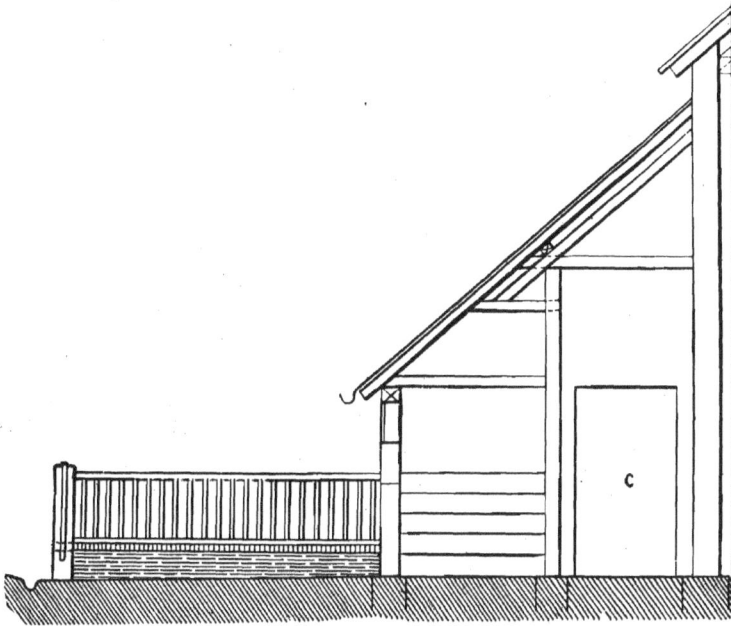

Fig. 46.—Side Elevation of Piggeries at the Colonie Agricole de St. Maurice, Lo're-et-Cher, France.

C, Passage at back of Sties.

convince the reader, it is only necessary to examine carefully the plans by which it is illustrated. The building is isolated, and forms one side of a large range of buildings erected in quadrangular form, the other side being occupied by a *bergerie*, or sheep-house, and a *vacherie*, or cow-house. The interior is about 33yds. in length by 5yds. in breadth. The walls are built of stone and mortar, the angles being of open

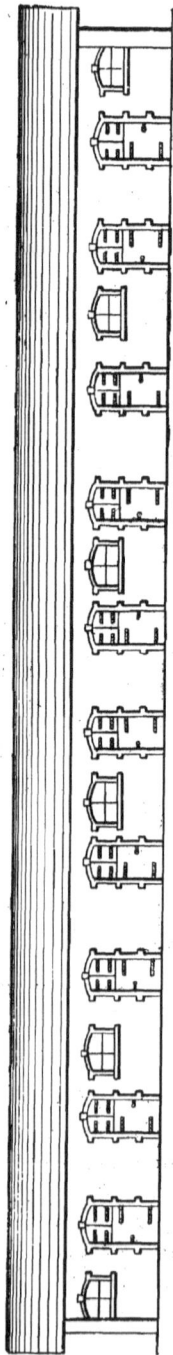

FRONT ELEVATION.

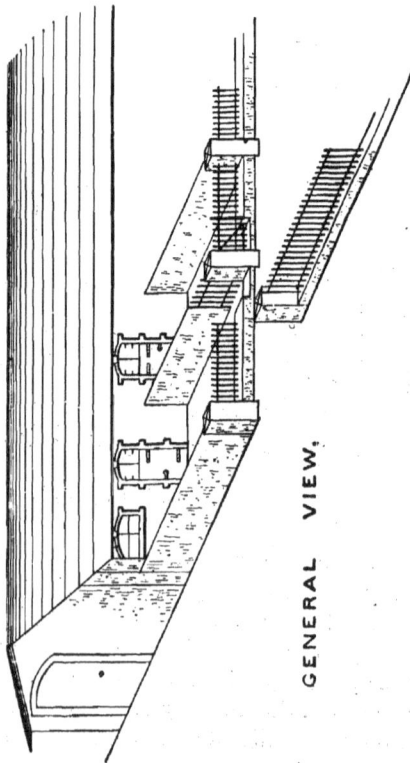

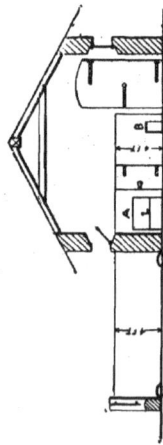

CROSS SECTION.

A. Feeding troughs with swinging shutter.
B. Passage for young pigs, with swinging door.

GENERAL VIEW.

Fig. 47.—Front Elevation, General View, and Cross Section of M. Louis Léouzon's Piggery at La Poule. Loriol, Drôme. France.

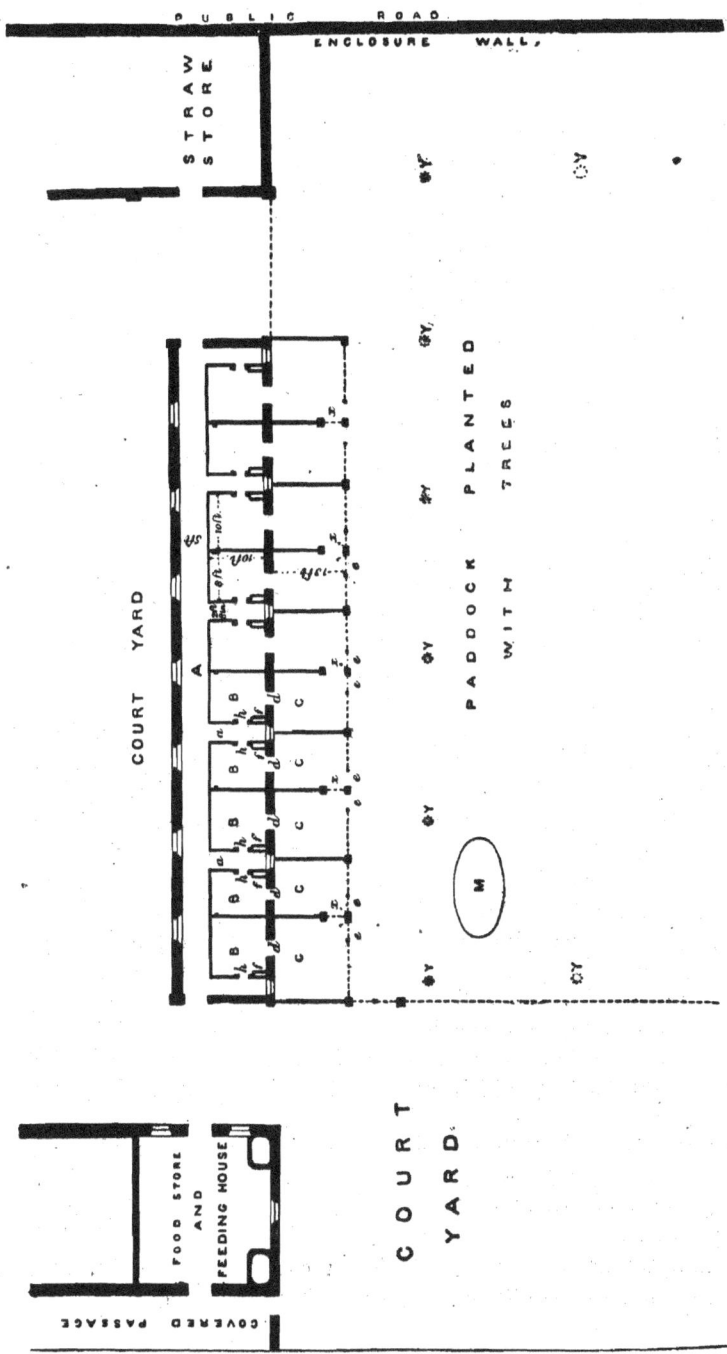

Fig. 48.—Ground-plan of M. Louis Léouzon's Piggery at Loriol, Drôme, France.

A, Passage; *a,a*, Side Passages; B,B, Sties; C,C, Outside Courts; *d,d*, Doors leading into Courts; *e,e*, Gates leading into Paddocks; *f,f*, Feeding-troughs; *h,h*, Doors leading from Side Passages into Sties; M, Bath; *x,x*, Doors between two Courts; Y,Y, Trees.

brickwork, giving to the buildings a charming appearance. The façade faces the east; for I regard this aspect as being the best. Nothing is more conducive to the healthy atmosphere of a sty than the first rays of the rising sun. They chase away the objectionable humidity which has arisen in the night, and prove a real source of enjoyment to the animals.

"There are ten inside courts. Of these, five are for sows, and are 10ft. in width, being provided with small openings to admit the young pigs into the adjoining compartment, where they can feed without the presence of the sow. These openings are provided with sliding doors. The service is conducted from a longitudinal passage, in close connection with the straw store and the apartment in which the food is prepared and kept. Adjoining this, too, is another apartment, in which roots and potatoes are stored. In this way, the feeding and management are conducted with a minimum of trouble and time; and I can also with convenience—as I intend to do—run a small tramcar for the transport of the food and litter.

"Generally speaking, piggeries in France are double, *i.e.*, they have one range of sties on each side of a central passage. I have not adopted this system of construction because, if one range is properly exposed to the sun, the other is quite the reverse, and it is my desire to give all the pigs the advantage of an eastern aspect. Ventilation is also as considerable as possible, and can be regulated at will by means of numerous windows and doors, the upper portions of which are independent, forming shutters which can be opened, although the doors may be closed. Every door opens upon the outside courts, each sty having a court to itself; thus the animals are enabled to exercise in the open air and to bathe in the small basins, which are filled with water during the summer.

"Adjoining the buildings is a large enclosure of grass, planted with fruit and other trees. I find this of the greatest utility, as it enables me to give liberty to sows with their pigs, and especially to young growing stores. The piggery is also shaded by a number of fine trees, which, in our warm climate, are generally in good condition. The roof is tiled; and, when the

weather is severe in winter, we generally use a large quantity
of straw, which is so arranged around and above the sties
that each forms, as far as possible, a warm chamber. I have
especially endeavoured to avoid humidity and the accumulation
of dirt; and I think nothing is more important than to provide
a clean and healthy bed for the pigs.

" I find some difficulty in determining where to place the
trough, which is naturally surrounded by moisture from the
feeding of the pigs. I could place it in the exterior wall,
and serve the food from the outside; but this is a bad system,
especially in winter, when the food, being given scarcely
lukewarm, rapidly cools and sometimes freezes. Then, again,
the attendant invariably fails in doing his work thoroughly
if the weather is very cold or wet, as he hurries through the
feeding much too rapidly.

" Another bad system, which is too common in England, is
to place the trough in the outside court. This, however, is
still worse, for, in addition to the inconveniences enumerated,
it has the disadvantage of compelling the pigs to go outside
to feed, notwithstanding the nature of the weather. I believe
I have solved this problem in the most satisfactory manner, by
the arrangement of a small passage of service between every
two sties. The food is carried *en masse* in pails upon a small
car : this is drawn to the opening of each small passage, when
it is served to the pigs in each sty. At the end of each
of these small passages is a window which lights every sty,
and more particularly the troughs; doors are also provided
for entrance into each sty. Such a system, I believe, does
not exist in any other piggery. The troughs are built of brick
and cement. They are not expensive to make, they can be
kept clean, they resist the teeth of the animals, and they
are 31in. in length.

" The floors of the sties are of concrete, and have an in-
clination towards the outside doors; thus the drainage from
the pigs is carried away under the doors by a gutter, which
leads into the liquid manure tank. The walls between the
sties are 3ft. 7in. in height; they are constructed of bricks
placed upon their edges, the thickness being $2\frac{3}{4}$in., and both
sides are cemented. It will thus be seen that the structure
is very light and simple. As a general rule, too little space

is given in the sties, and I have therefore endeavoured to obviate this fault by making mine more spacious.

"I have dared to call my new building a model piggery, because I have been at considerable trouble in avoiding all faults and uniting all possible qualities in its construction "

American Piggeries.

The following is a description of the piggery of Messrs. C. Street and Son, of Hebron, McHenry County, Illinois, shown in Figs. 49 to 52:

The building is 30ft. by 64ft., with posts 9ft. high. At one end is a water-tank, supplied from a well, and a stove,

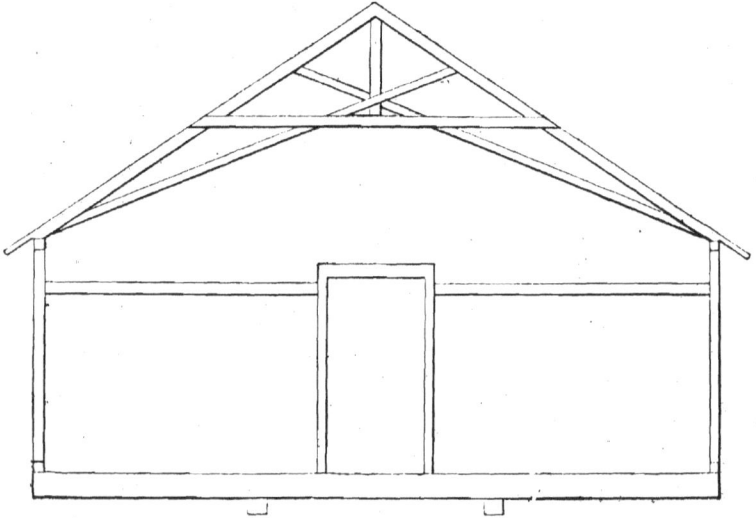

Fig. 49.—End-section of Messrs. C. Street and Son's Piggery, at Hebron, McHenry Co., Illinois, U.S.A.

used for heating purposes. At the other is the corn crib. A passage, 4ft. wide, runs down the centre of the building, upon either side of which are six pens or sties, each having an inner apartment in which the pigs sleep. This apartment is 6ft. by 6ft., and the sties are 8ft. by 13ft. The

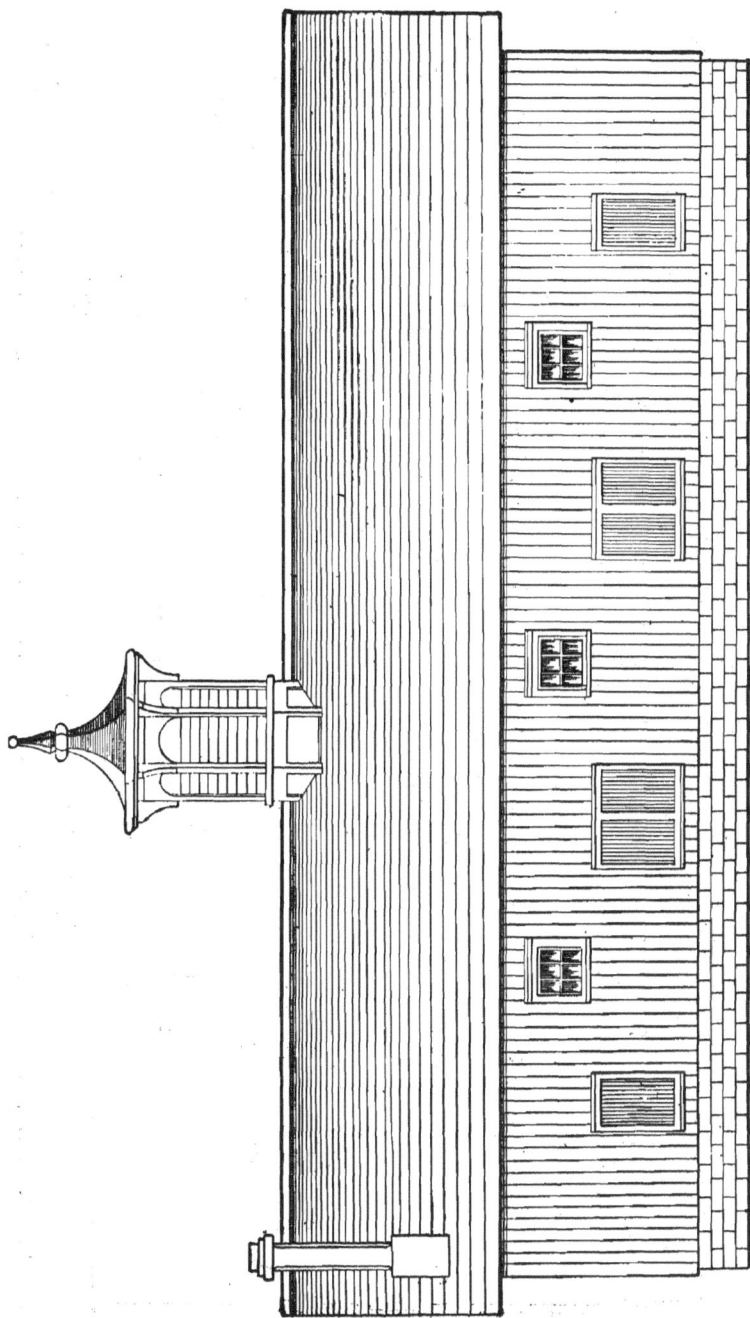

Fig. 50.—Side Elevation of Messrs. C. Street and Son's Piggery, at Hebron, McHenry Co., Illinois, U.S.A.

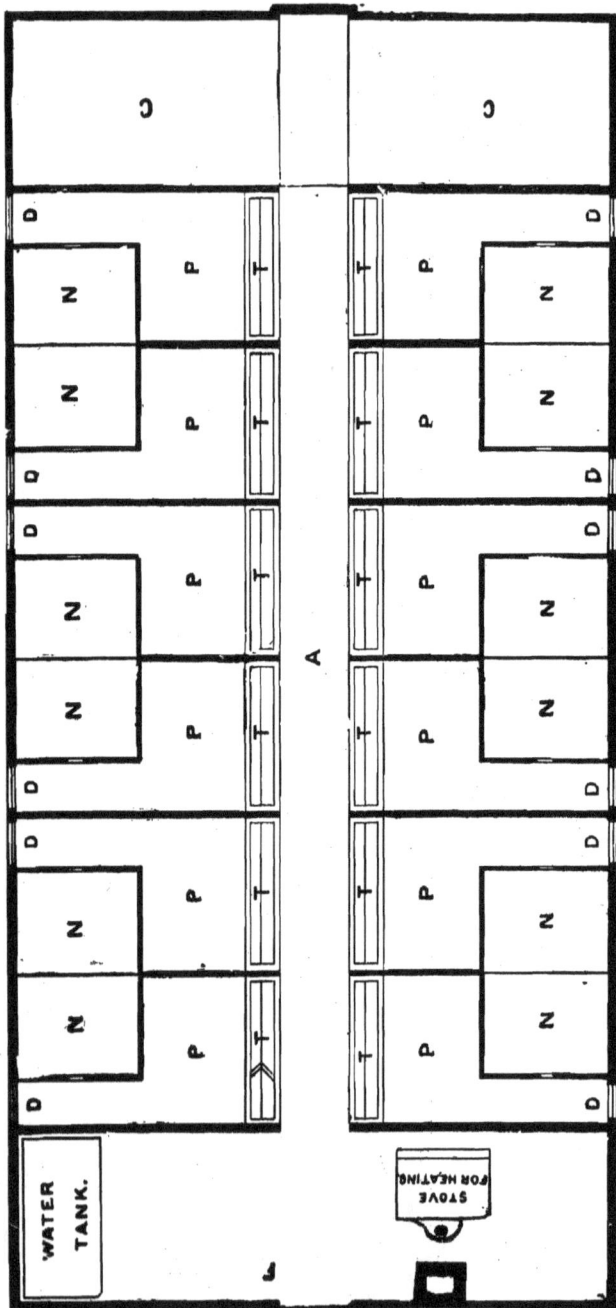

Fig. 51.—Ground-plan of Messrs. C. Street and Son's Piggery, at Hebron, McHenry Co., Illinois, U.S.A.

A, Alley; C,C, Food Stores; D,D, Doors; F, Front Hall-way; N,N, Sleeping-places; P,P, Pens; T,T, Feeding-troughs.

floor of the sleeping apartment is raised 2ft. above the floor
of the outer court, and the end of the same apartment connecting the small doors is enclosed with swinging partitions.
The front is open, and every other partition is so arranged
that it can be fitted for either a large or a small pen as
occasion may demand. The doors referred to are 2ft. by
3ft. 6in. in size, and all round the inside there are 1in. boards
to a height of 4ft. Over the head of the pigs (some 7ft.
from the floor of the sty) the bedding is stowed in winter,
as this is found to supply warmth to the animals; while a
door permits of its being shot into the sleeping apartment
when desired. The outside of the apartment enclosing the
bedding is made of 12in. boards and 2½in. battens.

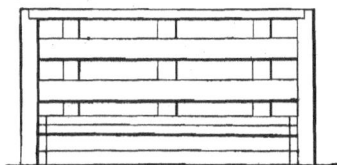

Fig. 52.—Swinging Gate over Trough in Messrs. Street's Piggery.

In connection with the ventilation arrangements is a slide
which, by means of a cord, can be shut or withdrawn, as is
found necessary. There are two windows in the end of the
building and three in the side. Three pairs of rafters, stayed
in the manner shown in Fig. 49, one in the centre and one
between this and the ends, keep the building in its proper
shape. A rib set into the posts helps to relieve the bearing
upon the joists. The height of the partitions in the sties is
3ft. 6in., and the swing partitions hanging over the troughs
have a latch which, as the partitions swing back, drops, and
effectually prevents the pigs from getting to the food as it is
being placed before them. These partitions are made of 4in.
boards, with a space between each.

Messrs. A. C. Moore and Son, the famous Poland-China
breeders, of Canton, Illinois, have an admirable piggery, which
is 120ft. long by 25ft. wide. The building is shown, both in
plan and in elevation, in Fig. 53. It is erected upon a
foundation of stone pillars, built to a height of 2ft , and at a

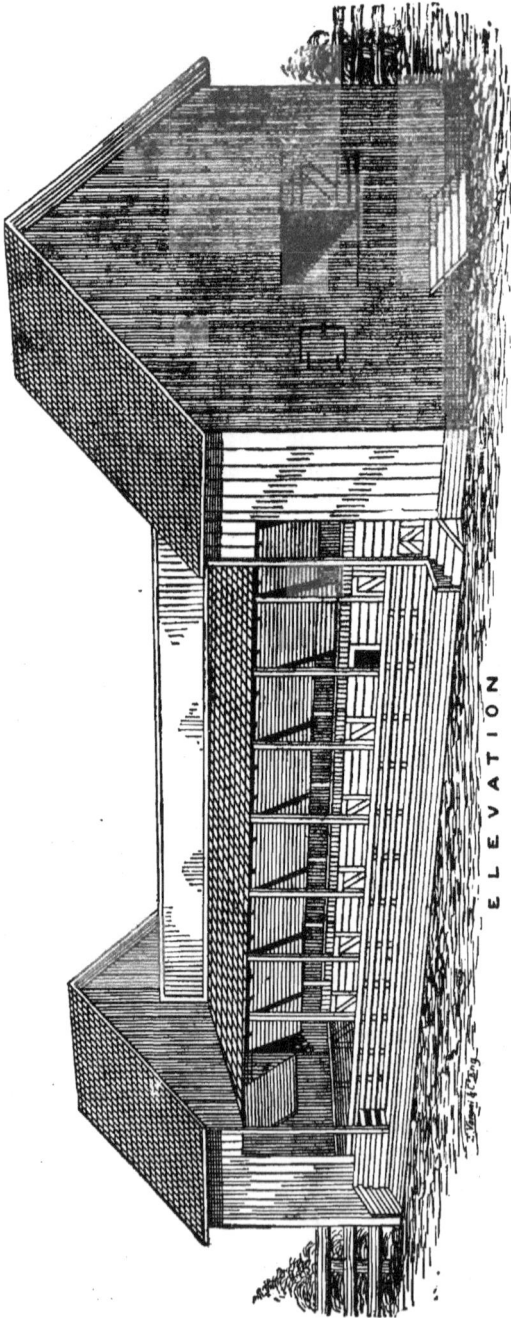

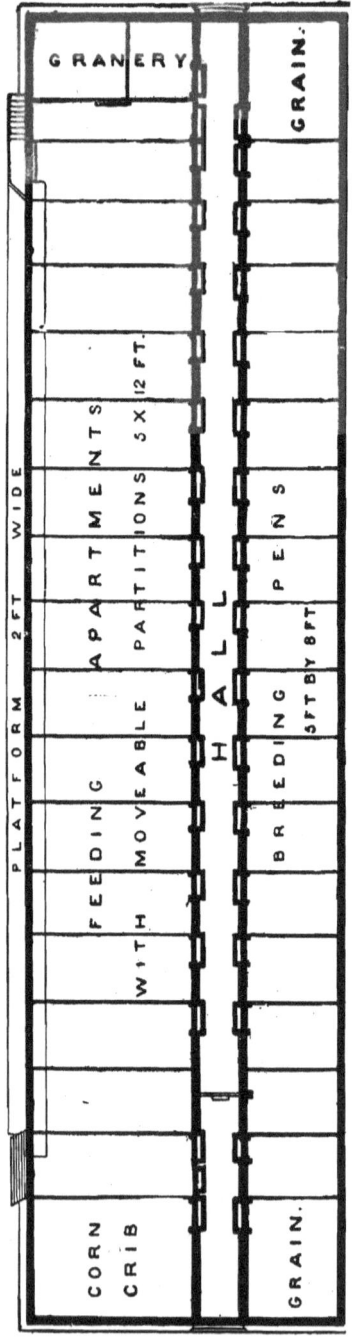

ELEVATION

PLATFORM 2 FT WIDE

GRANERY.

GRAIN.

FEEDING APARTMENTS

WITH MOVEABLE PARTITIONS 5 X 12 FT.

HALL

BREEDING PENS

5 FT BY 8 FT

CORN CRIB

GRAIN.

distance of 6ft. apart. Upon the pillars are placed three rows
of oak sills, 10in. by 10in., and upon these sills are mortised
oak rafters, 2in. by 8in., which reach from the outside to the
centre sill. There is also a piece of oak timber laid length-
wise under the centre of these rafters, and blocked up with
stone, in order to make them more solid.

The piggery, although a single building, is divided into
three parts. The two end portions are much higher than the
long range of sties in the centre, and are provided with an
upstairs floor. The buildings face the south, and the hind
part of the central building is 3ft. high at the back and
10ft. in the front, so that a 16ft. rafter covers it, and makes
the roof about half-pitch. The two end buildings are 15ft.
by 25ft., and are 12ft. high at the eaves, the roof being
provided with the same fall as that in the centre. The
apartments of these buildings are used for straw on the
upper storeys, and as food-rooms and granaries below. A
passage, 4ft. wide, runs the whole length of the piggery, on
the northern side of the pen; the remaining 8ft., upon the
southern side, is divided into breeding pens, 5ft. wide, with
a door in the front of each pen. The partition upon the
south side of the passage is boarded up half-way, the other
half being covered in by light doors, hung upon hinges, so
that they can be drawn back and fastened, to permit the
entrance of the sun into the back pens, or shut, to protect
them against foul weather.

The floor in front (which is about 12ft. by 100ft.) is also
divided into compartments 5ft. wide, with movable partitions
and doors in the rear, which are exactly opposite similar doors
in the north or breeding pens. These doors can be opened
and fastened together, when they form a partition for each
separate sow, so that she can come out of her pen to feed,
and thus keep her bedding clean. The front range is not
covered with a permanent roof, inasmuch as the pens are
required for young pigs, which must have plenty of air and
sun. The sows and pigs are kept in these compartments until
the youngsters are three weeks or a month old, when the
movable partitions are taken away, and the various females
turned out together upon the grass, after each pig has been
labelled upon the ear. When the young pigs are old enough

to feed, they are allowed to enter these sties as they choose; but the sows are kept away, and fed elsewhere. The floors are all of pine, which is less slippery than oak, and the sows are not so liable to strain themselves upon it. Water is raised by wind power into an elevated tank, and can be run into any part of the piggery.

The following description of a large piggery, which is stated to have had as much careful thought bestowed upon it by a practical breeder and farmer as any similar establishment in America, is described by Mr. Coburn (now one of the most prominent agricultural authorities in the United States) in " Swine Husbandry."

The building is 100ft. long by 30ft. wide, built of first-quality pine, upon stone foundations, and arranged with a view to the utmost economy of time and labour in feeding and care of the stock. There are fourteen pens on each side, divided by movable partitions, so that two or more pens can at a time be thrown together as one. Each pen is furnished with a fender in one corner, to prevent the young pigs from being overlaid and smothered by the sow. Through the centre of the building is a passage 12ft. wide, along which runs a wooden track with a truck or car for carrying barrels of food from the steamer and food-rooms. Each of the troughs extends through the partition between the pens and the passage, so that food can be poured into them from the outside without interference from the animals within. This is quite a novel and somewhat practical feature which has many advantages. All the pens open into outside yards, the gates between them forming, when open, a passage, through which the animals can be readily removed from one portion to another, and manure wheeled out to the compost heap. Fresh spring water runs through all the yards on either side of the building, and extensive clover pastures are accessible from the north, east, and south.

The owner of this piggery raised hogs by the hundred, and claims for his establishment that it economises labour and affords excellent protection to a large number of animals, giving warmth in winter and shelter and ventilation in summer. By opening the large doors at each end of the building, and the fourteen small doors on each side, the freest ventilation is secured in both directions; the interior walls of the pens are, of course, but a few feet high, and the space above them is open. In its owner's opinion, the abundant clover pastures adjacent, and the strong, never-failing springs, supplying the purest water, are among the most valuable additions to this structure, and they are prime necessities to the success of any other swine-breeding establishment.

The plan and elevation of a first-prize American piggery are illustrated in Fig. 54. In this building, which is 36ft. by 16ft., there is an additional extension for the pens of 14ft., also by 36ft., the height of the back being 8ft. The sills and joists of the pens are intended to be of oak, pine or spruce being used for the remaining work. There is a cornice, with a projection of 18in., and the roof is covered with timber of good quality, and provided with a ventilating cupola, as shown. Each pen, with the yards in the rear, is furnished with a sliding door. Doors are also provided in the partitions between the pens, upon the left side of the building, as shown in the end elevation, and in the main floor, as shown on the ground-plan. This floor is intended to be used for dressing and scalding, and should be matched, thick paint being used to make it watertight. The cellar-way below should be ceiled inside, and filled in with coarse mortar. The feeding-troughs, which are extremely simple and equally useful, are made of hard wood, preferably oak, the flap door over the troughs swinging backwards and forwards for convenience in filling, and fastening back with a button. The cellar, which is below the main floor, together with the foundation walls, are of common stone laid in mortar, the cellar-bottom being grouted and, if necessary, provided with a drain. A cistern is provided beneath the swill-tub, which is well cemented, there being an opening outwards, which is closed with a door. An iron pipe is connected with the furnace, to carry off steam, &c., this running through the loft over the main building. The outside is well painted with two coats of paint, both for preservation and appearance. The entire cost of the building is estimated at £136, but it may be erected for considerably less if the cellar and cistern are omitted.

PEN 11½ × 12½ PEN 11½ × 13½ PEN 11½ × 13½

TROUGH

FURNACE

HEARTH KETTLE

PUMP

SWILL TUBS

UP

14 FT.

16 FT.

36 FT

GROUND PLAN.

TROUGH OPEN.

PEN.

TROUGH CLOSED.

PEN.

DOOR

TROUGH

FRONT OF TROUGH

END ELEVATION

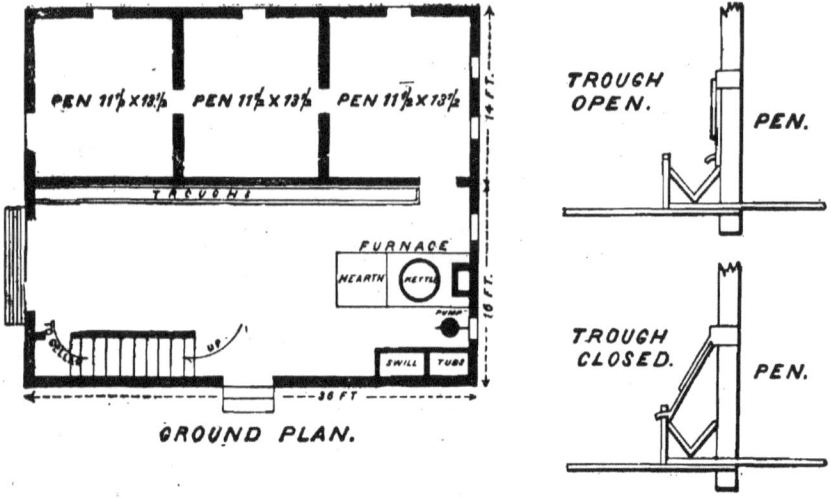

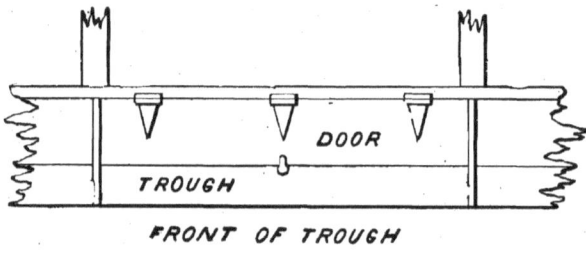

Fig. 54.—A First-prize Plan and Elevation of an American Piggery.

Fig. 55 is a plan of an extremely useful and simply made building, 48ft. in length and 30ft. in breadth, erected upon 12ft. posts, 6in. by 8in. in size, placed 16ft. apart. It was built for Mr. W. W. Ellsworth, of McHenry County, Illinois, U.S.A. *g, g* show the doors leading into the pens, while the line at *a* shows the wind-break partition by the doors. The enclosures *e, e,* are the sleeping-places, and *c, c,* are rails, 6in. high, intended to keep togther the straw upon which the pigs lie. The feeding-floors of the pens are shown at *d, d,* each pen being 8ft. broad. The lines *b, b,* show the wooden partitions between the pens, the doors leading from the passage (*f*) in the centre into the pens being shown at *i, i.* This passage is 6ft. wide, and is provided with a door at each end, that at *j* being double, and opening to the full width of the passage; while the other (*h*) is a small door for the attendant. The troughs are placed at *k, k,* the food being put into them from the central passage. The sleeping-floors are raised 2in. from the ground, in order that they may be kept perfectly dry for the pigs to sleep on. This building may be constructed entirely of timber, the roof being felted and tarred; or, if preferred, the latter may be covered with corrugated galvanised iron. In either case, if substantially made, it will answer equally as well as a much more costly brick or stone building.

In Fig. 56 the grass runs, or lots, connected with this piggery are shown. These may be as large as possible, and the larger the better, if the pigs can be at all times provided with clean grass. The end runs, which are larger than the others, are preferable for a number of growing pigs, which require more space, while the smaller ones are better adapted for sows with litters. The whole are furnished with wooden or iron palings, according to the discretion of the proprietor; but, in every case there should be gateways through which either a tramcar or a cart can be driven.

Mr. A. R. Cohoon, of Minnesota, tells us that he has found the buildings illustrated in Fig. 57 admirably adapted to his requirements. It will be noticed, however—as he has pointed out to us—that the elevation shows only five pens, whereas there are six, as will be seen by a reference to the ground-plan. The cooking-room (*a*) on the ground floor is 16ft. by

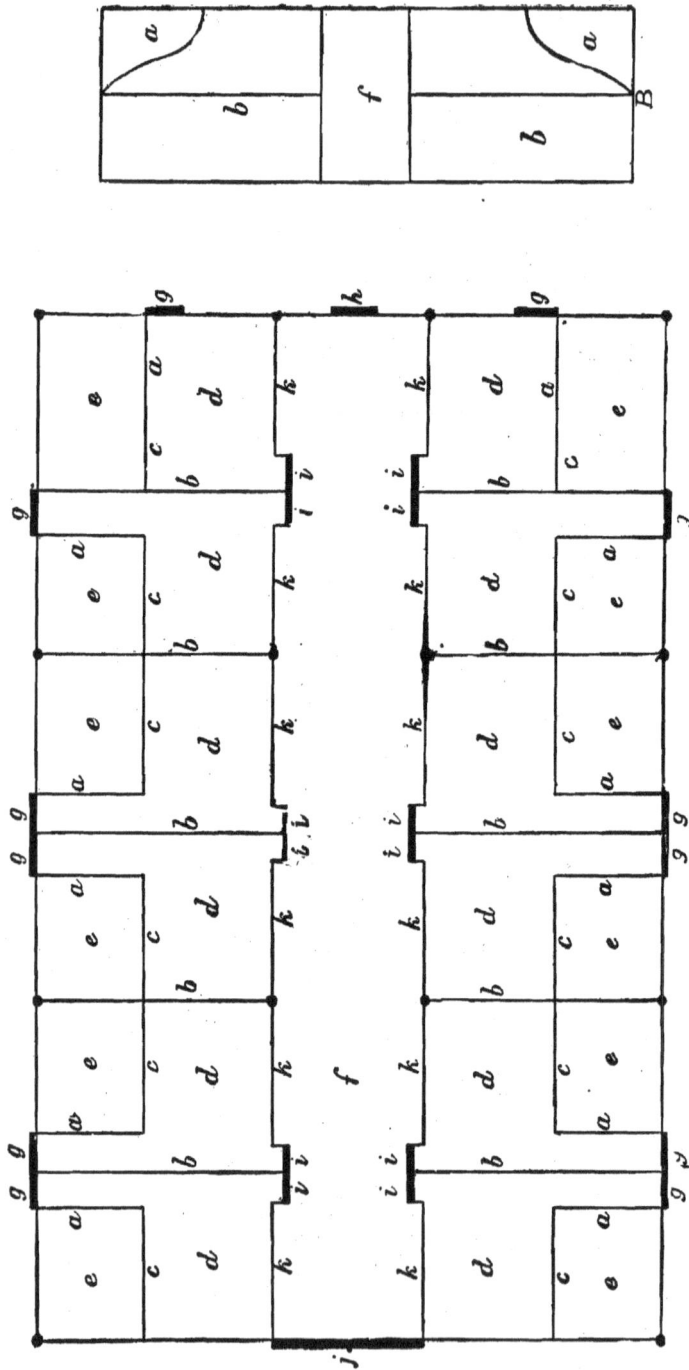

Fig. 55.—Ground-plan of Piggery owned by Mr. W. W. Ellsworth, of McHenry Co., Illinois, U.S.A.

a,a, Wind-Break Partitions by Doors; *b,b,* Partitions between Pens; *c,c,* Rails to hold Straw; *d,d,* Feeding-floors and Alleys to (*g,g*) Doors; *e,e,* Sleeping-places; *f,* Alley; *g,g,* Doors for Hogs outside; *h,* Back Door in Alley for Man; *i,i,* Doors for Hogs in Alley; *j,* Double Door in Front; *k,k,* Feeding-troughs; Posts in Building (6in. by 8in.). B. Diagram of Pen to Upper Floor, to show Partition (*b*) and Wind-Break (*a*) by Door, and Alley (*f*).

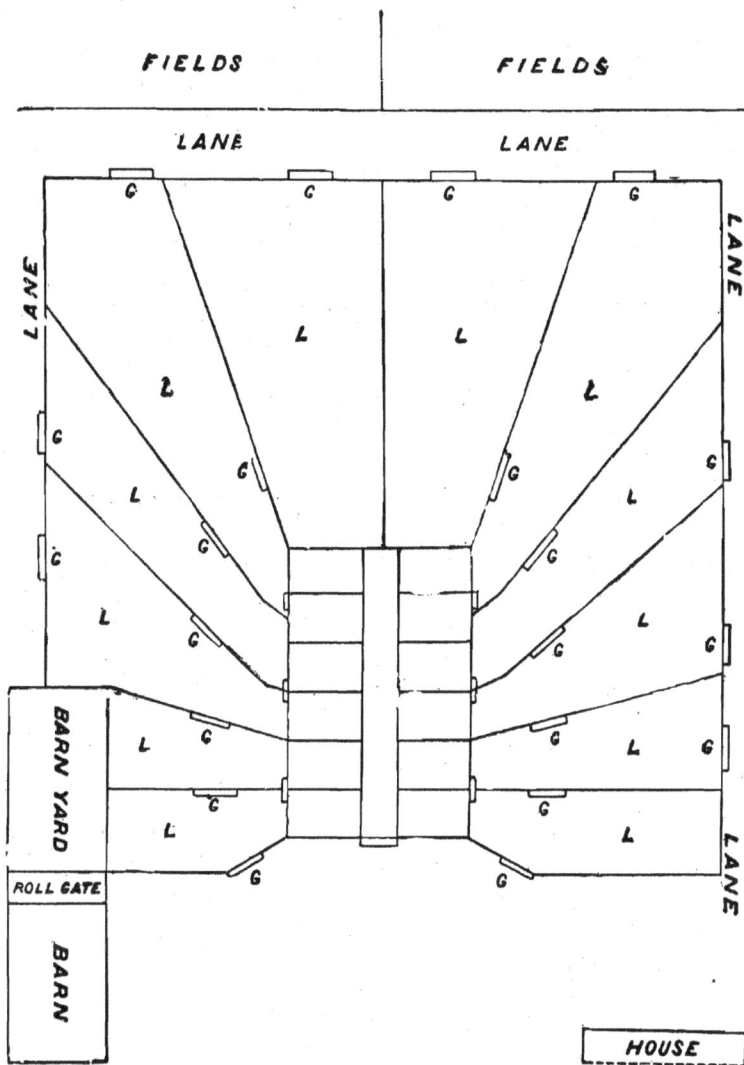

FIELDS   **FIELDS**

LANE   LANE

LANE

LANE

BARN YARD

ROLL GATE

BARN

HOUSE

Fig. 56.—Plan of Piggery and surrounding Grounds owned by
Mr. W. W. Ellsworth, of McHenry Co., Illinois, U.S.A.

L,L Grass Runs connected with Barn-yard and Pasture.   G,G, Gates through
which to drive trams to draw manure.

10ft. The bran or meal-bin (*b*) is 16ft. by 6ft., and has a
sliding door fixed at the bottom, so that the contents may be
allowed to fall into a tub beneath. The scalding-tub for food
(*c*) has a capacity of 40 bushels; it is sunk in the earth to a
level with the gravel floor of the cooking-room. *d* is a steam-
generator, and *e*, horse gear, connected by a belt with a
chaff-cutter in the upper floor; this is used for cutting hay,
which is mixed in the scalding-tub with the meal and roots,
the whole being steamed together; it is eaten with relish
by the pigs, and suits them admirably. *f* is the passage,
which is 4ft. wide, and *g*, *g*, are the doors, 2ft. wide. *h*, *h*, are
elevated pens, 4½ft. wide by 8ft. long. These pens are 3ft.
higher than the main floor, and as their bottoms are made
tight, there is no leakage; they are used for feeding young
pigs apart from the sows.

The run-ways for pigs (*i*, *i*) are 8ft. long by 14in. wide, and
16in. up and down. The lower end is made smaller by
tacking on a board, leaving sufficient room for those pigs
to go above which the breeder wishes. The run-ways are
fastened, at the upper end, by leathern hinges, to the
elevated floor; the lower end is fastened down by a prop
running from the upper side of the run-way to the joist
above, and driven under tightly while the run-ways are used.
When not in use, the lower end can be elevated to a point
between the upper joist, and there fastened by a pin running
through the joist into the run-way.

The nests, or sleeping-places (*j*, *j*) are 8ft. by 5½ft. Each
is raised 2in., so that it can be kept perfectly dry. The side
near the outer door is planked 2ft. high, as a sort of wind-
break, while a ledge, 2ft. by 4ft., is put in front, to keep the
litter in place. The front of the nest is enclosed by dropping
boards, 8ft. long, into grooves provided for the purpose,
thus protecting a sow about to farrow from other pigs in
the same pen. The nest-beds are arranged directly under the
large windows, in order to give the little pigs the advantage
of the early sun. They are also provided with adjustable
fenders, by using a plank, 6in. wide, fastened to the side
of the nest, 8in. above the floor. These are intended to prevent
sows from crushing their pigs while small.

The main pens (*k*, *k*) are 16ft. by 10ft., including the rest,

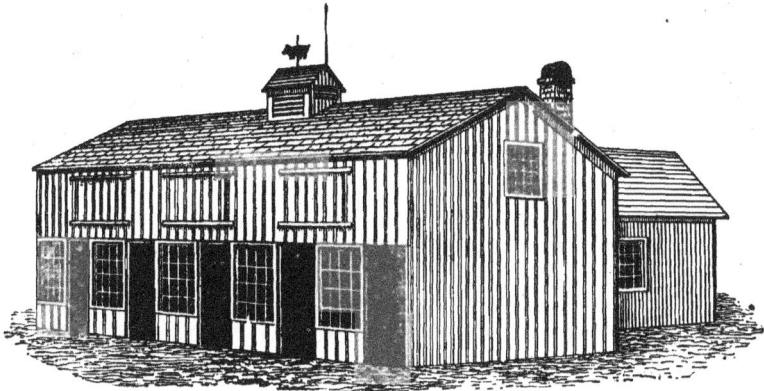

ELEVATION.

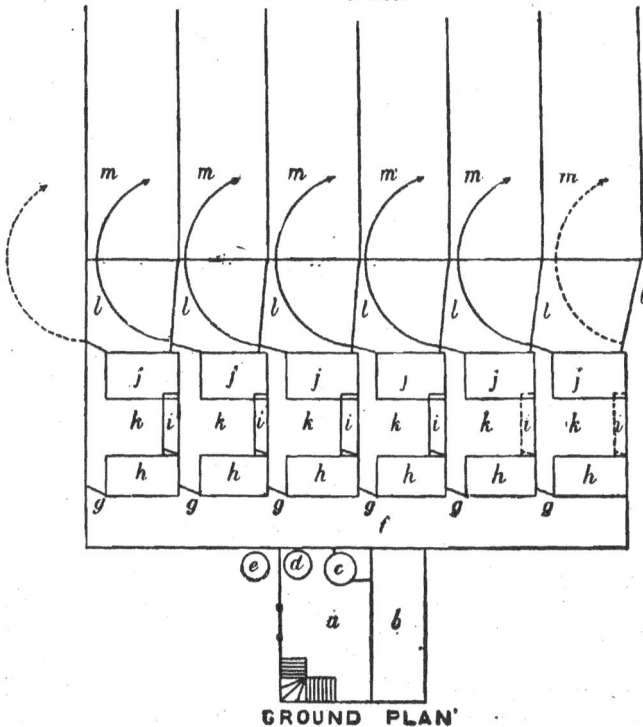

GROUND PLAN

Fig. 57.—Piggery owned by Mr. A. R. Cohoon, of Minnesota, U.S.A.

a, Cooking-room; *b*, Meal-bin; *c*, Scalding-tub for Food; *d*, Steam-generator;
e, Horse-power gear; *f*, Alley; *g,g*, Gates or Doors; *h,h*, Elevated Pens;
i,i, Run-ways; *j,j*, Nests; *k,k*, Main Pens; *l,l*, Gates; *m,m*, Yards.

and, with the sections of the partitions, are movable, so that
two or more pens may be thrown together. The doors leading into the yard are 6½ft. by 2ft., and are made in two
sections, the lower one being 2ft. 10in. high. They are
provided with hooks and hasps, so that they may be held open
or fixed close as the weather may require.

The windows in front are 8ft. by 3ft., occupying the whole
of the space between the doors. The gates (*l*, *l*), 10ft. wide
and 3ft. high, are hooked to the side of the building, and,
when swung parallel with it, with each gate hooked to the post
of the other, they provide a passage, permitting the house
to be cleaned out, or the upper storey to be filled with hay or
straw. They also form an outer yard, so that each lot of
pigs may be driven out of the house and kept together while
it is being cleaned. The gates can, too, be swung back and
hooked to the fence of which they form a part, so that two or
more yards may be thrown together. The yards (*m*, *m*), 50ft.
long and 10½ft. wide, are provided with tight fences, and are
well gravelled.

The loft, 60ft. by 20ft., is intended for the storage of hay or
straw. A box connects the lower storey with the ventilator.
This is provided with long rods, which are attached to the
upper end of the lattice blinds, closing them at will. The
height of the building to the eaves is 14ft., and its dimensions are 60ft. by 20ft. Its cost to Mr. Cohoon, including
fixtures and stone foundation walls, was £200.

Small Piggeries.

A convenient small piggery is shown in section at Fig. 58. It
is divided into two compartments, with a court to each, and is
adapted for a breeding sow and a few store pigs. The covered
building is 5ft. high at the eaves, and is provided with a stout roof,
preferably tiled, to keep out the rain. The depth from the front
to the back wall is 7ft. The walls are, in front, 4½in., and,
behind, 9in. in thickness. The front wall of the court is also
9in. in thickness, and the length of the court 9ft. The letters
aa show the floor or the outside court; *bb*, the floor of the
sty; and *c, c*, the drains in each. A partition runs from the

wall of the court to the rear wall of the covered sty, providing two compartments—that for the sow being 9ft. in width, and that for the store pigs 6ft. Doors are provided to admit the animals into their respective courts; while there are other doors in front of the courts for the admittance of the attendant.

Fig. 59 shows a ground plan of this building: *aa* is the sleeping apartment or cote for the pigs; while *bb* is their open yard, in which the feeding-trough, *d*, is placed, the door being shown at *c; ee* is the covered apartment for the sow, and *ff* the yard attached to it. The letters A and B indicate the section shown

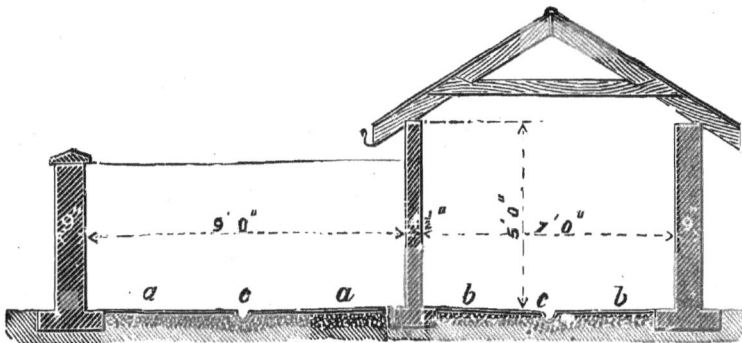

Fig. 58.—Section of a Convenient Small Piggery, divided into Two Compartments. The Ground-plan is shown in Fig. 59.

at Fig. 58. The Cheshire trough is filled by means of a shoot through the wall, shown at Fig. 60, in which *b* is the trough and *aa* the shoot. There is a surface drain both in the sties and in the courts, which are connected with the ordinary drainage system, these being made in either concrete or hard brick. The floors may, however, be arranged to slope to any given point with equal advantage.

The doors provided are shown at Fig. 61, and, like the gates in use at the Colonie St. Maurice, they defy the most cunning pig, it being unable to obtain any leverage to assist it in forcing them. The door *aa* is lifted by the holes at *bb*, when it slides in grooves, *c, c*. These grooves are shown at *d, d*, and the walls at *e, e*, in the plan. The letters *ff* show the door as it is fixed in the grooves, and *g, g*, are the walls at the side.

Another form of door is shown at Fig. 62, in which *aa* is the wall and *b* the door, the lower style abutting against the stone block (*d*), while *e* is the line of the floor. A part of the side elevation is shown at *f*, which is a portion of the wall *a*, *g* being the floor line.

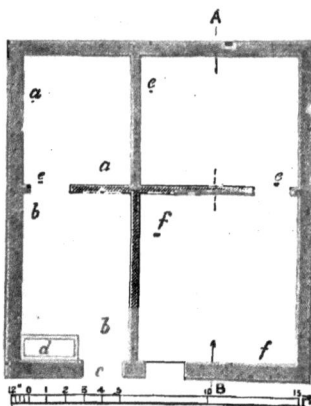

Fig. 59. — Ground-plan of the Small Piggery shown in section in Fig. 58.

Fig. 60.—Section of the Cheshire Trough and Shoot.

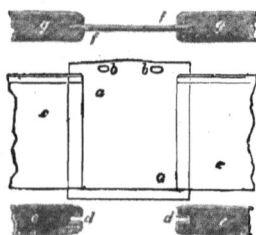

Fig. 61.—Elevation and Plan of an Ingenious Sliding Door.

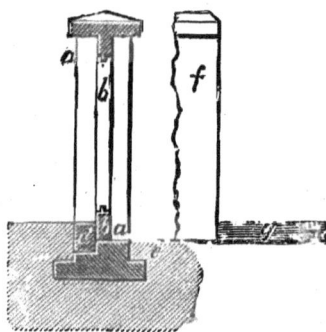

Fig. 62.—Section of another form of Door for a Pig Court.

There is another kind of floor which can be admirably adapted to a piggery of this kind, and which is shown at Fig. 63. The outside wall is seen at *aa*, and the back wall at *b*, this being of wood, where two apartments are side by side. It is mortised into a stone sill, c. An asphalt floor, dd, slopes towards the drain, g, in the centre. A wooden framework is shown at ee, the two end bars being 4in. by 3in. The remaining bars or spars, f, f, are about 3in. wide, with ¾in. spaces between each, to allow the manure to pass through as it is voided. This framework lies loosely upon the floor of the sty, or upon one corner, as may be found desirable. If the space beneath is sufficient in depth, say from 12in. to 15in., the solid manure will accumulate while the liquid will drain away, and the former can be removed from time to time by simply lifting the framework. There are various opinions as to the healthiness of this

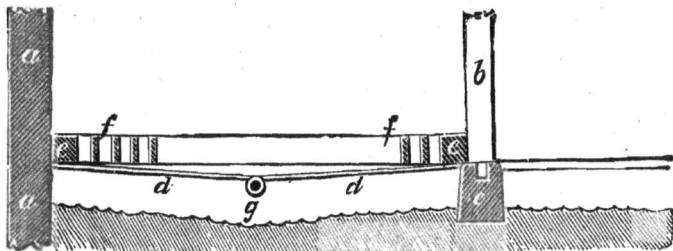

Fig. 63.—Section of a Sparred Floor.

system, which has been strongly recommended by many practical men; but it is our opinion that pigs can be kept in better health where the solid manure is entirely removed daily.

Figs. 64 and 65 show the walls of an outside court with an iron front, to which are attached a trough and a door of the same material. These are clean, convenient, and almost indestructible. The troughs are shown ready for the reception of food (Fig. 64), and closed with the food open to the pigs (Fig. 65). These arrangements are among the best of their kind, and are made at the St. Pancras Ironworks. The piggeries, covered and open, shown in Figs. 66 and 67, and made by the same company, are of the highest excellence; and that shown in Fig. 68, made by Messrs. Hill and Smith, of Brierley Hill, Staffordshire, is light, strong, and elegant, and is provided with one of the best-made troughs, resembling that shown at Fig. 26.

R

By the arrangement shown in Fig. 66, one roof, which is of corrugated iron, covers the inside as well as the outside courts,

Fig. 64.—Iron Front of a Court, showing the Trough Closed against the Pig.

Fig. 65.—Iron Front of a Court, showing the Trough Open to the Pigs.

both of which are well ventilated by apertures in the iron sides of each compartment. The outside court is reached by means of an iron door and the pigs are fed in the front of this court, the iron trough being opened and shut against them at will by an iron swinging shutter. If necessary, a number of these buildings can be ranged side by side, and the same

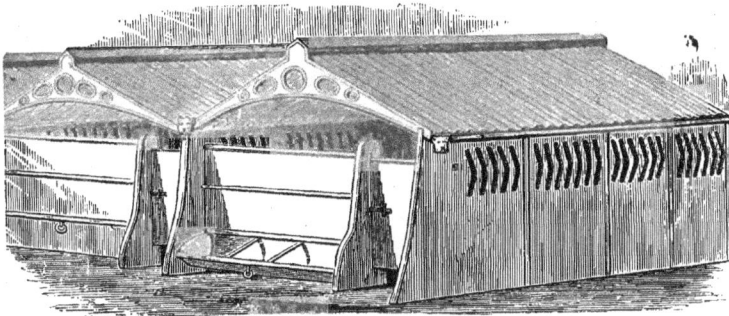

Fig. 66.—Covered Iron Piggery.
Showing the Feeding Troughs Opened for Filling and Closed for Feeding.

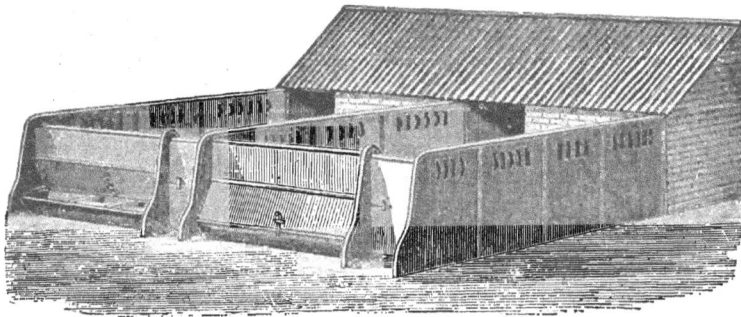

Fig. 67.—Double Piggery, in the old style, with Iron-walled Court.
The Troughs are shown Opened for Filling, and Closed for Feeding.

remark applies to the piggery shown in plan in Fig. 69. The drainage is managed by means of an open gutter, running down the centre of each building, as indicated in Fig. 69, the sides being higher than the middle. This gutter runs into a drain at right angles to, and outside, the piggery.

R 2

The building illustrated in Fig. 68 is adapted for pig-keepers who only require one sty, and who prefer that to be well made, of solid material, and of good appearance. It is almost entirely of corrugated galvanised iron, is well ventilated in the front

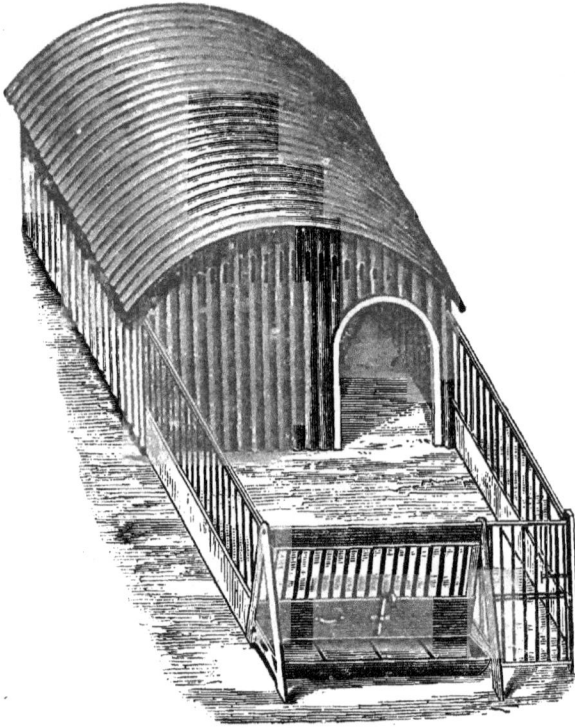

Fig. 68.—A Well-designed Modern Single Iron Piggery.

and at the sides, and is provided with an arched door leading into the outside court, the latter being surrounded by plain iron railings. In the front are a strong gate, of open ironwork, and a trough, made in modern style, and intended to shut backwards and forwards.

The double piggery, with iron-walled court, shown in Fig. 67, is erected in the old style. The main building is of brick, and

is substantially put together.   The roof slopes in the old-
fashioned way, and is made of corrugated iron.   The principal
features, however, are the improved iron walls and troughs of
the outside courts, all of which are solid and indestructible, this
material being better adapted for piggeries than perhaps any
other which can be named.

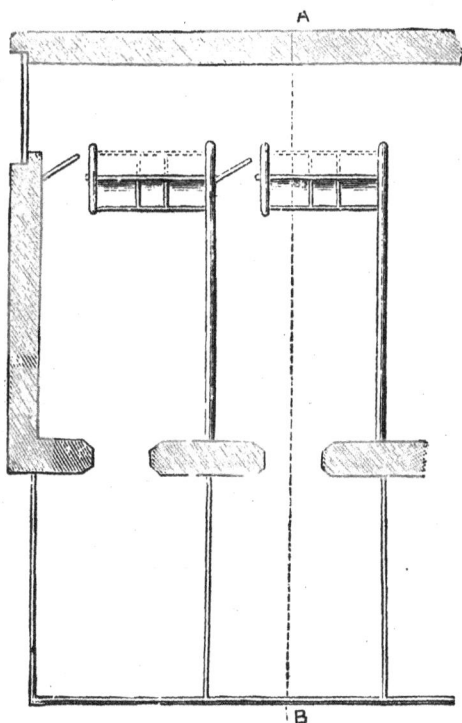

Fig. 69.—Ground-plan of Covered Piggery.

Showing Courts and position of Drain, AB.

A strong and convenient pig pen, suitable in cases where
a single pig is kept, is shown at Fig. 70. It is easily made
by any person who is handy with tools, and is very in-
expensive.   The door shown in the front, *abcd*, is made by
fastening inch boards to the quartering at B, the ends of which

are let into the corner posts of the pen, so that the door
is enabled to swing. It is shut to by the bolt at c, which can
be obtained ready made from almost any large ironmonger.
The trough is fixed across the front of the pen, its right-hand
end being at that part of the pen shown at D. When it is
necessary to feed the pig, the bolt is drawn, the door pushed
back, and the bolt shot again on the other side of the
trough, which can then be filled without hindrance from the
pig. The door is then shut, and the animal has undisturbed
possession of its food. If the roof is tiled, it will be found
convenient to have one or more glazed tiles. If, however, it

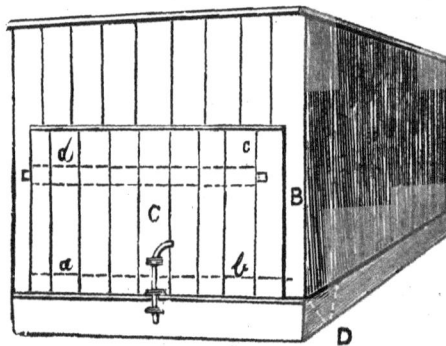

Fig. 70.—A Convenient Pen for a Single Pig.

is boarded, slated, or covered with corrugated iron, it will
be necessary to have a ventilating window in the front.
Ventilation can be provided over the door or on each side.

It may frequently be found convenient to adopt the following
combination of a door and feeding trough, for particulars of
which we are indebted to the *Rural New Yorker*. The trough
is made of two boards, 8in. and 10in. in width respectively,
which are nailed together, and may be of any required length.
At either end is a piece of board, some 2ft. in length, which
forms a stand for the trough, also an upright, made of 4in. by
4in. boards, which is extended to form a stud to the side of
the pen. In each of these uprights, at a distance of about
2½ft. from the ground, a hole is bored with a ⅝in. centre-bit.

Another board, 4in. by 4in., is then used for connecting the
uprights. A hole is bored into the ends of this by a $\frac{1}{2}$in.
centre-bit, and a piece of $\frac{1}{2}$in. round iron is next driven through
the uprights into the ends of the horizontal board, forming,
in this way, a very convenient, cheap, and durable hinge. In
the latter, at suitable distances apart, spaces are made for
the insertion of three or four 2in. by 3in. boards, which run
at right angles, extend to a little below the top of the
trough, and are sufficiently high to serve as a wall for the pen.
The middle one is made to extend above the others, to form a
handle. Boards are then fixed to these upright pieces, these
constituting a part of the wall of the pen, which continues
above to a proper height. In this way, by a simple move of
the handle, the partition is made to swing over either one
way or the other, and the food can be placed before the pigs
without fear of its being upset in the process. For the sake
of convenience, any spring that will answer the purpose
may be so fixed to the top of the pen and the top of the
swinging door as to keep the latter in position.

Movable Sties.

The only remaining class of pig-house to which it is necessary to refer is a movable sty, which, however, is not much
patronised in this country; farmers often preferring to allow
their pigs to roam about the grass fields at will, rather than
to confine them to the yard or sty, although at times they
would undoubtedly do much good upon the arable land, and
obtain most of their living.

A movable house for a number of pigs must necessarily be
large; but, if four old cart-wheels can be obtained, a building
can be easily and cheaply erected, which will be adapted to
the practice of moving the pigs about the farm, and will, at
the same time, restrain their liberty. In summer weather
—and this is the only period when the plan should be adopted
—no floor is required; but the walls of the building must
be brought close to the earth, to prevent the pigs from
getting out, which they certainly will do if they have half
a chance. A fixed trough should be fastened across one end,

and so arranged that it is always upon the level. There should be a swinging door, to enable the feeder to fill the trough; or, if preferred, one end may pass through the wall, in which case it can be fitted quite as well as if entirely exposed. A movable house, built upon four waggon-wheels, to cover six or eight store pigs, could soon be erected, and could easily be moved about the fields by a single horse.

CHAPTER XIV.

FOODS AND FEEDING.

IT is believed by a very large class of people to be one of the simplest things in the world to feed a pig; and the bare idea of conducting a system of feeding upon anything approaching scientific principles is to them the height of absurdity. There is, however, no better method of testing the truth of what science, combined with practice, has taught us than the conducting of an experiment with pigs of a similar race and age upon two systems of feeding—the one being in accordance with the common and, we might almost add, ignorantly applied method, and the other consistent with the principles which a thoughtful study of the subject has laid down.

In pig-feeding there are many things to consider, more particularly now that it is most difficult to feed them with profit, in spite of the lower prices of grain. Foods used for the purpose are numerous, and, in many senses of the word, extremely good; but, in order to obtain remunerative and rapid results, a feeder should make himself acquainted with their constituents, with the relative quantities required for the purpose of sustaining the animals, for maintaining their growth, and for producing fat. Just as there are certain foods, admirable in their way, upon which a human being could not long exist, so are there foods which, when given judiciously, are useful to pigs, but which, without the addition of other substances, would not only fail to increase their weight, however large the quantity given, but prove insufficient to maintain a healthy condition.

There is no domestic animal that feeds so rapidly as the pig, and probably none that so imperfectly masticates its food. This being the case, it is still more necessary that the food should be of a digestible nature, or that it should be given in such a form as will make the smallest possible demand upon the digestive organs of the animal, and enable it to appropriate the nutritive portions with a minimum of loss. The sole end of pig-breeding is to obtain a large quantity of meat, of a good quality, in as short a space of time as possible. This cannot be attained without breed and suitable feeding, nor can we expect quality without a combination of good feeding and management.

In consequence, perhaps, of the common custom of keeping pigs in sties, many persons are under the impression that they need neither exercise nor green food; and, indeed, it is often forgotten that the pig is a grazing animal. It is a fact, however, that there is no domestic animal upon the farm that pays better for liberty upon the pasture; and it may be laid down as a practical truth that the pig-breeder and feeder who grazes his breeding stock, as well as the young stock intended for fatting, will meet with a much larger meed of success than those whose animals are regularly confined to the sty, no matter how carefully they are managed or how admirably they are fed.

It has been said that, as the object of a pig's life is successful fattening, if he fails in that, his life is a failure. Providing the animal is of a good breed, and is healthy, such a result could scarcely be brought about, except by bad feeding and management. It is, nevertheless, a fact that, in the past few years, many practical breeders have lost money by their pigs, notwithstanding the care with which they have fed them; but, as a general rule, however bad the market may be, there are always persons who, pursuing a practical system, are able to make a successful return where their neighbours lose money upon every pig that they put up to feed.

Dr. Detmers, who was commissioned by the Missouri Board of Agriculture to investigate hog cholera, in remarking upon the hygiene of the pig, says that the organism of a domestic animal is composed of about fifteen to twenty elements, or undecomposable constituents of matter, united in numerous organic compounds. A constant change of matter is taking place, and a part of these elements, in the form of organic compounds, is constantly wasted and carried off by the processes of secretion

and excretion. The organism, therefore, in order to remain
healthy, and to maintain its normal composition, must receive,
from time to time, an adequate supply of those elements contained in suitable or digestible organic compounds, so as to cover
the continual loss, and, if the animal is young, to produce
growth and development. The simplest way to introduce these
elements into the animal organism is to give food that contains
them in nearly the right proportions. For instance, calcium
is present in water in the form of lime. One important element
—oxygen—enters the organism, also in large quantities,
through the lungs; but all others have to be introduced
wholly, or almost wholly, in the form of food. Almost all
kinds of food, however, milk perhaps excepted, lack some important elements, contain others in insufficient quantities, and
still others in greater abundance than is required. Therefore,
if such a kind of food be given exclusively—maize, for instance,
which is destitute of some of the mineral elements, and contains
only an insufficient quantity of the nitrogenous compounds so
important to the animal organisation—irregularities and disorders in the various organs will be the unavoidable results.

In support of the above theory, we may quote an experiment
made by Professor Knapp, who fed a thrifty, vigorous pig,
twelve weeks old, upon dry maize and water. Within three
weeks there were indications of fever, the limbs subsequently
became stiff, the skin dry, and the animal was extremely costive
and its appetite bad. During the fifth week there was great
weakness in the hind quarters, with swelling of the sheath, retention of urine and costiveness, and the appetite was still bad.
He then changed the diet to wash and cooked bran, and in three
weeks the animal was apparently well.

This alone, were other arguments not forthcoming, would be
a sufficient proof that feeding solely upon a particular article,
however good it may be, is highly inadvisable; and in practice
it has been found that a mixed ration not only suits the constitution of the pig very much better than any other, but goes
much further, especially if it is given in a warm state.

Experiments have also shown another fault in the system of
feeding entirely upon one class of food. The digestive organs
of the animal are unequal to the extraction of the whole of its
nutritive properties; and, as in the case of man, it has been

found that at least 50 per cent. is sometimes passed through the system into the manure without having benefited the animal in the slightest degree. Boussingault, whose reputation as a scientist places his opinion in a very high position, found from a number of experiments made with pigs: (1) that pigs fattened upon a mixed ration contained more fat than that which they had received in their ration; (2) that pigs fed exclusively upon potatoes produced no more fat than was contained in their food; (3) that foods which, given alone, have not the faculty of developing fat, acquire that faculty in an astonishing manner when fat is added to them, although fat given alone produces inanition; (4) that fat-producing rations, which themselves only contain a minimum of fat, are always rich in nitrogenous properties.

There are certain principles in feeding pigs which are of great simplicity, and which should be regarded with extreme care, whether equal care is taken in the selection of the food or not. There is nothing more important than that pigs should be fed regularly, and whether the supply be given twice, three times, or four times daily, it should invariably be at the same hours. It is not necessary, in feeding, that the trough should be filled, although it is a common supposition, especially in the country, that a man is a liberal, and consequently a good, feeder if a considerable quantity of meal is found remaining in the trough when the animal has satisfied itself and has lain down to sleep. It is a fixed rule with all who have mastered the question, to give no more than can be properly eaten. Like every other beast, a pig prefers fresh food, whether in a sweet or in a sour state, from the swill-tub. The trough, also, should not be filled from one end when a number of pigs are confined in the same sty. In such a case, the strongest will always get nearest to the feeder and obtain the largest quantity of the thickest food, leaving the thinnest to find its way to the bottom of the trough, where it is greedily consumed by the weakest pigs, which really need the best.

It has been frequently urged by old feeders that good food is comparatively thrown away upon fatting pigs if they are not groomed or kept in a thoroughly clean state. Some persons have gone so far as to say that a daily grooming is as valuable as an additional quantity of food; and that, moreover, it frequently answers to oil the coats of the pigs, both to prevent annoyance from vermin and to promote a feeling of satisfaction in the animals.

Water, again, is most necessary in all cases where pigs are confined; and if it is absolutely pure, so much the better for their health. There is, however, an additional reason why it should be provided in sufficient quantity. In spite of the supposed preference of the pig for mud, he invariably enjoys a bath in clean water, when it is provided for him in a basin in his sty; and there is no doubt that cleanliness thus promoted assists the proper assimilation of the food.

The feeding of pigs may be classed under two heads, one of which we may describe as feeding for a particular purpose, and the other as feeding upon a particular system.

In the former case, we have, first, the feeding of the boar and the breeding sows; second, the feeding of the litter subsequent to weaning; third, the feeding of young stores; fourth, fattening porkers; and fifth, fattening bacon pigs. In each of these instances there is a distinct *régime*, which is varied according to the custom of the country and the facilities at the hands of the feeder.

In the latter case, the systems of feeding are divided as follows: First, upon the produce of a cheese dairy; second, upon the produce of a butter dairy; third, upon the waste of a corn and sheep farm; fourth, upon the waste of hotels, public establishments, distilleries, &c.; fifth, upon a combination of garden produce and purchased food, as practised by cottagers and private individuals; and, lastly, feeding for exhibition purposes.

Feeding Systems of Various Breeders.

We addressed the following questions to some of the principal pig-breeders in England and America, and the answers are given below:

(1) What do you consider the best food for an in-pig sow?

The late Mr. James Howard, M.P., said: "Dry corn or bran, with a few raw mangels, turnips, or potatoes, diminishing the roots, especially the mangels, as the sow approaches farrowing, as the young are almost certain to be born dead if the mangels are freely given." Mr. Sanders Spencer says: "Vegetable food, with a little corn, beans, peas, or maize, until within a week of farrowing, when randan should be substituted." Colonel Platt: "Steamed Indian meal and a little sharps, mixed with water." Mr. Benjafield: "I prefer a good grass run, with bran, barley meal, and grains, mixed." Colonel Walker-Jones prefers crushed wheat, potatoes, swill, and bran scalded. Mr. James Robertson gives mangels, grass, whey, skim milk, and sharps. He also gives these foods to the boar. Mr. Frederick Coate says: "In-pig sows should not be in too high condition. Mine always do well on grass during the day, with a few beans, peas, or any dry corn, and a swede or two, morning and evening. I treat my boars in the same way." Mr. W. W. Ellsworth, of Woodstock, Illinois, prefers grass with a little maize, or, if confined to a yard, a mixture of oats, corn, and shorts.

(2) What is the best food for the boar?

Mr. Howard: "Dry food once a day; any kind of corn, oats, barley, beans, or peas, and moist food daily, such as barley meal, but not many roots." Mr. Spencer: "Roots and a little corn." Colonel Platt: "Any kinds of slops when he is in use." Mr. Benjafield: "I serve working boars once a day only, and give barley meal in their wash, with a few beans in winter." Colonel Walker-Jones uses Indian meal, bran, potatoes, and scalded swill in which a little raw wheat meal is mixed. Mr. Ellsworth uses oats, maize, and shorts, with grains once a day in warm weather.

(3) What is the best food for the sow and litter?

Mr. Howard: "Bran, randan, barley, pea, or wheat meal, mixed and steamed." Mr. Spencer: "Randan or sharps." Colonel Platt uses the same meal as for in-pig sows, with the addition of a little sharps and some buttermilk. Mr. Benjafield gives bran, grains, barley meal, and Thorley's food, mixed. Colonel Walker-Jones uses bran or grains and barley meal, with scalded potatoes and a little wheat meal mixed. Mr. Robertson prefers sharps, grains, skim milk, wheat, and a little barley or wheat and flour. Mr. Coate says: "Bran or pollard mixed with water, unless milk can be had, is the best diet for the first week or two. Barley I consider too heating, and consequently not good for milk-producing. I never use anything but bran or pollard and water. If the weather is fine, the sows run into the yard for a few hours daily, after the first week." Mr. Ellsworth says: "The sow and litter should have dry corn in the early morning, ground maize and oats, one half shorts, with about an eighth addition of oil meal mixed with milk, if it can be obtained. I would also give the young pigs, separate from the sows, some soaked maize in the afternoon, when they are from four to six weeks old."

(4) What is the best food upon which to wean the litter?

Mr. Howard: "New milk is the best, when it can be afforded, together with dann, barley, or meal, increasing the quantity of solid food as the young pigs get accustomed to the change of feeding. The food should be steamed, and the milk added afterwards." Mr. Spencer: "Randan and skim milk." Colonel Platt: "The same as for No. 3." Mr. Benjafield uses barley, maize, wheat, and peas, ground together, with maize or peas given whole once a day. Colonel Walker-Jones: "Scalded Indian meal, bran, a few boiled potatoes, and a little raw wheat meal." Mr. Robertson: "Skim milk, a mixture of barley or wheat flour, and a few peas." Mr. Coate: "Barley meal, mixed with milk or water, and seasoned occasionally with some condiment, such as Thorley's. It adds to the appetite, and keeps the pigs healthy." Mr. Ellsworth: "I prefer soaked maize in the morning, oats and maize finely ground, one half, and shorts one half, with about one-sixth part of oil meal; the whole to be mixed with milk, if it can be obtained."

(5) What is the best food for young pigs from twelve to twenty weeks old?

Mr. Howard: "Steamed bran, dann, barley, wheat, and maize meal." Mr. Spencer: "Randan, skim milk, and a little mixed meal." Colonel Platt: "Same as for No. 4." Mr. Benjafield prefers barley meal with a few whole maize or peas, and a good yard or field to run in. Colonel Walker-Jones: "The same as for No. 3, with a little swill." Mr. Robertson: "Barley meal, skimmed milk, or whey, which I also give to fatting pigs." Mr. Coate: "This depends upon circumstances. If I intend them for store pigs, they will require very little food, if turned into a meadow. A few beans, peas, or maize, or a little barley, with a few roots, would be sufficient." Mr. Ellsworth: "The same as for weaners." Mr. "Phil. Thrifton," the well-known American authority: "Maize and oatmeal in equal proportions, cooked together, and then thinned with skim milk, make an excellent diet for young pigs. Wheat bran and middlings, particularly the latter, may also be used with maize meal. Oats are the best and safest of all grains to be given, with or without cooking, at this age. When the pigs are older, other grains can be used. Peas ground with maize or oats, mixed with middlings, and all cooked together, may be used to great advantage; also cooked potatoes. The last-named should be well mashed and thoroughly mixed with cooked meal, and the entire ration afterwards thinned with skim milk. If cooked potatoes only partly broken up are given in bulk with the meal, the pigs are apt to gulp them down too fast, and so to overload their stomachs. They then become trough-sick, and throw up part of what they have eaten. Pigs sometimes do this with other foods, when given in bulk, particularly if they have been allowed to become very hungry. If their food is reduced to a liquid state, there is less danger in this way, and then, having drunk to their satisfaction, threshed oats or any other grain may be given them to crack and eat at their leisure. Oil meal in the proportion of one part to six parts of maize meal is recommended for young pigs. Barley and rye meal are also given. Variety in all diet is advisable, but no great or sudden change should be made in the feeding of pigs at this age."

(6) What is the best food for fatting pigs?

Mr. Howard: "Barley and wheat meal, with milk in all cases, if it can be afforded. Sugar or treacle may be added to steamed food with advantage." Mr. Spencer: "A few roots and meal, made from barley, wheat, and a few peas, beans, or maize." Colonel Platt: "Plenty of good oatmeal, scalded Indian meal, and sharps, mixed equally with buttermilk." Mr. Benjafield: "Any good mixture of ground corn, mixed with whey or milk; for example, five sacks of barley, one of peas, one of maize, and one of wheat, ground and mixed together, and the meal given slightly warm in cold weather." Colonel Walker-Jones: "Pea meal, Indian meal, and bran scalded. A little raw wheat meal and oatmeal to finish." Mr. Coate: "I know of nothing for quick and safe fatting like barley meal, mixed occasionally with a little linseed meal, with water or milk." Mr. Ellsworth: "Soaked maize in the morning, slop food at noon, mixed according to No. 4, and at night two-thirds maize and one-third oats, finely ground and mixed with either water or milk to form a thick paste."

(7) Do you recommend, as pig food, beans, peas, wheat, and barley? If so, please say in what order, and whether whole or ground?

Mr. Howard: "The foods named make the best of feeding when ground and mixed together, weight for weight, and afterwards steamed." Mr. Spencer says: "I prefer them ground and in the order given." Colonel Platt prefers all given as meal. Mr. Benjafield thinks they would pay better if cooked. Colonel Walker-Jones prefers peas, wheat, and barley, ground. Mr. Robertson: "Wheat, a small quantity; barley, unlimited, ground into meal; beans or peas whole, but in small quantities." Mr. Coate: "For store pigs running in a meadow, a few beans or peas, or barley or maize, twice a day. For fattening, nothing beats barley meal." Mr. Ellsworth: "We do not raise peas for hogs in America, as maize is much cheaper, and, when mixed with oats, we prefer it to peas. We never give beans to hogs; nor will they eat them. Barley makes a very good food, but hogs prefer maize. Wheat and oats, ground together, we prefer; but peas are undoubtedly good."

(8) What is your opinion upon cooked food?

Mr. Howard: "Cooked food is far better than uncooked. A less quantity is used when steamed, and it mixes readily with water, and divides further than uncooked food." Mr. Spencer does not like cooked food. As he sells about 350 pigs each year, he finds that they do not thrive after the change and journey when they have been fed upon cooked food. Colonel Platt uses plenty of cooked food. Mr. Benjafield thinks all pigs pay better upon cooked food. Colonel Walker-Jones thinks it much more digestible. Mr. Robertson says: "The indiscriminate use of cooked food is objectionable." Mr. Coate thinks cooked food is very good; but he has never tried it. Mr. Ellsworth: "We like to use cooked food, and to give it in the middle of the day. Whole grain should be given at least once a day where cooked food is used." Mr. "Phil. Thrifton": "I would cook enough during winter to give one warm meal every day. Usually, however, maize is so cheap, and labour so dear, that it does not pay to cook much."

S

(9) What is your opinion of mangels, turnips, and potatoes as pig foods?

Mr. Howard: "The potato is the best of the roots named, either steamed or raw." **Mr.** Spencer uses a great quantity of each. Colonel Platt says potatoes may be given steamed and mashed, but he does not recommend turnips. **Mr.** Benjafield: "I do not like mangels, excepting in the spring, and then not for in-pig sows. Potatoes are always good." Colonel Walker-Jones: "I like potatoes as a feeding stuff, and a few mangels for a change; but I find that pigs do not thrive well upon turnips, and I consider that they are too acid for young pigs." **Mr.** Robertson: "I am favourably disposed to the use, sparingly, of cooked mangels, potatoes, and swedes salted down and used with ground barley." **Mr.** Coate says: "A few of either of these roots every day can be recommended for store pigs." **Mr.** Ellsworth: "We like mangels as a substitute for grass. Turnips we do not like. Potatoes boiled with maize and oatmeal make a good fattening food." **Mr.** "Phil Thrifton": "They may be profitably used, by way of variety, in the diet of pigs."

(10) What do you consider to be the best system of fatting for exhibition?

Mr. Howard: "To force pigs from birth, which can be done for show animals while with their dams. Three months before being exhibited ball feeding (*i.e.*, the various foods chosen made up in the form of balls) should be commenced, and pushed on three times a day—the oftener the better. Barley flour with new milk, sugar, or treacle may be used with advantage." **Mr.** Spencer: "The best of food, and plenty of exercise, often given." Colonel Platt: "Good oatmeal and whey, mixed with buttermilk or sweet milk." **Mr.** Benjafield: "Good mixed meal, with condimental food, regular feeding, and a warm sty. Never leave any food in the trough. Give the pigs a good grooming every day." Colonel Walker-Jones: "I use Indian meal, pea meal, bran, oatmeal, swill, and milk, especially buttermilk, for all my young ones." **Mr.** Robertson: "This is a delicate question, and much depends upon private judgment; the pigs ought to have exercise to keep them on their legs, to be fed regularly, and with care as to the quantity of the food. Keep them cool by sufficient ventilation, and **give** suitable medicine when required." Mr. Coate: "A great deal of judgment is here required. Pigs intended for exhibition in the classes under six months, and under nine months, must be kept well from the time they are weaned; but with the older pigs of twelve and up to eighteen months it is different. These latter may be carefully guided along, in order to have them just fat enough upon the day of the show, instead of a month beforehand, as is frequently the case, when they get ' off their legs,' and are very bad in their breathing. In this way their lungs are soon attacked, and they often die very suddenly. The pigs should not be entirely confined for more than three months; they should be well kept, but be let out for a run every day for a few hours, and then be shut up and given as much food as they will eat. Barley meal and linseed meal, mixed with milk or water, is the best food, without exception." Mr. Ellsworth: "Keep the pigs upon grass, if possible. Use soaked maize in the morning; No. 4 food in the afternoon; and oats and maize, ground, at night."

The best system of feeding the boar and the breeding sows is clearly detailed in the replies given above. We do not endorse the opinions of those who are too fond of mangels and turnips for sows; but we cordially agree with all that has been said as to the value of grass, having found in practice that, where sows have access to a good meadow, they thrive upon a minimum quantity of food, and give the least possible trouble to their owners.

With regard to young pigs, we prefer to feed them away from the sow, three times a day, at from five to six weeks old, upon a mixture of grain, meal, bran, and skim milk. A little wheat, and a few handfuls of grass or pulped potatoes in addition, will be found of the greatest value. They learn to eat very rapidly, and can be weaned at eight to nine weeks; but, if it is desired to give them a first-rate start, they may be allowed to remain with the sow every night until they are twelve weeks old, after which, if they are fed upon a mixture of boiled meal, bran, and roots, they will put on flesh in the most rapid manner possible. Having had eight or ten weeks upon this diet, they may be finished upon barley meal, potatoes, and skim milk, which is the best known combination—a fact that has been proved by the well-conducted Wiltshire experiments.

Potatoes are a first-rate food for young pigs, but they are improved by pulping and afterwards cooking.

Some pig-feeders are over-fond of giving the young pigs their meal mixed too thinly, in fact, in a sloppy form, than which nothing can be more objectionable.

Boiling is undoubtedly advantageous, for it not only assists digestion, but it makes the food go further, although feeding upon raw meal, without doubt, saves time in fattening. When meal is used to any extent—and it may be advantageously composed of more than one kind—it should be mixed twenty-four hours before it is consumed, and, when ready, it should be scalded and salted. The object of mixing so long beforehand is to encourage fermentation, when what is believed to be lactic acid is formed, which greatly assists assimilation. When young pigs are intended for stores they may be preferably fed upon tares, clover, or lucerne during the summer months, a few whole beans or peas being given morning and night.

Unquestionably store pigs are grown to the greatest perfection when they have plenty of liberty, and they should be permitted to graze during the whole day in summer, and be allowed to have the run of the yard in winter.

Porkers and Baconers.

In fattening porkers for the market, much less exercise is necessary, but it is advisable to allow the animals a certain amount of liberty, in order to maintain a healthy condition. At all times, however, they should be provided with salt in their food, and with coals in the sty, in case they are unable to find sufficient in the short time they are at liberty. Porkers, when well fed, will put on from ½lb. to 2lb. of live weight per day, according to the breed, and they have been known to put on as much as 2½lb. Their feeding for the large markets is, in many instances, a most profitable occupation, but it requires judgment and good management, in order that the pigs may come in at the season when pork is in demand, thus realising a reasonably good price. It is especially necessary, for this class of business, to maintain a good quality of meat, and, consequently, a good breed; but the animals should not be fattened too heavily, nor grown to a large size; indeed, the younger they are, the better in every way for the producer. Dairy-fed porkers are chiefly in demand in the markets of London, and there is no doubt that the preference of the public is a correct one, for pork that has been produced from a milk diet is invariably most delicious, and, judging by the slight demand for skim-milk in many districts, it must be profitable to the grower.

There are a variety of systems for the conversion of pigs into baconers, but differences of opinion exist as to whether the animals should or should not have a certain amount of liberty every day. If they have a roomy court attached to their sty, we scarcely think that absolute liberty is necessary; but in this case they should be furnished with a little green or succulent food, with coal and salt, both of which are requisite to maintain health.

It is a common custom in some parts of the country to commence to fat a pig intended for bacon upon carrots. After a week or two have elapsed, meal food (barley meal is generally used) is commenced, the quantity given being gradually increased until the end of the period of fattening approaches, when the volume of food is gradually diminished, although the quality is maintained. A bacon pig is sometimes fatted to such an extent that it is barely able to rise to feed; but it should be the invariable custom of the feeder to insist upon the pig taking a little exercise every time it is fed. In some parts of France it is common for feeders to finish off a bacon pig with maize, and in others chestnuts are used for the same purpose, this food being very generally grown in the South of France.

A pig a year old, which is intended for conversion into bacon at Christmas, may be allowed to graze upon a good pasture at the commencement of spring, or, what is still better, it may be fed upon clover or lucerne, both of which are admirable preparatory foods. The meal feeding may be confined to an ordinary allowance, morning and evening, and it will be found that this system not only has an excellent cooling and medicinal effect, but also invigorates the pig, and is in every way most economical, paving the way for the production of a much larger quantity of flesh than if the animal were shut up in a sty or a small yard, and fed upon a large quantity of more condensed food. If this method be continued until the autumn, the pig will have been gaining flesh during the whole period, and be

ripe for finishing off during the next two or three months, when
the green food should be gradually diminished, and meal and
milk, or whatever food may be given, as gradually increased,
until the maximum point is reached. We believe that the
assertion that nothing is gained by fatting a hog over ninety
days is quite true, and that, if it is in a fit condition at the
commencement of that period, any food given at its com-
pletion is almost, if not entirely, thrown away.

It is the custom in America, when turning out pigs to grass
or clover lots, to cease grain feeding entirely, inasmuch as it is
argued that, if a condensed food be given, it is more costly,
whereas the animals themselves eat less grass, the very food
of which it is necessary for them to have their fill, in order
to develop their frames, to produce lean flesh, and to bring
them into condition for subsequent fattening. Mr. Moore says
that, if a limited quantity of maize is given, it only makes
"fools" of the pigs, for they expect it at the usual time, and
stand waiting at their feeding-place until it is supplied.
The consequence is that, as it is less than they require, they
continue to worry themselves so long as the attendant is near.
He argues that no maize should be given, when the pigs will
evince no desire for it, and remain perfectly contented on what
is their most profitable diet, viz., grass.

Fattening on Purchased Foods.

The question is often asked, Does it pay to fatten pigs upon
purchased food? This depends in a great measure upon the
demand. If a price of, say, 4s. 3d. to 4s. 6d. a stone can be
obtained, we are of opinion that it pays very well, but not
unless the pigs are of an improved breed, and managed upon
an absolutely perfect system. To feed carelessly is only to
throw money away; and when pigs are put up to fatten neither
trouble nor food should be spared until they are finished. We
cannot do better than counsel those who make a practice of
fattening pigs for sale to use the weighing machine with
regularity, and to see for themselves exactly what quantity of
pork or bacon they obtain for a given quantity of food. In
feeding upon purchased food, much depends, too, upon the
judgment of the buyer. Maize, for example, varies considerably

in price according to the season of the year. If bought at a high price, it may possibly mean a loss upon the transaction; and the same may be said of barley. Potatoes, again, offer many advantages to the pig-feeder, for their price has of late been so extremely low that nothing but the correspondingly low price of pork should prevent a purchaser from using them with a profitable result. It is, however, possible that a person who is a liberal and judicious feeder may, by some other deficiencies, let his profits slip through his fingers. The animals may perhaps be badly housed, and as warmth is equivalent to a certain amount of food, the sties should be, especially for the fatting pig, as comfortable as possible. Moreover, the feeder should not attempt to obtain a profitable result from the common, long-legged, long-eared, and long-snouted animals, which unfortunately exist in too large numbers.

The following rations for the feeding of pigs are based upon those drawn up by Heuzé; but, making allowance for certain differences and conditions in French and English feeding. we have made some few alterations:

RATIONS FOR YOUNG PIGS.

	lb.			lb.
Cooked potatoes ..	$2\frac{1}{2}$		Cooked potatoes ..	4
Skim-milk } ..	$10\frac{1}{2}$		Barley-meal	$\frac{1}{2}$
Wash .. }			Wash	7
	13			$11\frac{1}{2}$

These rations may be improved by substituting skim-milk or buttermilk for wash, or by mixing a small quantity of oilcake meal with the barley-meal; and, naturally, it may be given in larger or smaller quantities, in accordance with the race or destination of the pigs.

WINTER RATIONS FOR BREEDING SOWS.

	lb.			lb.
Potatoes	6		Potatoes	3
Carrots	1		Carrots	5
Barley-meal	1		Swedes	7
Wash	18		Barley-meal	3
	26		Wash	15
				33

WINTER RATIONS FOR BREEDING SOWS—*continued.*

		lb.			lb.
Potatoes	..	9	Potatoes	..	7
Barley-meal	..	1	Grains ..	..	7
Cake	..	¼	Bran	..	1
Bran	..	¼	Barley-meal	..	½
Wash	..	11	Wash	..	7
		21½			22½

WINTER RATIONS FOR BOARS, AND FOR SOWS WITH YOUNG.

		lb.			lb.
Potatoes	..	9	Potatoes	..	5
Barley-meal	..	1	Cooked maize ..	..	2
Cake	..	½	Pollard	..	2
Grains ..	..	7	Carrots ..	..	5
Bran	..	1	Wash	..	14
Wash	..	15			28
		33½			

SUMMER RATIONS FOR BREEDING SOWS.

		lb.			lb.
Barley-meal	..	2	Potatoes	..	3
Bran	..	1	Barley-meal	..	3
Green clover	..	12	Cooked nettles	..	9
Wash	..	18	Wash	..	14
		33			29

		lb.			lb.
Potatoes	..	5	Potatoes	..	3
Barley-meal	..	2	Pollard	..	2
Cooked nettles	..	9	Skim-milk	..	5
Skim-milk	..	7	Cake	..	¼
		23	Green clover	..	14
			Wash	..	7
					31½

SUMMER RATIONS FOR BOARS, OR FOR SOWS WITH YOUNG.

			lb.					lb.
Pollard	..	..	3	Bran	..	..	..	1
Vetches, or lucerne	..	9		Pollard		..	..	2
Skim-milk	..	..	7	Cake	..	..	..	1
Wash	..	..	9	Wash	..	..	..	16
			28					20

			lb.					lb.
Pollard	..	..	2	Potatoes	..	..		9
Potatoes	..	..	7	Barley-meal	..	..		2
Cake	..	..	1	Acorns	..	..		2
Green clover	..	..	9	Mangels	..	..		4
Whey	..	..	4	Skim-milk	..	..		5
Wash	..	..	9	Wash	..	..		9
			32					31

Heuzé quotes Parent, who states that he found by experience that the following quantities of different foods were respectively required in the production of 100lb. of live weight:

			lb.				lb.
Cooked rye	..	..	416	Bran	..	..	820
Barley-meal	..	..	480	Cooked potatoes		..	2000
Cooked buckwheat	..	568	Cooked carrots	..	..	2340

The Various Foods.

MAIZE is perhaps, of all foods, the one most largely used in the feeding of pigs, for, among the millions annually raised in America, the great majority are fattened upon this food. Maize contains 60 per cent. of carbo-hydrates, $8\frac{1}{2}$ per cent. of albuminoids, and $4\frac{1}{2}$ per cent. of fat. Comparing its weight with that of other commonly used foods, it is the cheapest in the market, and, when it can be obtained as low as 25s. per quarter, as is often the case, we are surprised that it is not still more largely used, its price being considerably less than $\frac{5}{8}$d. per lb. It has been shown by experiments that one bushel of maize adds 15lb. of live weight to a well-bred pig, whereas a bushel of barley, which is a much lighter food, adds 10lb. The late Dr. Voelcker said that maize contained about 6 per cent. (it is really about 4½ per cent.) of ready-made fat; but it has been proved, as in the case of other fattening foods, that, in order to utilise the whole of this fat, a nitrogenous food, such as milk or peas, should be added to or mixed with maize.

In America, maize, more especially in the ear, is the principal food used in the fatting of pigs, although it appears to be generally given without a complete knowledge of its value; farmers, on account of its cheapness, using it without any discrimination, and seldom taking the trouble to add any nitrogenous diet to it. The best American authorities, however, are of opinion that it is not an economical food when used exclusively, and that it tends to produce disease and to depreciate the inherent properties of good herds. Mr. Coburn says that Western farmers who use whole maize generally estimate that ten bushels will produce 100lb. of pork; but he is of opinion that, when fed in a different form, and in conjunction with other food, it will make much more, this fact having indeed been frequently and fully demonstrated by careful feeders, both in Europe and in America. Mr. Coburn, who has pursued the question of maize feeding very extensively, adds that he does not wish to argue that maize is not a suitable food for swine, or that it is not the best single fat-producing material for the money in the world for general use; but he desires to enforce the fact that variety is essential to perfect health and development in all animals, and that a single article of food is satisfactory to none, not even to the pig.

With regard to the capacity of maize to add weight, there appear to be many well-established instances in which as small a quantity as 4lb. has produced 1lb. in weight. Professor Miles, of the Michigan Agricultural College, obtained this result. In the year 1868 he fed six pigs, of the age of ten weeks, for five periods of four weeks each. The amount of maize meal required to make 1lb. of live weight was, in the first four weeks, 3.81lb.; in the second, 4.05lb.; in the third, 4.22lb.; in the fourth, 5.24lb.; and in the fifth, 5.92lb.; the average for the whole of the periods being only 4.66lb. In feeding maize for

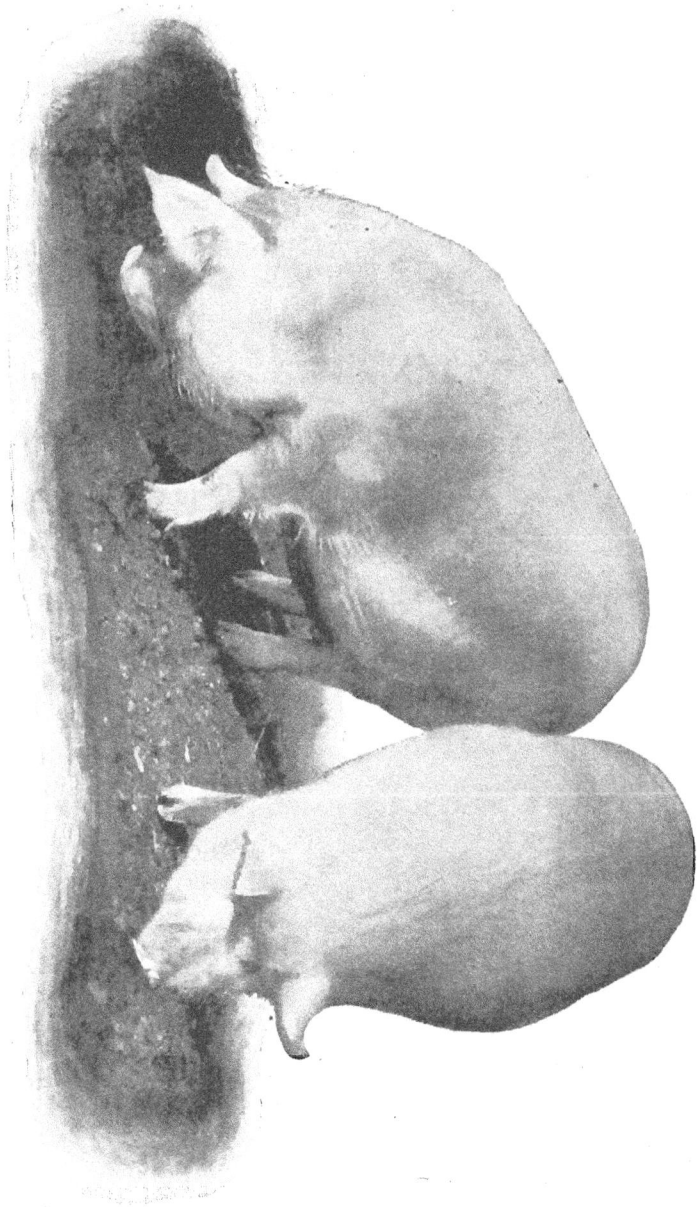

LARGE WHITE SOWS.

Bred by Mr. Alfred Brown. Sire, "Wimpson I."(Herd-Book No. 7289); dam, "Wimpson Princess" (45, 108). Winners of First Prizes, Norwich and Birmingham, 1903.

Well-bred Large Whites are good breeders and mothers, and the young attain considerable weights at an early age. They have thick bellies and provide a large proportion of lean meat; their hams are broad, full, and

weight, however, almost everything depends upon the quality as well as the age of the pigs. A mixture very largely used in America, especially where considerable numbers of pigs are kept together, is composed of maize meal, ground oats, and bran. This, given in a thin state, is highly esteemed. Maize meal is also much used with whey, particularly when roots are given in addition.

With regard to the produce of the maize plant used in a cooked form, there has been much discussion. Mr. Johnson, of Marysville, a practical feeder of authority, says: "I do not believe that grinding and cooking will pay under all circumstances. If cooked food is to be used in a paying way, it should be fed with great punctuality, three times daily, at least; and the feed, during cold weather, should have a temperature of between 95 and 110 degrees. This will cause the pigs to fill up without getting chilled, and they will go to their sleeping quarters feeling comfortable, and remain at rest until their appetites call them forth for their next meal. Warm food in winter I regard as essential for fattening pigs, for they are slow feeders, and remain a long time at the feeding trough, if not driven away by being chilled. Maize and maize cob meal are ground, thoroughly cooked, and seasoned with salt, and, when ready, wheat shorts are added. The shorts and cobs act very beneficially, the cobs keeping the food in the stomach in a porous condition, and giving the gastric juices full chance to hasten digestion. The shorts supply the wants of the bone and muscle to a great extent, and this, I think all pig-breeders will agree, is very necessary, as all our fattening pigs are young, and should be kept growing rapidly at the same time that they are fattening. I have fed dry maize to some of my pigs, on a platform, twenty feet from the sleeping quarters (in order to keep it clean), and once a day I gave them all the bran or shorts they wanted. This diet suited them better than one consisting entirely of maize; but I found that, on many cold days, they would not gain at all, and I am satisfied that they deteriorated instead. It is, however, the reverse since I began cooking, and a steady gain during cold and warm weather is achieved."

Messrs. Moore and Son, the Illinois breeders previously referred to, use maize in large quantities for their pigs. For

the young pigs, a large boarded floor is provided, which is
sheltered from the sun and rain, and on which the feeding
troughs are placed. These are filled with maize, which has been
previously soaked for twenty-four hours in barrels sunk in the
ground. The water used in the soaking process is afterwards
mixed with ground oats and maize, to which bran or pollard is
added, and this mixture is given to the larger pigs.

A very able essay was, some time ago, prepared by Mr.
Joseph Sullivant, an officer of the Ohio State Board of Agriculture, in which, having dealt very comprehensively with the
feeding of pigs, he showed that, in feeding value, 100lb. of
maize equals:

	lb.			lb.
Barley	103	Carrots		721
Beans	103	Buttermilk		508
Rye	117	Fresh milk		863
Oats	118	Red clover		665
Buckwheat	122	White clover		665
Cotton cake	117	Timothy grass		298
Linseed cake	119	Lucerne		593
Peas	106	Cabbage		1018
Potatoes	360	Skim-milk		721
Mangels	665	Turnips		1235
Parsnips	618			

It is an undoubted fact that, in feeding whole maize to pigs,
a large quantity passes through the system entirely unused,
and it is believed by American feeders that ground maize
produces a better result, to the extent of a third, than the grain
given whole. Judging by the results of numerous experiments,
it would seem that one bushel of maize meal should furnish 12lb.
of pork, but that the same amount of maize given whole and
uncooked seldom yields so much as from 8lb. to 9lb.

Some years ago, Sir John Lawes made a valuable experiment,
in succession to Professor Miles, to test the value of maize
meal. He selected thirty-six pigs, which were divided into
twelve pens. They were from nine to ten months old, and
averaged 143½lb. each in weight, being an exceedingly even
lot, and differing very little in size. The lot of three fed upon
maize meal, to which we need only refer, consumed 362lb. each,
or nearly 6½lb. daily, for eight weeks, gaining during this

period 79¾lb., or nearly 1¼lb. daily, which would be at the rate of 12lb. per bushel of maize. Strange to say, this result was very similar to those obtained by Professor Miles and others, who had made analogous experiments. Sir John Lawes found —what should be clear to all feeders of pigs—that there was a very rapid decrease in the consumption of food as the animals fattened, and that, while they consumed less food, a larger quantity of it was required to produce each pound of increase. Thus it appears that, as a pig approaches perfection of fatness, it is quite possible to produce each additional pound of pork at a cost which exceeds its value. Upon the same point, Professor Miles remarked that the best return is obtained by liberal feeding during the early stages of growth. To this Sir John Lawes added, that the larger the proportion of nitrogenous compounds in the food, the greater the tendency to increase in frame and flesh; but that the maturing or ripening of the animal—in fact, its fattening—depended very much more upon the amount in the food of certain digestible, non-nitrogenous constituents.

WHEAT has not in the past been considered one of the most suitable foods for pigs, chiefly, perhaps, on account of its high price; but it has been tried by many practical feeders of our acquaintance, principally in the form of meal, although in every case, without exception, it was found unsuitable, and was quickly stopped. We were ourselves induced to make a trial with a ton of wheat meal, which cost a shade less than ¾d. per pound, but our experience was not such as to warrant its continuance, although, when mixed with other foods, we have every reason to believe that it is a valuable article of pig diet. We have invariably recommended the use of whole wheat for young pigs at weaning, and, with the exception of peas, there is no better food which can be given to them in small quantities between meals. They pick up the grain with great eagerness, it keeps them well employed, and they feed well upon it.

An experiment was made in America upon the feeding of cooked wheat which is worth quoting, the result being sufficient to warrant a repetition by some of the more enthusiastic of our own countrymen : "On the 4th of August, fifteen hogs were put up, weighing 2400lb., and fed upon 5¼ bushels of cooked wheat the first week. On the 11th their weight was 2600lb.;

thus showing a gain of 13⅓lb. to the hog, being nearly 2lb. a day. The next week they received 6 bushels of the cooked wheat, producing an increase of 215lb., or 14⅓lb. to the hog, being a gain of over 2lb. per head per day. The third week they consumed 10 bushels of cooked wheat, resulting in a gain of 260lb., or 17⅓lb. a head, or 2½lb. a day. The fourth week 11¼ bushels of cooked wheat were given, the gain being 320lb., or 21⅓lb. a head, or a fraction over 3lb. a day each. The hogs were then sold and taken away. They gained, in four weeks, 995lb. on 32½ bushels of wheat. By this manner of feeding the owner received a good price for the wheat, as the hogs were sold at £1 14s. 4½d. per 100lb."

BARLEY.—As a food for pigs there is perhaps nothing so largely used in England as barley, which is generally given in the form of meal. It is, however, too little understood by purchasers of pig food that the barley-meal common in the market is a most inferior article, containing a lot of refuse material, although it is usually sold at almost as high a price as they would pay for good barley in their own neighbourhoods. Boiled barley has been frequently recommended, and, where cooked foods are given, it is an admirable change, especially when given in conjunction with milk; but, upon the whole, barley-meal is to be preferred, and with some reason, as the results universally obtained from feeding upon a good sample are such as to justify its continued and still more extensive use.

An experiment upon barley feeding was made in France which lasted an entire year. A crossbred pig was used for feeding. At the commencement it weighed, after castration, 19lb., and at the finish of the trial, 309lb. The animal consumed 1854lb. of barley, and the experiment showed that 1 kilogramme (2⅕lb.) of live weight was produced for every 13⅓lb. of barley; that 1 kilogramme of carcase weight, net, was produced for every 13lb. 9½oz.; that 1 kilogramme, without bone, was produced for every 14lb. 2oz.; and that, for every 225lb. of live weight, 1343lb. of barley was consumed.

OATS are not so common a pig food as barley, but they are exceedingly valuable in two forms—given whole to young pigs, or in their ground state mixed with water. Strange to say, there is a general deficiency of knowledge with regard to ground oats and oatmeal. The oatmeal commonly consumed in

Scotland and England is a much too expensive food for stock; indeed, it is less valuable than the ground oats principally sold in Kent and Sussex, and prepared by millers who use special stones for the necessarily very fine grinding. In ground oats the whole of the oat is used, the husk being ground almost as fine as the interior of the grain. This is an exceptionally fine food, and we are not sure but that, could it be obtained in all parts of the country, it might be considered the best of all meals, not only for pigs, but for every kind of stock.

BUCKWHEAT is another grain of high value for pig-feeding, but one not sufficiently known and valued in this country, where its price is monstrously high. It is largely cultivated in the North of France, and can be obtained from the port of St. Malo at a most reasonable price in large quantities. Indeed, some years ago we received quotations which were considerably less than the current prices in the London markets; and, as is well known, the freight to this country is exceedingly low. Buckwheat is largely used in France, and, as it is of a very oily nature, is highly useful for fattening pigs. It is, however, stated to be liable to cause congestion when given in undue quantities.

BEANS are not very largely used for pig-feeding, but they are exceedingly valuable, both in the ground and in the whole state. The chief value of bean-meal is that it forms a useful constituent in a mixed diet, providing the requisite amount of nitrogen. Whole beans have lately been in general use among pig-feeders of a high class, who give them to adult pigs in small quantities during the summer months, and to young pigs, between meals, after weaning. They are not suitable for giving in large quantities; but a handful at a time is found to maintain the condition of the animal, and, in fact, to be a sufficient dry diet when it is grazing throughout the day. Beans are sometimes given in a boiled state, and there are few foods which are more relished by young pigs than Egyptian beans, boiled.

PEAS are equally as good as beans, and are much more largely used, as there is a great preference among certain classes, in almost all parts of the country, for finishing porkers and bacon pigs upon pea-meal. Whole peas are also largely given to young pigs, and, when care is taken that the peas are old, the animals generally thrive remarkably well upon them. When given in conjunction with a more fattening food, they are exceedingly good, and could scarcely be replaced. Pigs generally thrive well when turned upon a field which has been occupied by beans or peas. In America, however, the latter are comparatively unknown to the large pork-producers, and it is questionable whether they will ever be appreciated as they deserve to be, where maize can be grown in such enormous quantities, and with comparatively such little trouble. At the same time, if our experience in this country is of any value, they certainly produce more lean than maize, and, consequently, a much firmer and more highly appreciated flesh. It must, nevertheless, be remembered that, weight for weight, they are considerably dearer than maize, and this is one reason why they are not in such general use, especially among small pig-keepers, who are compelled to look at every shilling laid out upon their pig.

Acorns and Beech Nuts, which more nearly resemble grain than the green foods given to pigs, are certainly of value, but it is questionable whether either food has the property of producing meat in anything like the degree that corn feeding has. Meat produced from acorns is unusually lean, and this fact suggests that the value of the acorn is greater when mixed with a fat-producing food. Both foods, however, have an astringent property, which is very undesirable. M. Magne has suggested that, if the acorn is allowed to germinate, it becomes a better food, having lost the tannin which it naturally contains. For store pigs fed upon a green diet, acorns and beech nuts are useful, and may be given morning and night, a few handfuls at a time.

Grass.—The value of grass, especially if clover is present, as a food for pigs can hardly be over-estimated. It contains all the elements necessary for the development of the growing animal, and for the production of bone and muscle, while at the same time it materially assists in counteracting disease, whether from unhealthy surroundings or from careless grain-feeding. Its cost, moreover, is so small that it cannot be too highly recommended.

We have advised the system of turning young and store pigs upon grass, and we cannot do better than refer to a system of feeding which is only second to that of grazing. What is known as "soiling," or the carrying of green crops to the animals of the farm in their stables, has long been practised in connection with cattle, but it is most uncommon in Great Britain as regards pigs. Where, however, a pig-owner is unable to turn his swine upon a pasture, he cannot do better, if it is within his power, than to cut them either clover, lucerne, vetches, or Timothy grass, and to bring it to them in their yards. When given at stated periods, green and fresh, it is almost as good as feeding upon the pasture, and we question very much whether the system causes more waste. The analytical value of these grasses is shown in the table given on pp. 292 and 293, but we should prefer lucerne, which, when cut young, and before the stems have had time to become hard and fibrous, is one of the very best foods which can be given to stock. Tares or vetches we do not like so well, as they cause a loose state of the bowels, which requires special feeding to check. Timothy grass, or even rye grass, may be used with advantage, the former being one of the best grasses known, as well as an abundant cropper, and we are only surprised that it is not more generally cultivated in this country. Instead of being added in small quantities to mixtures for laying down pastures, Timothy should be sown with two or three other strong-growing grasses, and laid down into permanent or semi-permanent pastures.

It has been estimated that 15lb. of clover will make 1lb. of pork, and, upon this basis, Mr. Coburn has constructed the following table:

	Gross Product per Acre.	lb.	Pork per Acre. lb.
Wheat	15 bushels	900	225
Barley	35 ,,	1680	420
Oats	40 ,,	1320	320
Maize	40 ,,	2240	560
Peas	25 ,,	1500	375
Green Clover	about 5½ tons.	12,000	800

If these figures are correct, an acre of clover is worth as much as three and a half acres of wheat, or two and a half acres

T

of oats, **for** manufacturing into pork. It should, however, not
be forgotten that pigs that have made their first growth upon
such concentrated food as grain have stomachs too small for
them to be successfully fed upon such bulky food as the artificial
grasses, and they cannot, therefore, eat sufficient of them, or
properly use what they do eat, to produce a maximum quantity
of pork. If the pigs are allowed to graze in summer, or are
provided with green food in an open yard, and roots in winter,
they will seldom take any harm.

In estimating the value of clover or grass for pigs, we may
quote from the essay contributed by Mr. Sullivant to the Ohio
Agricultural Report, in which he goes very closely into the
matter. It is, of course, impossible to estimate accurately
what quantity of pigs a given number of acres will support, as
much depends upon their quality and upon the crop the land
carries.

Mr. Sullivant says: "We will assume, to begin with, that
one acre, with a good set of Timothy and clover occupying the
ground in equal proportions, will give a product of 12,000lb.
during the season. We think this a moderate estimate, for the
reason that it requires less than 1oz. of green food per month
from each square foot during five months of pasturage. Suppose the average weight of the hogs, when turned on to grass,
to be 125lb., and that it be the fact, as has been frequently
stated, that an animal requires 3 per cent. daily of his live
weight in dry food, or its equivalent in green food, to keep him
in a growing and fattening condition, then 7½lb. of grass and
clover will be consumed by one hog daily from May to October,
or 153 days—or 1147½lb. during the whole period. Thus it is
evident that the acre of grass and clover will support as many
hogs as 1147½ is contained in 12,000lb. (the product of one
acre), or 10½ hogs nearly!

"But we prefer to base our calculations upon the supposition
that it requires 1⅓lb. of corn to maintain a hog of 150lb. in
condition alone, and, of course, it requires a corresponding
portion of green food to do the same thing. Suppose, therefore, that it takes 6.75lb. of clover to equal 1lb. of maize, then
1.33lb. of maize (the amount necessary to keep the hog in condition) requires 9lb. of green clover, or an equivalent, to supply
the daily waste in the animal organism, and, of course, an

additional amount is necessary to increase the hog in weight; and, if we take the increase at ½lb. daily, then 6.75lb. more of clover is needed, or 15.75lb. altogether. As the feeding value of Timothy grass, however, is, in relation to clover, as 298lb. is to 675lb., a less amount, or 11lb., will suffice than if feeding clover alone; but, as some is wasted and trampled down, we think that a daily allowance of 15lb. to each hog is none too much.

"Fifteen pounds of green food—which we have determined as the ration to sustain the hog and to fatten him ½lb. daily—are contained in 12,000lb. (the product of one acre) 800 times, and would support one hog for 800 days, or 5⅕ hogs for 153 days, or five months, from May to October—the period of pasturage. Omitting the fraction, our five hogs increasing ½lb. daily for 153 days, we have a total return in pork of 382½lb. from one acre of Timothy and clover."

CABBAGES are valued by some persons as a food for pigs; but they contain a very large percentage of water, and, although they are a healthy article of diet, of which a pig can scarcely consume too much, a little care is needed at first to observe that they do not cause any ill effect.

POTATOES.—One of the most valuable of all foods for porcine stock is the potato. It has been shown by Boussingault that the pig requires 9lb. daily for every 100lb. of his live weight, 8lb. of which is necessary to maintain his existence. A breeding sow, however, unless when she were suckling her young, would only require 7lb. The first-named ration may be described as one providing for the growth of the animal, and the second for the mere maintenance of the system. We have referred to the systems of pulping and boiling potatoes. Some years ago a number of experiments were conducted in Denmark, with a view to ascertaining whether a better return could be obtained by the use of boiled or by raw potatoes. Ten young pigs from the same litter were divided into two lots, at the age of ten weeks, one lot being fed upon boiled, and the other upon raw potatoes. Each lot received, in addition, 2½lb. of barley meal, which in one instance was given in a boiled state, while in the other the barley was only bruised. In four weeks the increase in the weight of the pigs that had been fed upon boiled food was found to be 173lb.; whereas in the other case it was only 115lb. It was also shown that pigs are greatly improved by grooming; and, in order to test between pigs fed for pork, which had been well groomed in one case, and not cleaned in another, six were selected, and divided off into threes, with the result that each of the regularly cleaned animals showed an increase of 30lb. more in weight than the dirty ones.

In an experiment made by Boussingault in the feeding of pigs upon potatoes, he found that a pig weighing 135lb., which consumed 16lb. of potatoes in twenty-four hours, returned in—

		lb.	oz.
Excrement (about)	.. ..	2	15
Urine (about)	.. ..	6	13
Total	.. ..	9	12

So that the actual amount of food absorbed was 6lb. 4oz. At the same time, it was found that the animal consumed in—

		lb.	oz.
Carbon	.. ..	1	7
Oxygen	.. ..	2	11
Oxygen drawn from atmospheric air	.. ..	13	4

Another experiment may be quoted, in which twelve pigs were fed upon potatoes, and divided into two lots, the first lot comprising six hogs, and the second six gilts. The former were fed upon cooked, and the latter upon raw, potatoes. At the end of a fortnight the feeding was changed to beans, cooked and raw. In another week, seeing that the gilts did not eat the soaked beans with avidity, the potatoes were given as before to each lot in addition, and upon this *régime* the experiment was continued. The result was as follows:

	Hogs.	Gilts.
Weight on July 1st	243lb.	180lb.
Weight on Oct. 12th	792lb.	429lb.
Increase	549lb.	249lb.

Turnips are an admirable diet when judiciously given. We have known many thoroughly practical feeders to use them largely in the winter with good effect. At the same time, numerous instances could be quoted in which they have caused disease, weak litters, and even abortion. Like all roots, turnips should be minced as small as possible for pig-feeding, and mixed with the meal at least twenty-four hours before they are given. When used in this way, they are believed to be as nourishing as when cooked.

Mangels.—Many feeders prefer the mangel to the turnip, and we remember one case in which a large quantity of mangels returned 26s. a ton by feeding to pigs. It is a mistake, however, to use them early in the winter, as they are of at least twice as much value when given in the spring. Pigs should not, of course, be fed wholly upon mangels, any more than upon any other roots; but, when an unusually large quantity is given, any ill effect may be checked by the addition of a few beans to the daily ration.

Carrots have long been used for pigs, although they are too rich for breeding animals. Numerous experiments have been made with them, and it has been shown that, in some instances, they have returned as much as 30s. a ton by being converted into pork. They are an admirable winter food, if given in medium quantities, but should always form part of a mixed diet.

Parsnips are also a succulent food very much relished by pigs, but the same remarks that have been applied to other roots apply equally to these.

Jerusalem Artichokes, which are largely grown in France and the United States, are not so much known as a food for swine as they deserve to be. We have seen them grown in very considerable quantities in Touraine, where the stalks are frequently preserved in the silo. They may be given either raw or cooked, but preferably cooked, and few foods are more relished by pigs. Mr. Coburn quotes numerous instances of successful feeding upon artichokes. He cites one case in which forty Poland-Chinas, with their young, were kept upon an acre of artichokes, without other food, for nearly half a year. In same parts of America it is the custom to turn pigs into an artichoke field, allowing them to dig up the tubers for themselves.

There is one decided advantage in connection with the artichoke: it seldom requires re-planting, for the same ground bears a crop for several successive years, a certain number of tubers invariably remaining in the soil—sufficient, in fact, for the following crop, after the produce of the year has been lifted.

PUMPKINS are considered to be of value where they can be grown, although their production in this country is very small. Before being given to pigs, they should be cut open, and the seeds removed, as these are injurious to the bladder. The pumpkin, however, is a mere adjunct, and is not of such feeding value as the turnip or the mangel; but, in conjunction with grain or meal feeding, it may be recommended, more especially to those persons who possess unusual facilities for growing them.

MILK.—We have already referred to the feeding value of milk, which cannot be too highly esteemed for pigs. At the same time, it ought to be clearly understood that milk in any form is too valuable for use as pig food, and that producers of it should use every effort to sell it in some form for human consumption. New milk, unless employed for the purpose of preparing pigs for exhibition, is out of the question, and should not be treated as a pig food; but skim-milk, where its sale is impossible, may always be used with the greatest advantage. It contains nearly 10 per cent. of solid matter of high value, and, when mixed with a fat-producing substance, can scarcely be excelled.

Whey, which is the residue of the milk used in the manufacture of cheese, is not saleable as food; yet the yield of one cow is generally estimated to be of the value of £1 per annum when employed as a pig food. Its principal ingredient is sugar; it should be kept in tanks, as far from the cheese-room and as near the piggery as possible. Whey is much better when it has been mixed with meal, and allowed to remain from two to three days to ferment. Barley meal and oatmeal are both valuable for this purpose, and, in winter time, minced or steamed roots, with bean meal or lentil flour, may be added with great advantage.

The following is the average of eighteen analyses of whey made by Dr. Voelcker:

	Natural State.	Calculated Dry.
Water	93.02	—
Butter (pure fat)..	.33	4.80
Albuminous compounds	.97	14.00
Milk, sugar, and lactic acid ..	4.98	70.18
Mineral matter (ash)	.70	11.02
	100.00	100.00

It will thus be seen that water forms much too large a percentage. This being the case, the only practical method of decreasing it is to add a sufficient quantity of dry food, which should preferably be of such a nature as will make up for the deficiency in the whey, it having been robbed of its fatty and nitrogenous properties. Foods containing nitrogen in some form should be added, and here bran or pollard, pea or bean meal, or ground oats, will be found extremely valuable. One of the best combinations of dry foods for mixing with whey is the following:

								lb.
Pollard ..	..	..	..	..	..	..	..	3
Bean meal	..	..	..	..	..	..	..	½
Maize meal	..	..	..	..	..	..	..	2
								5½

This should be added to every six gallons of the liquid. Upon a diet of this description the pigs will do remarkably well; but the food should be cooked, or steamed as we have seen it upon some farms in Cheshire, where a large engine is in daily use. The whey-vat, which, in these cases, is filled by gravitation from the milk-room, is fixed in the feeding apartment, and is so arranged that a powerful jet of steam can be turned into it at any moment.

BREWERS' GRAINS, DISTILLERY WASTE, and FLESH FOOD.—Of brewers' grains and distillery waste as foods for pigs we need

say little. The former is useful, when given in small quantities
and mixed with other foods, to sows suckling their young; but
for breeding sows it is not so good, and, if given in excess, is
liable to cause abortion. We can neither recommend refuse
materials of this kind nor flesh feeding, whether it be animal
matter from a slaughter-house or the flesh of the horse. If
either be given, however, the food should be invariably cooked,
for, in addition to the fact that pork made from flesh is always
of inferior quality, flesh feeding is at all times liable to set up
disease.   Some authorities have declared that cooked flesh,
when given with a large proportion of vegetable or milk food,
stimulates the growth of young pigs, although we cannot but
think that it is calculated to impair their disposition, and
materially to alter the most valuable characteristics of our
improved breeds.

HOTEL WASTE is, when it can be obtained, one of the most
valuable foods that can be given to pigs. It naturally contains
a certain proportion of animal food; but this is generally of
such a nature and quality that its introduction into the diet of
the pig is of little moment. We could quote instances of pig-
feeding upon a very large scale, which have been conducted in
connection with a considerable use of this kind of food, but
the success that attended the work was materially affected by
its cost. The profit of this system of pig-feeding, therefore,
depends entirely upon the price paid for the food. When used,
the waste should invariably be mixed with a large proportion
of bran or pollard, and be allowed to ferment for at least
twenty hours.

The Value of Cooked Food.

Mr. Coburn quotes a number of American experiments with
cooked food. In one instance, a feeder selected four pigs
weighing 245lb. each, and four others weighing 170lb. each.
One was taken from each lot, and put in a pen by itself. The
other six remained together, and were fed upon cooked maize,
gaining 15lb. to the bushel consumed. The two pigs kept by
themselves were fed upon raw maize, and only gained 10lb. to
the bushel.

Another gentleman, a breeder of the Chester White pig, states that he has used cooked food for his pigs for many years, and that, with a steamer he has fixed up, he can cook from one to two hogsheads at a time. He believes that there is great economy in cooking, and that one-fourth of the quantity of grain is saved. The mixture used is generally two parts of maize to one part of oats, ground together, and with this a large quantity of whole maize is given before it becomes hard and dry in the autumn. It is found preferable to allow the cooked food to become almost cold before it is given to the pigs. This gentleman experimented with two sows from the same litter, No. 1 weighing 292lb., and No. 2 280lb. The former was given 176lb. of unground cooked maize, consuming 2 bush. 5 galls., and gaining 36lb.; while No. 2, fed the same length of time upon raw maize, of which it consumed 3 bush. 3¼ galls., only gained 30lb.

Messrs. Sisson, an eminent Illinois pig-breeding firm, write as follows: "We have been cooking food for hogs for some years, and we state, as the result of our experience and observation, that, in the great hog and maize-producing States, cooking food for hogs, *generally*, will not pay; still, there are times and circumstances which will make cooking, to a limited extent, profitable.

"We do not think it profitable to cook maize or meal for hogs whenever they can have access to good, tender grass, and when the temperature is such that maize can be soaked in water. Soaking will then answer every purpose; but winter, when there is no grass, and dry maize is the principal food, is the time that cooking will pay, if ever.

"Hogs need something besides dry maize (it is too concentrated), something with more bulk; and, to meet this requirement, we do some cooking. If a mixture be made of maize and oatmeal, middlings and bran, finished up with potatoes, well cooked, and fed in connection with dry maize, we think the advantage will be very apparent.

"It is not absolutely necessary that this should be fed more than once a day, but *pigs* especially should have enough, once a day, to fill up and properly distend the stomach. In speaking of pigs, we mean those six months old or more."

One of the most practical sets of experiments conducted in

America to ascertain the actual value of cooking food was
made by Mr. S. H. Clay, a famous breeder of Berkshires, and
well known as the winner of the Grand Thousand Dollar Prize,
at the Chicago Exhibition, in 1871. Mr. Clay made these
experiments, as we are informed by Mr. Coburn, to settle in
his own mind the question as to what extent, and under what
circumstances, cooking food could be profitably followed. The
result was that Mr. Clay obtained from each bushel of maize,
when ground and cooked, as much pork as from three bushels
given without cooking or grinding. In this way, 40 bushels of
maize, which had been cooked and ground, yielded a better
result than 100 bushels of the dry food. Another advantage
was that the cooking process enabled the feeder to put 50lb.
upon one pig, while another, of exactly similar breed and size,
gained only 10lb. upon dry maize in the same space of time.

The first experiments were made with six hogs, which were
twelve months old, and weighed from 240lb. to 285lb. each.
They were, first of all, fed for twelve days upon cooked maize
meal mixed with water, so thin that they could drink it. At
the end of this time their weights were as follow:

No.			Original Weight. lb.	Weight at End of the Twelve Days. lb.	Increase. lb.
1	..	..	..	255	294	39
2	..	..	..	285	318	33
3	..	..	..	240	290	50
4	..	..	..	240	276	36
5	..	..	..	265	290	25
6	..	..	..	245	282	37

An alteration was now made. Nos 1 and 2 were fed upon
boiled maize, consuming 7 bushels, and gaining respectively 50lb.
and 52lb. in thirty days. Nos. 3 and 4 were fed upon boiled
meal, given thin, consuming 4¾ bushels in the same period,
and gaining respectively 30lb. and 50lb. Nos. 5 and 6 were fed
for the same length of time upon dry maize, consuming 7¼
bushels, and gaining respectively 10lb. and 32lb. In this way
Nos. 1 and 2 yielded 14½lb. of pork to each bushel of maize;

Nos. 3 and 4 yielded 16½lb. for nearly the same quantity of food; and Nos. 5 and 6 yielded 5¾lb. A second change was now made, and Nos. 5 and 6 were given cooked meal for twenty-six days, of which they consumed 234lb. They gained respectively 40lb. and 34lb. Nos. 3 and 4 were fed for the same period upon dry maize, consuming 6½ bushels, and gaining 34lb. and 10lb. respectively. Nos. 1 and 2 were kept upon their original diet of boiled maize, the result being as successful as before. Nos. 5 and 6, in this second trial, produced 17¾lb. of pork to the bushel, or three times as much as they had done in the first trial; while Nos. 3 and 4 yielded only 6¾lb., a tremendous falling off from their return when fed upon the boiled meal. During the first twelve days, No. 5 gained 25lb.; but, when fed upon dry maize for thirty days, the same animal increased only 10lb. in weight, although it had consumed the enormous quantity of 202lb. of maize. When, in the second change, it was fed upon boiled meal, it only consumed 117lb. in the twenty-six days, and actually gained 40lb. in weight. Again, in the twelve days, No. 4 gained 3lb. daily, and, during the following thirty days, it consumed 135lb. of meal, and gained 50lb. in weight. The same animal, however, after achieving this splendid result, when put upon its second trial with dry maize, consumed 182lb. in twenty-six days, and showed a gain of only 10lb. in weight.

Various kinds of apparatus are used in America for cooking food, more especially upon the large pork-growing farms in the West; but that adopted by Mr. Clay is believed to be equal to any yet invented, whereas it costs considerably less than the patented machines commonly known. It consists of a box, 2ft. wide, 6ft. or 8ft. long, and 18in. to 24in. deep, made of 2in. hard-wood plank, and somewhat wider at the top than at the bottom. The bottom is of heavy sheet iron, nailed firmly to the sides and ends. The box rests on brick or stone walls, sufficiently high to give plenty of fire room underneath. (A trench in the ground might serve in place of walls.) The front of the fireplace has a door of sheet or cast iron, with a damper by which to regulate the heat of the fire. The door is of sufficient size to permit the use of refuse pieces of wood found about the farm, or from the wood-pile. At the rear end, a chimney, or suitable escape for smoke, is constructed; for this purpose large-sized stove piping answers well. In making the box, thick white lead is carefully spread on the bottom edges before nailing on the iron bottom; this makes it less likely to leak. After setting the box on the walls, earth is banked up against them, this extending to the sides of the walls in such a way as to prevent the escape of smoke and sparks through them. For drawing off the contents of the box, a sliding gate, with a tin spout underneath, is fixed in the front end. A cover, made of 1in. pine, cut on a bevel with the flaring sides of the box, should fit inside, instead of on the top, and have some sort of handles at each end for convenience in lifting. A few strips of wood at intervals on the bottom, and upon them a false bottom, with numerous small perforations, is desirable, as it will prevent meal or any other fine food from burning beneath. Whenever the box is emptied, it should be cleaned out under the false bottom, and, if emptied of food when there is a fire below, some water should at once be poured in the space, to prevent injury to the pan. With such an arrangement as this for boiling maize, either shelled or in the ear, potatoes, turnips, pumpkins, beets, &c., and with cheap fuel, and feeding the mass when cold, or but moderately warm, it is believed that almost any farmer can secure a fair compensation for the time and labour expended in cooking a considerable proportion of the food for his pigs. When it is found more suitable to soak than to cook the maize, the box can be used for this purpose; while, for heating water, and scalding pigs at killing time, it is of considerable use.

In dealing with the question of the manufacture of pork, American writers have confined themselves more closely to an investigation of the cost of its production from the feeding of maize, their common food; and, as Americans keep pigs in very much larger numbers than we do in Great Britain, we have not followed this point up so closely as they have. We may, however, inquire, in a few words, what quantity of food is necessary to maintain a pig of average weight, and to produce each pound of additional weight which it adds to its carcase. Assuming the mean weight of the pig to be 150lb., Mr. Joseph Sullivant says that the amount of carbon necessary will give the average amount required daily for the whole life of the pig. Thus, 28 grains, multiplied by 150 (the amount required to

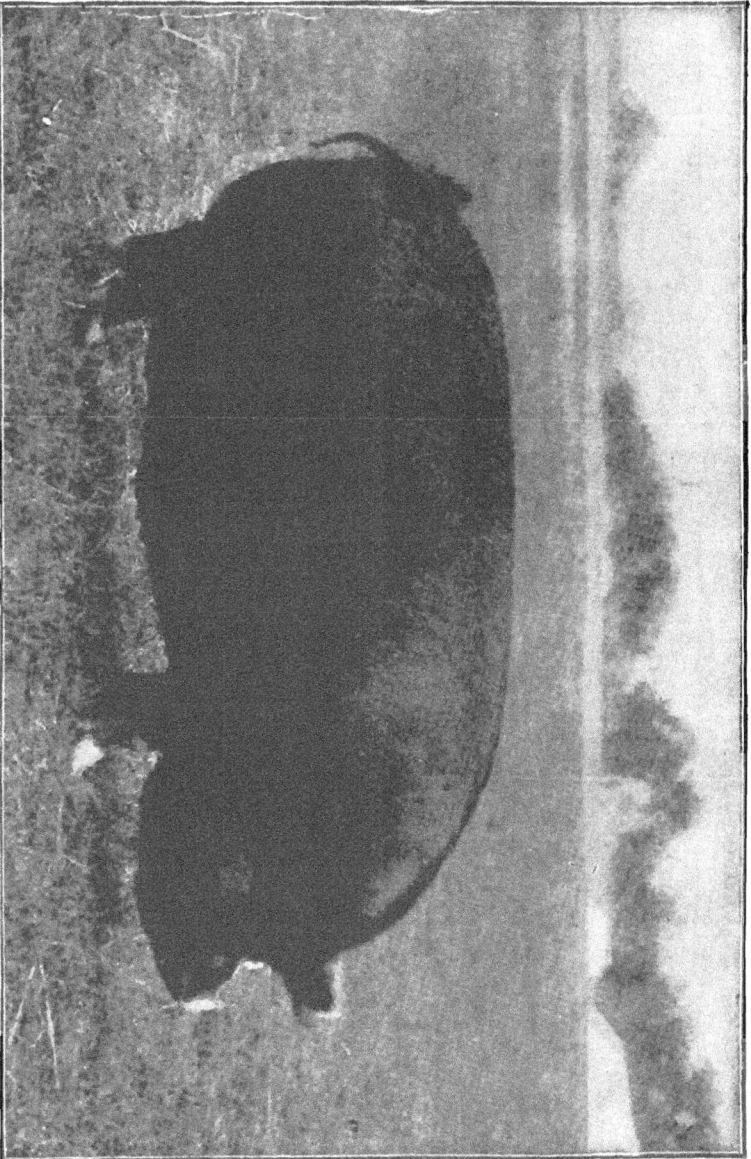

BERKSHIRE SOW.

The Hon. Claud B. Portman's "Manor Empress Queen" (Herd Book No, 8660), bred by Mr. Arthur Hiscock, jun.
Winner of First Prize, Royal Agricultural Society's Show, 1904.   .   .   .   .

sustain a man of the same weight), gives 9 ounces of carbon, which can be obtained from 22oz. of maize, and which would keep the animal in normal condition. Directly, however, we proceed to add additional weight to the pig, we require, over and above this quantity of carbon, 16oz. of muscle and fat-forming material, which can be obtained from 2lb. of maize, containing about $3\frac{3}{25}$oz. of the flesh-former, and $13\frac{3}{25}$oz. of carbon. By this method of reasoning, therefore, $3\frac{3}{8}$lb. of maize constitutes a sufficient daily ration for a pig required to make 300lb. of pork in 300 days. From these figures, it will also appear that every bushel of maize contains a trifle more than $16\frac{1}{4}$lb. of pork.

Effects of Temperature on Fattening Pigs.

We have shown that warmth is of especial importance in the feeding of pigs, and that, where they are kept at a low temperature, it is necessary to increase the quantity of food in order to maintain the warmth of the body. As a general rule, feeders pay little regard to a matter of this kind, and, rather than take the trouble of providing warm piggeries or sties for the protection of the animals during the winter months, they give an additional quantity of food at an infinitely greater cost.

In the winters of 1880-81 and 1882-83, Professor Shelton, of the Kansas State Agricultural College Farm, made a series of experiments, which are the most complete of their kind within our knowledge. They were undertaken with a view to ascertaining the actual effect of cold upon fattening pigs, and to establishing facts that would be a useful guide to the swine-breeder. In one set of experiments, in 1880-81, ten well-bred Berkshire pigs were selected, the dates when they were farrowed being as follow:

PIG.				PIG.			
No. 1.	Farrowed	April 12, 1879.		No. 4.	Farrowed	July 4.	1879
„ 5.	„	„	„	„ 6.	„	„	)
„ 7.	„	„	„	„ 8.	„	„	,
„ 2.	„	July 4	„	„ 9.	„	..	,
„ 3.	„	„	„	„ 10.	„	March 20	,

The three dates represent three different litters,

Previous to the commencement of the experiments the pigs
had been kept upon pasture, receiving, in addition, a few
heads of maize daily. They are stated by Mr. Coburn, whose
authority is undoubted, to have been a remarkably uniform lot,
and of excellent quality. In both sets of experiments the pigs
were divided into two lots, the first (Nos. 1 to 5) being kept in
separate pens, in the basement of a warm stone barn. The
second lot (Nos. 6 to 10) were kept in an open yard, surrounded
by a 5ft. board fence, having no other protection except straw
for beds, which was furnished as was found necessary; like the
first lot, they were divided into pens, in order to maintain the
proper apportionment of food.

In the first set of experiments, the animals in Pens 1, 2, 5,
6, 7, and 10 were fed exclusively upon maize, while those in
Pens 3, 4, 8, and 9 received a ration of about 2lb. of bran per
day, in addition to the maize. In a short time the pigs left a
portion of their bran, and finally refused it altogether, when
it was discontinued. In the second set of experiments, maize
was used alone, the animals receiving all they could eat, and
great care was taken to prevent waste, or the possibility of any
of the pigs receiving less than they required. They were fed
at 8 a.m. and 4 p.m., the food being weighed on every occasion.
At the 8 o'clock feed the temperature was noted by the ther-
mometer, while each pig was weighed at the end of every
week, so that there is reason to believe that the experiments
are most trustworthy.

The table on page 287 shows the weight of each pig at the
commencement of the trial, the total increase, the increase per
cwt., and the average increase per cwt., in the experiment
made in 1880-81. The even quality of the pigs is conclusively
shown by the remarkable uniformity in their increase in
weight.

The table on page 288 shows the general results of the
experiments.

It appears that 100lb. of increase in Pens 1, 2, and 5 was
obtained at a cost of 515lb. of maize, while 100lb. in Pens 6, 7,
and 10 cost 548lb., showing a loss of 33lb. of maize per cwt.
of increase when compared with the pens in which the same
feed was used in the barn, whereas the 418lb. of increase in
Pens 6, 7, and 10 showed a loss of 138¼lb. of maize. Again,

DATE.	PIGS KEPT IN WARM PENS IN BARN.					PIGS KEPT IN OPEN PENS IN THE YARD.				
	Pen 1.	Pen 2.	Pen 3.	Pen 4.	Pen 5.	Pen 6.	Pen 7.	Pen 8.	Pen 9.	Pen 10.
	lb.	lb.	lb.	lb.	lb.	lb.	lb.	lb.	lb.	lb.
November 1st, 1880	272	240	258	275	226	244	229	249	252	285
,, 8th ,,	281	257	267	294	238	253	239	260	259	292
,, 15th ,,	296	266	285	309	251	263	245	269	278	313
,, 22nd ,,	313	282	297	325	273	287	259	292	293	330
,, 29th ,,	331	304	319	338	289	304	275	310	308	352
December 6th ,,	349	328	336	357	305	323	288	317	320	362
,, 13th ,,	365	339	356	376	321	347	306	331	338	387
,, 20th ,,	389	359	373	396	340	356	321	339	346	392
,, 27th ,,	400	371	390	409	351	373	336	348	355	403
January 3rd. 1881	413	381	399	422	359	382	346	357	356	398
,, 10th ,,	424	394	410	429	374	384	357	363	356	407
,, 17th ,,	435	404	424	439	382	401	366	372	369	409
Total Increase	163	164	166	164	156	157	137	123	117	124
Increase per cwt.	50.90	68.30	64.30	59.60	6.00	64.30	59.80	49.30	46.40	43.50
Average Increase per cwt.	63.90lb.					52.20lb.				

taking Pens 3 and 4, we find that 100lb. of increase was obtained from the consumption of 481¼lb. of maize and 70¼lb. of bran; whereas, in Pens 8 and 9, 100lb. cost 577¾lb. of maize and 83⅓lb. of bran. There was considerably more variation in the pens in the open yard than in those under cover, and the loss throughout the entire term of feeding was most distinct. It was also noticed that the most docile of the pigs bore the severity of the weather better than the others, and yielded the largest return; they passed much of their time, when the

	Total Increase.	Total Maize Consumed.	Total Bran Consumed.	Maize Consumed for each 100lb. of Live Weight.	Bran Consumed for each 100lb. of Live Weight.	Maize Consumed for each 1lb. of Increase.	Bran Consumed for each 1lb. of Increase.
FEED : MAIZE.	lb.	lb.	lb.	lb.	lb.	lb.	lb.
Pens 1, 2, and 5, in the Barn ..	483	2487.50	..	22.03	..	5.15	..
Pens 6, 7, and 10, in Open Yard	418	2291.00	..	21.64	..	5.48	..
FEED : MAIZE AND BRAN.							
Pens 3 and 4, in the Barn ..	330	1589.00	232.00	21.09	4.13	4.81	0.70
Pens 8 and 9, in Open Yard ..	240	1386.50	200.00	19.82	4.14	5.77	0.83

temperature was at its lowest, in a condition which, Mr. Coburn states, closely resembled hibernation, seldom feeding more than once a day, and lying perfectly still during the remainder of the time.

This experiment tends to show that bran, given in conjunction with maize, is not of such value as has been supposed. In Pens 1, 2, 5, 6, 7, and 10, for instance, in which maize was exclusively used, 901½lb. of increase in weight was obtained from 4778½lb. of maize; whereas in Pens 3, 4, 8, and 9, in which maize and bran were given in combination, 2975lb. of maize and 432lb. of bran yielded 570lb. of increase. In other words, 100lb. of increase in Pens 1, 2, 5, 6, 7, and 10 cost 530½lb. of maize; while 100lb. of increase in Pens 3, 4, 8, and 9 cost 521¾lb. of maize and 75¾lb. of bran. Thus, it appears that 8.42lb. of maize had a feeding value equal to that of 75.78lb. of bran.

The table given on page 290 shows the weight of each pig at the beginning of the experiments of 1882-83, together with the weight at the close of each week, the total increase, the total increase per cwt. of each animal, and the average gain per cwt.

The figures on page 291, which it is necessary to quote, show the average temperature of the weather during each week of the experiment, together with the total food consumed, the food for each 100lb. of live weight, the total increase, and the quantity of food consumed for each pound of increase in each experiment.

From this table, it appears that Pens 1, 2, 3, 4, and 5 produced 583lb. of pork from 2870lb. of maize, and that Pens 6, 7, 8, 9, and 10 produced 490lb. from 2819lb. of maize. Thus, each pound of pork made in the warm pens cost 4.9lb. of maize, whereas each pound made in the cold pens cost 5.7lb. of maize. Dealing with the question from another point of view, it is found that, in the warm pens, one bushel of maize produced 11.3lb. of pork, while in the cold pens, one bushel produced only 9.7lb.

Here, then, is a clear answer to the question as to the effects of temperature, for a set of carefully conducted experiments like the above proves incontestably that, in every bushel of maize consumed, a quantity equal to the manufacture of 1.6lb. of pork was required to maintain the warmth of each animal. During the three weeks of the lowest temperature in the winter the result was still more striking, for while 136lb. of increase was obtained from the consumption of 823lb. of maize in the barn, the pigs in the outside pens consumed 740lb., and only yielded 33lb. of increase. Thus, finally, during the most severe weather, every pound of pork produced in the barn cost 6.05lb. of maize, whereas outside the same quantity cost no less than 22.42lb. of maize.

DATE.	WEEK OF EXPERIMENT.	PIGS KEPT IN WARM PENS IN THE BARN.					PIGS KEPT IN OPEN PENS IN THE YARD.				
		Pen 1.	Pen 2.	Pen 3.	Pen 4.	Pen 5.	Pen 6.	Pen 7.	Pen 8.	Pen 9.	Pen 10.
		lb.	lb.	lb.	lb.	lb.	lb.	lb.	lb.	lb.	lb.
Nov. 27th, 1882..	Beginning of Experiment	252	211	223	214	214	200	227	196	237	209
Dec. 4th, „ ..	First	269	222	235	228	228	218	251	216	237	209
„ 11th, „ ..	Second	275	244	257	253	251	234	270	236	271	232
„ 18th, „ ..	Third	293	253	268	260	258	244	278	249	288	246
„ 25th, „ ..	Fourth	309	271	285	275	295	253	289	241	299	242
Jan. 1st, 1883..	Fifth	315	279	291	285	277	263	299	259	315	261
„ 8th, „ ..	Sixth	324	282	305	290	287	262	295	269	321	266
„ 15th, „ ..	Seventh	343	301	319	311	296	278	319	276	338	270
„ 22nd, „ ..	Eighth	346	308	330	322	301	283	323	272	344	275
„ 29th, „ ..	Ninth	356	317	337	334	316	283	334	289	360	286
Feb. 5th, „ ..	Tenth	373	330	347	346	322	288	337	283	363	282
Total Increase ..	..	121	119	124	132	108	88	110	87	126	78
Total Increase per cwt.	..	48.00	50.00	55.60	61.60	50.40	40.00	48.40	44.40	53.10	38.20
Average Gain per cwt...	..	54.20lb.					45.90lb.				

Week of Experiment	Pigs kept in Warm Pens in Barn					Pigs kept in Open Pens in the Yard				
	Average Weekly Temperature, Fahrenheit.	Total Feed.	Feed for 100lb. of Live Weight of Animal.	Total Increase.	Pounds of Feed for 1lb. of Increase.	Average Weekly Temperature, Fahrenheit.	Total Feed.	Feed for 100lb. of Live Weight of Animal.	Total Increase.	Pounds of Feed for 1lb. of Increase.
		lb.	lb.	lb.			lb.	lb.	lb.	
First	38°	227	19.8	68	3.30	31°	243	21.9	87	2.80
Second	33°	298	24.2	98	3.00	22°	323	26.9	92	3.50
Third	36°	324	24.8	54	6.00	21°	331	26.0	62	5.30
Fourth	42°	336	24.2	101	3.30	29°	332	25.3	19	17.50
Fifth	32°	310	21.5	12	25.80	15°	322	11.8	73	4.40
Sixth	21°	277	18.8	41	6.80	5°	268	19.0	16	16.70
Seventh	29°	289	18.9	61	4.70	13°	279	19.2	68	4.11
Eighth	19°	272	17.1	37	7.30	12°	248	16.6	16	15.50
Ninth	27°	263	16.1	53	4.90	15°	249	16.3	56	4.40
Tenth	20°	274	16.8	58	4.70	2°	224	14.4	1	221.00

Table showing the Composition of Pig Foods.

FOODS.	WATER.	DIGESTIBLE CONSTITUENTS.		
		Albumi-noids.	Carbo-hydrates.	Fat.
CORN AND SEED:				
Maize	14.4	8.4	60.6	4.8
Wheat	14.4	12.0	64.3	1.2
Barley	14.0	8.0	58.9	1.7
Oats	14.0	9.0	43.3	4.7
Rye	14.2	9.9	65.4	1.6
Buckwheat	14.0	6.8	47.0	1.2
Peas	14.3	20.2	54.4	1.7
Beans	14.5	23.0	43.6	1.4
Lentils	14.5	23.8	49.7	2.2
Linseed	12.3	20.5	18.9	35.2
Acorns, fresh	55.3	2.0	30.9	1.6
ROOTS AND TUBERS:				
Potatoes	75.0	2.1	1.8	0.2
Mangels	88.0	1.1	10.0	0.1
Swedes	87.0	1.5	10.6	0.2
Artichokes	80.0	2.0	16.8	0.3
Beet	81.5	1.0	15.4	0.1
Carrots	85.0	1.4	12.5	0.2
Parsnips	85.5	1.6	11.2	0.2
MEALS AND VARIOUS:				
Wheat Bran	13.0	11.8	44.4	3.4
Barley-meal	11.0	11.6	34.5	3.4
Rice-meal	11.0	9.9	63.3	6.4
Grains	76.0	3.9	10.8	0.8
Malt Dust	10.1	19.4	45.0	1.7
Rape Cake	11.3	25.3	23.8	7.7
Linseed Cake	12.0	24.8	27.5	8.9
Cotton Cake, decorti-cated	11.2	31.0	18.0	12.3
Beechnut Cake	10.0	24.0	16.7	5.2
Palmnut Cake	10.5	16.3	55.4	9.5
Cocoanut Cake	9.4	18.2	47.4	11.2
Pumpkin	89.0	0.6	6.5	0.1
Skim Milk	89.6	3.0	5.2	0.7

FOODS.	WATER.	DIGESTIBLE CONSTITUENTS.		
		Albumi-noids.	Carbo-hydrates.	Fat.
MEALS AND VARIOUS— (*continued*) :				
Buttermilk	90.1	3.0	5.0	1.2
Whey	93.5	0.8	5.0	0.3
Horse Chestnuts ..	49.0	3.4	34.8	1.0
GREEN FODDER:				
Lucerne, Young ..	81.0	3.5	7.3	0.3
Sainfoin	81.0	3.2	8.0	0.3
Grass	75.0	2.0	13.0	0.4
Rye Grass, Italian ..	73.0	2.3	12.5	0.4
Timothy	70.0	2.1	16.2	0.5
Trifolium	81.5	1.5	7.5	0.6
Red Clover	83.0	2.3	7.4	0.5
White Clover	80.5	2.2	7.9	0.5
Maize	82.9	0.7	8.4	0.3
Tares	82.0	2.5	6.7	0.3
Cabbage	89.0	1.5	7.0	—

Food Constituents Consumed by Pigs.

. According to Dr. Emil Wolff, the following three tables show the chief food ingredients consumed by pigs at particular stages of growth:

At per 1000lb. of live weight, per day:

GROWING SWINE.

Age in Months.	Live Weight.	Total Dry Organic Matter.	Albumi-noids.	Carbo-hydrates and Fat.
	lb.	lb.	lb.	lb.
2 to 3	50	42.0	7.5	30.0
3 to 5	100	34.0	5.0	25.0
5 to 6	125	31.5	4.3	23.7
6 to 8	170	27.0	3.4	20.4
8 to 12	250	21.0	2.5	16.2

FATTENING SWINE.

Period.	Total Dry Organic Matter.	Albuminoids.	Carbohydrates and Fat.
	lb.	lb.	lb.
First period	36.0	5.0	27.5
Second period	31.0	4.0	24.0
Third period	23.5	2.7	17.5

The carbohydrate group includes sugar, gums, starch, and cellulose. The last-named, when consumed in young plants, is a valuable food, but is not easily appropriated by stock consuming old plants; it is practically the wall of the cells of which vegetable foods are largely composed. Albuminoids chiefly incude casein, albumin, and fibrin, all of which are well known to the general reader.

Per head, per day:

GROWING SWINE.

Age in Months.	Live Weight.	Total Dry Organic Matter.	Albuminoids.	Carbohydrates and Fat.
	lb.	lb.	lb.	lb.
2 to 3	50	2.1	.38	1.50
3 to 5	100	3.4	.50	2.50
5 to 6	125	3.9	.54	2.96
6 to 8	170	4.6	.58	3.47
8 to 12	250	5.2	.62	4.05

The table on page 295, prepared by the late Sir John Bennet Lawes, Bart., shows the food increase and manure of fattening pigs. It should be remembered, in dealing with fresh vegetable foods, that, although a pig may consume a large quantity, it does not necessarily obtain a large amount of nourishment, inasmuch as they contain a very considerable quantity of water. The following figures give some idea of the quantities of water:

Turnips, 92 per cent. ; turnip leaves, 88 ; mangels, 88 ; mangel leaves, 90 ; cabbage, 86 ; potatoes, 75 ; tares, 82 ; red clover, 80 ; grass, 75 ; sainfoin, 80.

Nature of Substance.	500lb. Barley-meal produces 100lb. Increase, and Supplies—				100lb. Total Dry Substance of Food Supplies—			Amount of each Constituent of Food Stored up for 100lb. of it Consumed.
	Its Food.	In 100lb. Increase.	In Manure.	To Respiration, &c.	In Increase.	In Manure.	To Respiration, &c.	
	lb.	lb.	lb.	lb.	lb.	lb.	lb.	lb.
Nitrogenous Substance ..	52	7.0	59.8	276.2	1.7	14.3	65.7	13.5
Non-nitrogenou Substance ..	357	66.0			15.7			18.5
Mineral Matter .	11	0.8	10.2	—	0.2	2.4	—	7.3
Total Dry Substance ..	420	73.8	70.0	276.2	17.6	16.7	65.7	—

CHAPTER XV.

THE COMMERCE OF THE PIG.

In this chapter—which must necessarily be somewhat comprehensive—we propose to deal with what we have termed "The Commerce of the Pig"—to refer to its various products, to compare the nutritive value of the edible parts one with the other, and to describe the systems of slaughtering and preserving, as well as the manufacture of some of the more famed delicacies which are prepared from the animal.

It has been stated that a larger proportion of the carcase of the pig is employed for some useful purpose than that of any other animal. The fat, in addition to its value as an article of food, is largely used in the preparation of perfumery. The skin has long been in demand for the manufacture of harness, more especially of saddles, and is converted into numerous smaller articles of general use, such as purses, pocket-books, and even shoes; while the bristles are commonly employed in brush-making, and have already been the means of creating a large industry.

It may here be remarked that pig-skin is covered with a number of small holes like pin-pricks arranged in series of threes, thus: .·. In one of these there is a small hair, while the other two holes have no hirsute growth. In the imitation article the holes are much blunter (being made by a machine) and the general appearance of the leather (usually calf) is, apart from the holes, of a more uneven and "crinkly" appearance. Pig-skin is much harder on the surface than other leathers, and takes a polish better than any other. It is so tough that the portion taken from the back of

an old boar will, even if undressed and exposed to all
weathers in its natural state, remain for years before it finally
decays.

It appears that, since the year 1853, when three and a
quarter million pounds of bristles were imported into England,
the annual imports have averaged two and a quarter million
pounds, the declared value of which very nearly approaches
half a million sterling. The trade is principally American and
English, although the large mass of the supply comes from
Russia. The swine which furnish the bristles are chiefly fed
upon the refuse of tallow factories, and it is believed that the
Russian product owes its quality to the special system of
feeding of a race of pigs which very closely resemble the wild
boar. The best samples of bristles come from Siberia, and
average from £7 to £8 the *pood* of 36lb. In all, Russia exports
bristles to the value of a million sterling per annum. France
also exports bristles to the extent of some two million pounds
per annum, and these bear a high reputation in America,
principally on account of their whiteness, their softness, and
their firm and elastic touch.

In the bristle trade, it is customary to look, first of all, to
length, which should be 6in., as those which are longer generally
lack toughness. In sorting, the bundles are divided into colours
of white, dark, and brown bristles respectively, the last
named being bleached, if necessary, by being placed for two
or three days in a saturated solution of sulphurous acid. The
bristles used by shoemakers are the highest in value, and some
idea may be obtained as to the range of prices when we state
that they have reached from 1s. to 7s. per lb. It has been
stated that the first manufactory established in connection
with pigs' bristles was that of a Frenchman; and it was also
a native of France who first invented a machine for the
extraction of oil from lard, and for the manufacture of stearine
candles from the refuse. This does not exhaust the useful
purposes to which portions of the pig have been put, for it is
now well known that prussiate of potash is made from the
blood.

It may, indeed, strike the casual breeder as a remarkable
thing that these industries should be conducted at all; but he
may not be aware that there are establishments in the world,

more especially in America and in Germany, where pigs are killed by the thousand, and where, if the waste material were not usefully employed, a considerable loss would result.

Great Pig-Killing Establishments.

In the year 1883, during a visit to Hamburg, we were shown over the large pig-killing establishment of Mr. J. C. Koopmann, when we were able to see what pork-making upon a very large scale was really like; while in 1893 we paid a visit to one of the largest establishments of a similar character in the Chicago stockyards. In Schleswig and Holstein very large numbers of pigs are bred by the smaller farmers as well as the large, for dairy work is part and parcel of their system, and, consequently, pigs form a considerable item in the economy of the farm. Vast numbers of these pigs find their way to the Hamburg markets, whence they are exported; and, as the price they command is exceedingly small, as compared even with English prices during bad seasons, it is no wonder that Mr. Koopmann has built up a huge business, and that he can compete upon advantageous terms with a country like this, where pork cannot be grown upon the farm at the price at which that from Germany is sold in the London markets.

At one of the German International Exhibitions, we noticed that the German farmer is making great progress with the best British pigs, and already the quality of his stock is improving, for, although it is a reflection upon ourselves, there is no doubt that in Germany there are far greater numbers of well-bred large Yorkshire pigs, of an exhibition type, than there are in England at this moment, although perhaps these are not bred to quite such perfect form as a few of the best to be seen at our shows. Mr. Koopmann's factory is not a model building; indeed, it immediately gives one the idea, which is quite correct, that, as the business extended, fresh buildings were added and equipped; consequently, there is nothing to describe but the system which is pursued, from the killing of the pig until it finally leaves for the London market.

Mark Twain once gave a happy description, in a few words, of the Chicago pig, which was, he said, "squeaking at one moment, and hanging, in quarters, in an ice-room, at the next." This would almost apply to the Hamburg factory, for the work is done in a marvellously expeditious manner. Here a thousand pigs are killed daily, and their meat is conveyed to the cold-room as follows: A large number of men are employed in the work, each man doing one particular portion, and that only; consequently, they are most skilful, and carry it out with lightning speed. The pigs, as fast as they arrive, are driven into the yard, where they are caught and slung up in a moment by their hind legs, the cross-piece upon which the pig hangs being fixed to a roller: these run along iron rails, extending from the killing-room, right through the building, to the packing-room upstairs. When the pig is first slung, it is pushed along to the man who sticks it, and, as this is his entire work—merely sticking a thousand pigs daily—he is able, as we are informed, and as we could see, to find out the right spot every time.

Passed along from the killer, the pig has to run the gauntlet of a dozen men, each of whom does some particular portion of work before it is ready for cutting into halves. It is dipped into scalding water, and immediately afterwards scraped roughly and dried. It then turns a corner, and enters the furnace room, the roller running on to a circular rail, upon which four pigs hang at a time in front of the furnace door. This door is opened, and the four pigs are run in, one at a time. Each is allowed to stay several seconds—sufficiently long, in fact, to destroy all the remaining hairs, and to give the apartment a somewhat agreeable odour. The pig is again run along the rails, when he is quickly opened, and the entire contents of his body taken out, one man only appropriating one portion; thus, one person takes the liver, another the heart, a third the bladder, and so on. Again pushed on, a man cuts off the feet; a little further, and the pig loses his head; still further, and the carcase is in an instant divided into sides. After this, the pigs are sent up an incline, one after the other, to a high storey of the building. Subsequently the sides are weighed and despatched into a cooling-room, the roller being sent down another way, at considerable speed, for use again. If any carcase appears to need examination, it is taken off the roller and placed upon a table for attention.

The livers of the pigs, of which there are, of course, a large number, are sold in Hamburg for sausage-making, and many of the intestines are also used for the sausage trade. All that is not edible in any form is sold for manure. The lungs are used for sausages also; so that, if there are many of these establishments in Germany, the enormous trade in sausages can easily be accounted for. The heads and feet of the pigs are sold in Hamburg. The fat is, as far as possible, taken from each pig, and rendered into lard in the building, the bladders being dried, blown, and used for this purpose; consequently there is a trade in lard in connection with Mr. Koopmann's business.

Although everything is perfect in its way, by far the most important machines in the building are those employed for compressing and drying the air, and afterwards expanding and cooling it. The one in work upon our visit was of 60 horse power, but there was another of 80 horse power used for the same purpose. The warm compressed air is cooled by means of a system of tubes, over which cold water is continually running. When cool, it comes into a cylinder, where it is expanded, and afterwards conveyed into the cooling-rooms, the system which it has undergone having completely dried it. These rooms have a temperature which is a little above freezing point, and the air in them is continually being changed, the fresh entering on one side, and that which has been used leaving on the other, whence it is again run through the machinery, re-dried, and cooled. The sides of pork hang here for from fifteen to eighteen hours, when they are ready for removal. If they are sent away more quickly, the outsides are certainly firm, but the insides are soft—in other words, the cooling has not sufficiently done its work.

There were four of these cooling-rooms on the occasion of our visit, each holding 1300 sides; and also two ice-rooms for salting the pork. Into all of these the sides are run upon wheels, and can, by an ingenious system, be placed on any row that is necessary. In one ice-room, which is a very large one, pork is packed in winter to about 8ft. in depth;

and there are also very large bacon-rooms at each end. In these the sides are packed upon each other in salt, and remain thus for fourteen days before being sent away. Through these rooms cold air is also passed. When the salted sides are ready for market, they are sorted, according to their quality and the proportion of lean and fat, and then packed for London. They are placed upon packing tables, and six or seven are put in each cloth, and well sewn up, all being marked with Koopmann's brand.

Care is taken in the preparation of the lard. It is melted in huge vats, and thoroughly strained, that which we saw being as clear as water. The best only is placed in bladders, the remainder being poured into barrels, and every particle of solid matter excluded. Much attention is also paid to the selection of the pigs, which are preferred of an equal size, and such as is generally attained in the country in about nine months. They must also have been carefully handled and driven, for, if this is not the case, the meat is spoiled, and, therefore, rejected. The majority come from Schleswig and Holstein, although many are from Mecklenburg and Denmark. The blood is all used for manure, and the largest quantity of this, as well as other manure produced from the pigs, Mr. Koopmann (at the time of our visit) used himself, for at one place he owned a number of acres of asparagus.

The system of slaughtering, however, is conducted upon a still larger scale at Chicago, and, as there is some little difference between the method of conducting the work in America and in Germany, we give the following brief description of a Chicago establishment:

About twenty hogs are driven at a time into a square room just large enough to hold them, when a man enters, closes the door, and, with a short, heavy hammer, knocks down one animal after another by a smart blow between the eyes. While this is being done, a second apartment is similarly filled, which he visits in turn. The stunned pigs are seized, one by one, by two men, and dragged through the inside door on to a platform, where they are stuck with a knife, the blood flowing into a receptacle placed beneath. After bleeding sufficiently long, they are slid into a huge vat of scalding water, in which the temperature is regulated. They

are pushed along by men who take care that every portion is
scalded, and reach the end of the vat in about two minutes,
ten animals being scalded at one time. At the end of the
vat they are hauled out by a pulley on to the scraping-table.
At each side of this are eight or ten men, all of whom perform
their particular work in the scraping of the pigs. Thus, one
removes the bristles for saving, another scrapes one side, while
a third scrapes the other. Next come the head and legs,
after which the carcase is shaved with sharper knives.

At the end of this table, a stick is thrust between the
tendons of the two legs, and the pig is swung upon a large
wheel above. This wheel is provided with eight hooks, about
4ft. apart, upon each of which a pig is hung. Here it is
washed with clean water and scraped down. A slight turn is
given, and the carcase reaches another man, who cuts it
down, and removes the large intestines; these are thrown
on a table behind, where other men are engaged in separating
the fat from them. Again the wheel is turned, and clean
water is dashed within the carcase. The pig is then carried
away, and hung in another part of the establishment, where
hooks are provided, from one to a thousand, as the case may
be; and here they remain until the next day, when they are
packed, after having been weighed and cut up. The whole
work is performed in the most rapid and clever manner.

Slaughtering.

Farmers and others who breed and feed pigs in this country
generally elect to sell them alive in preference to slaughtering
for sale as fresh pork by retail or by the carcase, or to selling
them in a cured form. One of the principal reasons for this
is that they receive an immediate and actual money value,
which they do not in all other cases. But, in the immediate
neighbourhood of large towns, it is the custom, at certain
seasons of the year, to slaughter pork for sale, especially in
London, where the price in the great market reaches from
4s. 3d. to 4s. 9d. per stone. In such a case, it is usual to
employ a local man who is an adept in the slaughtering of
pigs, and who generally does his work for a small fee. There
are farmers, however, who kill and dress their own pigs,

and pack them for the London market quite as well as the work can be done by a butcher.

In England there is but one general system of slaughtering. The pig, having fasted for a number of hours, is caught by a noose being slipped over the upper jaw. The rope is held by an attendant, who lifts up the head, exposing the throat as much as possible to the butcher, when the knife is immediately thrust in. The animal, having thoroughly bled, is then removed to the scalding-tub, passing through a process which is commonly understood in all parts of the country.

In America, however, it is quite common to stun the pig before it is stuck, and Mr. Coburn tells us that, when it is secured for sticking, it should be turned upon its back, no twist being allowed in its neck, so that the butcher is certain to divide the main artery without permitting the knife to penetrate either shoulder. It is not considered desirable to touch the heart, most persons preferring that the animal should die from loss of blood, and, if properly stuck, the blood will leap, as it were, through the wound in a powerful stream.

In some parts of America, a rectangular scalding-tub is used, this being made of two-inch deal, 6ft. long and 2ft. wide, and the ends rounded at the bottom. The ends and bottom are covered with a heavy plate of sheet iron, which projects about an inch on either side. This vessel is set upon two walls of brick, built 18in. high, one end being built up in the form of a chimney. The fire is lit underneath, so that, in a very few minutes, the water is ready for the reception of the pig. Upon one side, and level with the top of the tub, a platform is erected, which is some 6ft. wide and 8ft. long, and upon which the pigs are cleaned after scalding. At one end of this, an incline should be formed to facilitate the raising of the animals. Once upon the platform, they can easily be placed in the water by means of two ropes, 8ft. or more in length, which are fixed to the side of the platform nearest the vat, and upon which, with the aid of two men, the pigs can be lowered or raised, as deemed expedient. To avoid any possible injury to the skin or carcase from burning, a number of wooden strips should be placed across the bottom of the vat.

After scalding for a few seconds, each pig is pulled out of the water, to allow it to come into contact with the air for a short time, after which it is again immersed. As soon as the hair readily leaves the skin, the pig should be taken out of the vat and the remaining hairs removed. When this is done, the hind legs should be cut below the gambrel (hock) joint, so that both main cords may be reached, under which the gambrel should be inserted, this being made of wood some 2ft. in length, and notched to prevent the pigs from slipping off. Two posts with a cross-piece should be fixed near the platform, on to which the pigs, after scalding, can be hung and run along to the end, so as to be out of the way of the scalding operations. The next thing to do is to remove the intestines, after which, a piece of wood should be placed within the mouth, to keep it open and to facilitate the removal of the blood that yet remains in the body and in the neck.

One point, which must not be overlooked, after the performance of all this work, is to allow the pig to hang until it is quite cool in every part; otherwise, it will be found impossible to cure or preserve it, no matter how much trouble may be taken in endeavouring to do so.

In the abattoirs, or slaughter-houses, of Paris, the pigs are generally struck a violent blow upon the head before the throat is cut. This is done in order to diminish the suffering which they naturally undergo. The ordinary method of slaughtering is as follows: An assistant ties a rope to the right leg of the pig, behind, and accompanies the animal to the apartment where it is to be killed. The slaughterman then takes hold of the left ear and head, and turns the pig upon its right side, the assistant at the same time securely fastening the hind legs. The slaughterer next places his left knee upon the body of the pig, and keeps the head back with his right foot. This done, the throat is cut with a very sharp-pointed, narrow, and rather long knife, inserted in the direction of the heart, to sever the jugular vein. In addition to turning the knife round with as much rapidity as possible, it is sometimes found advisable to move one of the hind legs about, in order to disperse and hasten the emission of the blood. A third person receives the blood in a receptacle as it flows from the pig, and keeps it in constant agitation to prevent coagulation, and from it the fibrin is extracted.

When life is extinct the cleansing operations are commenced. In the North of France, and in the whole of Germany, it is the custom to burn or singe the pig. For this purpose, a quantity of dry straw is strewn over one side of the animal, and immediately set on fire. When the whole of this has burned away, the skin is thoroughly well brushed, and the parts that have still some hair upon them are again subjected to a flame, this time obtained from a few wisps of straw. The other side is similarly dealt with, and then the hoofs of the pig are removed. Sometimes the longer bristles are extracted before the singeing is concluded, these being used chiefly in the manufacture of brushes.

The more general custom, however, at all events in the South and West of France, and in Italy and Spain, is to scald the pig. The work is performed in a manner very similar to the scalding process already described, and therefore need not be again referred to; but we may state that, in the South of Europe, it is regarded as the best process for securing meat of high quality.

After singeing, the pigs are washed and scraped, upon a bed of clean straw, and, as in scalding, hung obliquely, and by their hind legs, upon a piece of wood made for the purpose, near to the wall. An opening is then made from the tail to the throat, the intestines are taken out and washed, the interior of the body thoroughly cleaned with cold water, all traces of blood removed, and the sides of the carcase kept apart by a piece of wood or any other suitable material.

In summer, in the South of France, however, pigs are often skinned as soon as they are killed. When the skin is taken off, the hairs are singed, and the former disposed of to the merchants who deal in such articles. At Hamburg, where this system has been adopted, the pig, as soon as it has been bled, and while it is yet warm, is singed for some minutes in a specially heated apartment, or "oven." The object of this is to cause the skin to rise somewhat, and to facilitate its being drawn off by the 'hand. The intestines are then removed and cleaned, and the large and small intestines separated while the whole are still warm.

X

Salting, Curing and Smoking

The systems of curing pork, whether for consumption as bacon, or as what is commonly called "pickled pork," are extremely numerous. We give examples of those which have been most successfully adopted in England, as well as in France and America.

One method is as follows: A little salt is sprinkled upon the meat as soon as the pigs are cut up, in order to expel the blood; the "kernels" at the same time being pressed out by means of the finger. When this is done, the bacon is allowed to remain for a day or two. It is then placed in portions in a deep square bin or tub, standing upon four legs, and having a groove along each side to carry off the brine which is formed. The best plan is to place a flitch of bacon at the bottom of the bin, with the rind downwards, covering it with coarse brown sugar and plenty of salt. Upon this flitch should be put another, and, at the top, the two hams, each part being dressed in the same way as the first, above mentioned. The salt should never be rubbed into or upon the bacon, as the tendency of this is to make the meat hard. The most important part of the work, however, is connected with the regular moving of the pieces of pork. Every third day, or thereabouts, the position of each piece should be changed, a further quantity of salt being also added if there is not sufficient left to rub in or to be absorbed. About a fortnight or three weeks afterwards—this time varying according to the size of the pigs—the flitches should be taken out of the bin, the hams still remaining; and, after each has been well drained, and the loose salt scraped off, they may be hung up in the kitchen or any other suitable apartment to dry, being subsequently removed to a cooler place. The plan of packing the bacon and hams in a chest, which is adopted in some parts of the country, cannot be recommended, as weevils are thus encouraged, which, of course, considerably affect its quality; at the same time, some splendid long-keeping meat is produced by this method.

A very good quality of bacon can be made by using common salt alone as a preservative, but it will be found much better to add a small proportion of saltpetre and bay salt (2oz. to 3oz.

MIDDLE WHITE SOW.

Sir Gilbert Greenall's "Walton Rose XIV." (Herd Book No. 12,002).

The Middle White is small in bone and large in flesh of the best quality, while it is one of the most docile races of pigs, and is second to none for maintaining high condition as a grazer. Its long, fine, and silky coat should be quite free from black hairs, and there should be no black or blue spots on the skin.

for a porker size), although, if too much of the former is
employed, it will make the meat hard.

The following is another system : The first thing to be done is
to cut off the head of the pig, the upper part being used for
manufacturing into potted meat or some other dish of an
agreeable description, and the under part, which is termed the
chap, for salting. The leaf of fat should then be removed, and
placed in a cool apartment. Next, the pig should be cut up,
and this may be done according to the taste or wish of the
curer. In curing, a stone table should be provided, in a well-
ventilated yet dry apartment, and covered with ground salt to
a quarter of an inch in thickness. On this should be placed the
flitches, hams, and chaps, rind downwards, and, above these,
twice as much salt as before, particular care being taken to rub
the salt well into the joints. Half a pound of pulverised saltpetre should also be sprinkled over the salt. The hams should
be examined every other day for three weeks, and, where the
salt is observed to have melted, more should be used. At the
end of this time the salt may be washed or brushed off, and the
bacon, hams, &c., hung up, in winter, in the kitchen, some
little distance from the fire, so that they shall not become
rancid. In spring, they should be sewn up in calico bags, and
placed upon a rack in a proper bacon chamber, where they will
be free from excessive heat and the depredations of the blue-
bottle fly, which so often causes damage to both fresh and preserved meat.

The following are American systems : In the first of which we
give particulars, the parts for curing, after being neatly cut,
are thoroughly covered with molasses, that from Porto Rico
being preferred. The entire surface of the bacon is then
sprinkled with salt and saltpetre, in the proportion of 9lb. of
the former to 4oz. of the latter, this being first mixed, and
heated to a very high temperature. When cool, the mixture
is rubbed into the bacon, great care being taken that no part—
more especially about the joints—shall be left unsalted. When
the hams are very large, a small opening is made, extending to
the bone at the joints, and this is filled with the hot salt. The
bacon is now placed in a suitable apartment, three or four
pieces deep, and allowed to remain for three days, after which
the salting process is repeated, and each piece laid in a cool,

dry place, where it remains for at least ten weeks; then, if deemed desirable, it is smoked.

By another system, when the meat is well cooled, the hams and shoulders are rubbed thoroughly with the best salt that can be obtained. Afterwards, the shank and bone are rubbed with a quantity of pulverised saltpetre. Next, a barrel is used, the bottom of which is well covered with salt. In this the hams and shoulders are placed, with the shank ends up, and the rind next to the barrel. The pickle pork is cut into strips of about 6in. in width, and, after the bottom of the barrel in which it is to be placed has been sprinkled with salt to a depth of half an inch, these strips are packed as closely as possible around the interior, with the rind outward, and, if necessary, the middle part is filled with one or two pieces of pork, cut for the purpose. As each layer of pork is packed, however, a large quantity of salt is sprinkled upon it, for the purpose of preservation. In two or three days a pickle is made as follows: A quantity of water is taken, and as much salt added as is sufficient to cause a potato to float. With this is mixed, for each ham and shoulder (and, in a like proportion, for the pickle pork), 1oz. of saltpetre, the whole being placed in a large kettle, boiled, and skimmed until it is clear. A $\frac{1}{4}$lb. of brown sugar is then added, in the same ratio as before, and the mixture stirred until the sugar has dissolved, when it is poured upon the meat in its hot state. The meat should always be kept weighted down, but the barrel should not be covered too closely or too tightly, or a certain percentage of the pork will be lost. The hams and shoulders are sometimes taken out of the barrel in time to have them smoked before flies appear in the spring. At this season of the year, the brine from the pickle pork should be poured off, boiled, and skimmed, so as to remove all blood drawn from the meat, after which the same brine may be again used to preserve the meat further.

In packing, if it is intended to use brine, the pork may be built up in the tub or barrel in layers; but it should always be remembered that, whether the meat is intended for home consumption or for sale in open market, it should be thoroughly cooled before it is packed.

Another good plan adopted in America is to sprinkle a quantity of salt over each layer, in the proportion of 8lb. to every 100lb. of pork, until the barrel is full. After this, pure water is poured in to fill up the interstices, and the whole is then completely covered. Where open barrels are employed, it will be found necessary to use a weight or weights of some description, to keep the pork under the brine.

Some persons prefer to prepare the brine for pouring over the meat, and a very good recipe, which will enable any reader to do this for himself, is that quoted by Mr. Coburn, as follows: For 100lb. of pork, take 4oz. of saltpetre, three pints of common molasses or 2lb. of brown sugar, and 7lb. of clean salt. When the whole is thoroughly dissolved, the brine is poured over the meat, which it will cover if it is properly packed. As there are often some impurities in the pickle, these may be removed by boiling, and then skimming the refuse which ascends to the top; but in this case the mixture should not be used until it is quite cool again.

Another good recipe, and one for curing the hams and shoulders without brine, is, for 150lb. of meat, to mix together 12lb. of fine salt, two quarts of molasses, and ½lb. of powdered saltpetre. This mixture should be thoroughly rubbed into the joints, and the operation repeated at the end of the first and second weeks, when the meat may be smoked.

A simpler way, in which any part of the carcase may be cured, is that mentioned by Mr. Coburn: A layer of salt of about half an inch in thickness is sprinkled on a platform, floor, or the bottom of a large box or cask, then a layer of meat, and above this a further quantity of salt. This is continued until the top is reached, when an extra covering of salt may be used.

The following method of home-curing bacon and hams produces excellent results:

When the pig is cut up, assuming that it weighs 9 or 10 score, place the sides and hams in a cool place (on a brick floor, if possible), flesh side uppermost. Mix 1oz. of powdered common washing soda with 2oz. of saltpetre, put on with a flour dredger, and allow this to remain until the following day. Then put a good layer of salt on, and place one side on top of the other, the rind side of the top one resting on the flesh side of the bottom one. The only rubbing required is just along the thick part of the back, and round the hocks and shoulders. Sprinkle a little more salt on every other day, and change the

sides about, using in all ½lb. of salt to every 20lb. of meat.
If the bacon is required for fairly quick consumption, take it
out of salt on the eighth day, wash over with clean water,
thoroughly scrape clean the rind side with a knife, and send
to be smoked; or if preferred white-dried, hang in a dry,
moderately warm place. If required to be kept a long time,
use ¾lb. of salt per score of meat, and allow it to remain
fourteen days before washing off. The hams should be treated
in the same way as the bacon for a week, and in addition
should be well rubbed with a handful of salt on the rind side
twice during that period. Put them into a pickling pan or
shallow tub, crush 2oz. of juniper berries, and put half on the
flesh side of each. At the same time get 2lb. of foots sugar
and put about half that quantity on also. Two days later give
each a little more salt. Mix the remaining pound of sugar
with ½lb. of salt and 1 pint of old beer, turn into the pan, and
baste the hams with this twice a day. Let each remain in
pickle one day for every pound the ham weighs, then wash off.
When thoroughly dried or smoked, tie up tightly in canvas
bags to keep the jumpers, or hopper flies, away, and store in a
dry place.

The salting of pork is a simple affair, and will generally
succeed under any intelligent system, if it is conducted with
care.

It will be noticed that the practice adopted under the various
systems described above is sometimes contradictory; but in
giving them we do not endorse either the one or the other as
the best.

Pork not intended for smoking should be stored in brine
before insects get a chance of depositing their eggs in it. In
all cases, a period of ten days is usually considered quite
sufficient to smoke a pig's carcase, unless the pieces are
unusually large and very thick.

A capital method of rendering salt pork more palatable is to
cut it into slices ready for frying, and half fill a crock or any
other suitable receptacle with the pieces cut. The whole should
then be covered with sweet skim milk, fastened down, and
placed in a cool apartment. In six or eight hours it will be
ready for use. In frying, the meat may be about half cooked,
and then rolled in flour and "browned." Two or three eggs

and a small quantity of pepper and salt mixed together, in which the meat should be frequently dipped, add to the delicacy of the dish.

Curing Hams.

The French system of preparing hams is as follows: The leg of pork is well pared with a knife, to give it a neatness and roundness of form and to remove excrescences, after which the skin is pricked with care, so as not to damage it, but to assist the brine in penetrating to the centre. The brine used consists of 11lb. of salt, about 1oz. of pepper, and 2oz. of saltpetre. With this the entire surface of the ham is rubbed. It is then placed in the salting-pan, and covered with the dry salt mixture, the outside of the ham being at the top. After about eight or ten hours it is bound with a thread, in order to compress it, and boiled in lightly-made brine, to which thyme, cloves, bay leaves, and basil are added. It is afterwards again soaked in brine, and, when it has remained from fifteen to twenty days, it is removed from the salting-vessel, and placed under a press for from ten to twelve hours to drain more perfectly. It is then suspended in a smoking-room until fit to eat. Sometimes, after the smoking, it is rubbed with wine lees and preserved in a dry place.

The famous hams of Hamburg, which are small but exceedingly good, are prepared as follows: The leg having been well washed in rain water, or, what is better, in brandy, is thoroughly salted with a mixture composed of 8oz. of salt, 2oz. of saltpetre, 1oz. of pepper, and ½oz. of powdered cloves. It is then put in a vessel lined with bay leaves and garlic, and covered with a clean cloth. At the end of twenty-four hours it is washed with cold water, and put for a fortnight into a tub of wine-dregs. It is then enveloped with thin paper, and hung in a chimney to smoke for a month or six weeks. It is next suspended from time to time in a small barrel, at the bottom of which is burned some juniper wood. After this second smoking, the ham is preserved in fine wood ashes until it is required for consumption or sale.

The Westphalian ham is larger and longer in shape, and always exceedingly well smoked. It is prepared in the winter, between November and March. In the first part of the process

of curing it is kept in salt to which a little saltpetre and bay
leaves are added, for five or six days. At the end of this time
it is entirely covered with very strong brine, and in three
weeks is removed, and plunged for twelve hours in rain water.
The ham is next carefully wiped with a clean cloth, and hung
in the smoking-room for a further three weeks, the wood used
in the smoking process being juniper.

With regard to the smoking of hams, Professor Nessler says
that their keeping qualities depend more upon a uniform and
proper system of drying than upon the amount of smoking
which the meat is made to undergo. Smoke of a high temperature, moisture, and the condensation of water are all injurious
when allowed to come into contact with the meat. With hot
dry smoke the surface is dried too rapidly, a crust is formed
which is filled with cracks, and, as there is a tendency on the
part of the fat to liquefy, the drying of the meat may be considerably retarded. Before smoking, and after removal from
the salt, it is a good plan to roll the meat in sawdust or bran,
or to cover it with either of these substances, as the crust
formed in the smoking process will not then be so thick. Moreover, if any moisture condenses upon the surface, it will remain
in the bran or sawdust, and the brown colouring matter of the
smoke will not penetrate to an undesirable distance. Warmth
of itself is not regarded as injurious to smoked meat, provided
moisture is not present and the air is not of too confined a
nature.

In some countries the smoked meat is kept in an airy place
in the shade of trees, this system being preferred to that of
standing it in cellars, which, although considerably cooler, are
at the same time much more damp. If a cellar is not dry, the
meat in it is certain to become mouldy, even though it is
covered with charcoal, ashes, sawdust, &c. A warm apartment,
too, is much to be preferred for the purpose of preservation
than one subjected to great variations of temperature, for in
the latter there is generally a quantity of moisture, which,
in condensing, must affect the quality of the meat; but probably the best place to keep smoked meat is in an apartment
which, while it keeps the meat dry, does not dry it entirely,
and for this reason the system of hanging it in a chimney is
very often practised.

In Yorkshire, the system of curing hams is not a very elaborate one. When the pig is cut up, the ham is well rubbed with saltpetre and common salt, after which it is packed in a tub and covered with a layer of salt. Here it remains for at least a fortnight, being generally examined once a day. At the end of this time a further quantity of salt is added. In another fortnight it is considered fit for removal, and, having been dried with a cloth, it is hung up in the kitchen for from six to eight weeks to dry. This is generally in the winter season. When removed from the kitchen ceiling, the face of the ham is usually washed with a mixture of lime and water, which prevents it from being attacked by flies. It is then hung in a cool and perfectly dry room, until required for consumption. Smoking is not practised in all parts of Yorkshire.

Where a number of hams are cured there is another system, by which each is rubbed well with a mixture of salt and saltpetre. They are then laid, with the shank end below, in the tub or pickling trough, so that each ham is on the slope. In three days they are taken out and again rubbed and put in as before, but in fresh positions. At the end of six days they are once more carefully rubbed and then packed flat and close together, the thick part of one ham being placed against the shank of the other, so that each is covered by the brine. In ten days they are usually ready for removal. Having been taken out of the tub and stood upright to drain, they are hung up for a week, after which they are taken to a warm drying-house. In Berkshire and some other parts of England, the wood used for smoking the hams is generally oak or ash.

The following is a plan of curing hams as well as bacon which has been followed with considerable success: When the meat is firm enough it is thoroughly rubbed with a mixture of 2lb. of salt, 2oz. of saltpetre, and 6oz. each of bay salt and sugar. A salting tub or pan is then partially filled with a quantity of brine in which half a gill of vinegar has been poured. The hams are packed in, being taken out and rubbed each day for at least four days. They remain in the tub for three weeks, after which they are removed, wiped dry, and hung in brown paper bags in the smoking-room, where the dust of ash or oak is used for the purpose.

Upon American farms that are distant from a town, it is

customary to erect smoke-houses, which are generally 7ft. or
8ft. in length by 6ft. in width, and 7ft. to 8ft. in height. They
are preferably fireproof, but are often constructed of wood, and
provided with a fireproof bin, which does not in any way interfere with the operation of smoking. This bin usually occupies
one-half of the floor, and is built of brick or stone. The fire is
built upon the other portion of the floor, and over it is suspended, at a height of 18in., a piece of sheet iron, 2ft. square,
which spreads the heat over a greater surface of the apartment,
and prevents the flames from reaching the joints hanging above.
Rails of wood are fixed across the room, being fastened to the
plates upon either side, and upon these cross-pieces are pegs,
upon which the meat is hung by means of pieces of string or
hooks. In the roof is a miniature chimney, which serves to
empty the house of smoke, when necessary, and to ventilate it.

M. Heuzé, in his work, gives an illustration of an ingenious
apartment erected for the purpose of smoking meat of all kinds
upon a large scale. In this case, the hearth or fireplace is
situated in a cellar, and the smoke ascending is allowed to enter
an apartment above through a trap, which is opened or shut at
will. As it reaches the apartment, it passes through a screen,
by which it is purified. In the chambers are two diaphragms,
which are merely movable boards, fixed together and resting
upon supports at the side, one diaphragm being fixed to one
side of the apartment, and the other to the other side, but
neither actually reaching across the chamber. The smoke is
thus driven by the current completely over the apartment,
finally reaching the chimney on the side opposite to the sieve,
where it is permitted to escape at will, a trap being fixed to
check it when it is found necessary. When the meat to be
smoked is suspended upon the hooks attached to the diaphragms,
the fire is lit, and kept going day and night, great care being
taken to select wood as aromatic in its smell as possible. The
woods preferred are generally juniper, rosemary, thyme, lentisk,
and lemon or citron; but, where these cannot be obtained, the
wood of the beech or birch tree is used. Humid woods produce
a thick smoke, and an odour resembling the taste or smell of
slightly burned animal or vegetable substances, which sometimes spoils the quality of the pork; but this effect is minimised
when the joints are covered with a cloth; indeed, it is believed

by persons experienced in the work, that a ham which is smoked when enveloped in this way is more agreeable in both taste and appearance.

"In France, it is customary, under ordinary circumstances," says M. Heuzé, "for large hams to remain in the smoking-room for from six to eight weeks, and it is found that moderate smoking continued for a length of time produces a superior result to an abundant smoking for a shorter period." The chief requirements in a smoke-house are, first, that it be absolutely dry; second, that it be not heated by the fire from which the smoke is made; and, third, that the meat be hung at such a distance from the fire that vapour thrown off is condensed before it reaches the meat.

Keeping Bacon and Hams.

Both bacon and hams as now prepared by professional curers are impregnated with a chemical preservative, chiefly boracic acid. In this way the keeping properties are improved. There is, however, a strong objection to the systematic employment of these materials, the constant consumption of which cannot fail to be detrimental to health. If pig-meat is thoroughly well prepared by salting and smoking, especially in the thicker parts of the joints, it may be kept for a considerable time when sewn in fresh calico bags or covers, and hung in a thoroughly dry, cool apartment. In the North of England hams are preserved in chests, and covered with oatmeal, and although this is an old-world practice, which in these days of quick consumption is a non-essential, it is quite certain that with age, well-cured meat kept in this way is among the best that is placed on the market. In preserving cured hams and bacon, the object in view should be to prevent blow-flies, &c., from depositing their eggs, and the attacks of even more objectionable pests, as well as to keep the meat where the access of heat or of a humid, and especially a warm humid, atmosphere is impossible.

One of the worst pests that can obtain a foothold in bacon and hams is a species of beetle often met with in houses, and known popularly as the Bacon Beetle, and scientifically as

Dermestes lardarius.　There is nothing repulsive about the beetle itself, which may be readily distinguished by a broad, yellowish-grey band across the wing-cases; but the grub is decidedly objectionable-looking (Fig. 71), resembling somewhat a very hairy caterpillar. These creatures by no means restrict their attention to bacon. On the contrary, scarcely anything comes amiss to them, from cork to the covering on jam-pots; they are, therefore, most unwelcome household pests. "Jumpers," too, affect bacon, and impart thereto anything but a pleasant flavour; they also are "undesirables." With a view to the exclusion of insect and other pests from bacon it is not an uncommon practice to sprinkle the exposed parts with pepper, and even with a special insecticide—one that is harmless to man, that is not likely to impart an objectionable flavour to the meat, and that at the same time is a preventative of such attacks as those referred to. Bellows are used for distributing the insect powder.

Fig. 71.—Grub of the Bacon Beetle (*Dermestes lardarius*). A most omnivorous Household Pest.

Minor Preparations of Pork.

In this country, although a large number of persons are occupied as pork butchers, yet we have no purveyors of food at all analogous to the *charcutier* of Paris, who prepares and utilises the produce of the pig in many different forms, laying his dishes in tempting morsels in his shop window as only a Parisian *charcutier* can.

It is quite unnecessary to describe the preparation of sausages and other dishes made from the pig which are already well known in this country, but the great reputation of the various French and German dishes is a sufficient inducement to us to introduce descriptions of a few of the principal preparations made in both France and Germany.

Sausages are made in France in different forms. The commonest are the round, which are similar to those of England, and the flat, which are called *crépinettes*. One system of manufacture is to mix together 2lb. of selected lean pig-meat with a small portion of the coarser meat and 1lb. of fresh pork without regard to the apportionment of lean. When the whole has been finely chopped and mixed together, and some salt, pepper, and spices have been added, it is introduced, by means of a funnel, into the skin, which is generally from the intestine of the sheep. When the skin is properly filled, the ends are closed with thread, and the whole is divided off into sausages of about the same length as those common in this country.

The flat sausage, or *crépinette*, does not keep so well as the other varieties, but is more delicious in its fresh state. It is made with the same mixture as that above described, but, instead of introducing the meat into a sausage-skin, it is enveloped in a piece of the caul of the pig. It is flat and oblong in shape, and is cooked upon either the stove or the gridiron. It was at one time forbidden to sell sausages in France between the first day of Lent and the 15th of December.

The French are not so famous for the large class of sausage as the Italians and the Germans, but there is one variety which they make extremely well. 4lb. of fresh lean pork and 2lb. of the fillet of beef are chopped and mixed together as intimately as possible. To this is afterwards added 2lb. of bacon, cut in small cubes the size of dice, and the whole is then seasoned with pepper, both whole and ground, with the occasional addition of a small quantity of garlic. The mixture is then introduced into a skin, which is usually taken from the calf or bullock. It is thoroughly pressed within this, in order that there may be no spaces left in the skin, and that the sausage may be thoroughly packed; indeed, great care is taken to prevent air from being imprisoned within. The two ends are next closed, and tied so that they shall form a rosette. The sausage is next put into brine, where it remains for from three to four days. It is then enveloped in a piece of oiled paper, and afterwards dried and smoked. It is subsequently covered with the lees of wine in which thyme, bay leaves, and other aromatic plants have been boiled. Finally, it is wrapped in a piece of thin, flattened metal or paper, and preserved in a suitable apartment. It is eaten hot or cold.

The famous sausages of Hamburg, which are known in all parts of Europe, and are made in large quantities, are manufactured from the lean flesh of the pig and the ox, chopped, and mixed together with fresh pork. The sausages of Bologna, which are almost as highly appreciated as those of Strasburg, are made with fresh pork, chopped and mixed with half the quantity of bacon, the whole being seasoned with salt, saltpetre, pepper, and herbs, and afterwards put in a well-closed bladder. The sausage is next steeped in brine for from ten to twelve days, after which it is hung in a smoking-house for four or five days. It is ready to eat after having been boiled for two or three hours.

In making German sausage, a quantity of the second quality of pork—that which is well interlarded with fat—is chopped up, but not so finely as in the previous instances. It is then seasoned with parsley, cinnamon, bay leaves, pepper—both in the grain and in powder—and spices. When the mixture is ready for filling, it is passed into skins prepared from the intestines of calves, and divided into sausages about 4in. in length. These are then exposed to smoke for a week, and, before being eaten, are boiled in water for half an hour. In some instances pea meal, the meal of French beans, and lentils are used for mixing with the sausage meat. Cochineal is also added by some makers, in order to produce the tint which is so well known. In the South of France, sausages of this kind are made with the addition of garlic.

The intestines or *chitterlings* of the pig are prepared in France as follows: Having been thoroughly well cleaned, they are pickled for from six to twelve hours in a brine flavoured with thyme, coriander, and bay leaves. They are then taken out, and the pieces of meat adhering to them are removed, cut up fine, and, with the smaller guts themselves, which are also cut into strips, are introduced into a larger gut, which when filled is tied at both ends. The whole is then cooked, great care being taken to prevent the skin from bursting. They are again placed in brine for three weeks, after which they are taken out to drain, and then dried with a clean cloth. They are next suspended in a smoking-chimney, where they remain for a fortnight. This preparation is eaten after being boiled, or first boiled and then fried; but the usual course is to boil it previously in water to which parsley, thyme, and sage have been added.

It may be here remarked that a good deal of trouble is involved in cleaning and preparing chitterlings for table. Still, when properly cleansed, they are appreciated in many British households, and even where they are not required for home consumption they have a real market value. The following in brief is the way to prepare chitterlings for either table or market; the great things being to prepare them as soon as possible after the pig has been killed, and on no account to break the "gut." Cut them into convenient lengths for manipulation, and clean out the food, taking care not to get any on the outside. Having done this, pour plenty of clean, cold water through them. Next take a stick with rounded ends and some 1½ft. long, and gradually, by the aid of finger and thumb, work the chitterling to the end, turning it (the outside thus becoming the inside). Again well wash in plenty of cold water, and allow the chitterlings to soak in salt-and-water for a day, or even for three days. At the end of that time, well rub them through the hands, plait them for convenience in cooking, and boil or fry. The "nightcap" portion, having been slit open and the contents removed, is best cleaned like tripe, either in lime-water or in salt-and-water.

The French are famous for the variety of *pig puddings* which they make, and which are composed of pork, generally mixed with some other kind of flesh. They are much more delicious than the black puddings which are common in this country.

The ordinary *black pudding* is made as follows: Twelve large onions, of medium size, are finely chopped, and cooked upon a slow fire, in some melted lard. Three pounds of the finest internal fat is then cut up in small squares, and seasoned with salt, pepper, parsley, and some finely-chopped onions. The two quantities are next mixed together, and afterwards added to about four quarts of blood, which has been maintained in a liquid state by means of heat. A clean sausage-skin is then taken, and fastened up at one end with a piece of thread. The maker blows in at the other end to distend it, in order to ascertain if there are any holes. A small funnel is next inserted, and the whole mixture passed into the skin, in a warm state. This must be very carefully done, in order that the more solid portions may not cluster together in lumps, and so spoil the pudding, and also so that the skin may not burst. When filled,

the open end is closed up, and tied with thread. The pudding
is next put into a saucepan of hot water, and carefully cooked
for from fifteen to twenty minutes. Great care is taken to
prevent the water from boiling, as the skin would then immediately burst. To ascertain whether the pudding is sufficiently
cooked, it is lifted out of the water by means of a long thread
left on the end of it in tying, and pierced with a pin. If the
blood is coagulated, the pudding is ready; if it runs through the
puncture made, the pudding must be further cooked. The
properly-cooked pudding is also firm when pressed by the
fingers. When taken out of the saucepan, it should be wiped
with a clean cloth, to remove the red scum which envelops it,
and afterwards rubbed with a small piece of fat to improve the
appearance. It may be eaten either cold or grilled upon a
lively fire, according to taste.

The *white pudding*, according to M. Heuzé, to whom we are
indebted for these particulars, was formerly made from veal,
capon, and pig fat, finely chopped, and mixed with milk and
spices. Now, however, it is made by mixing finely-chopped
portions of the fowl, pieces of bruised or pounded pork, and
some pieces of the choicest fat with milk, breadcrumbs, chopped
onions, butter, the yolks of eggs, cream, salt, chopped sweet
almonds, and spice. The spices commonly used, and which are
generally included under the term "spice," are ordinary pepper,
nutmegs, cloves, and cinnamon, all bruised, sifted, and mixed
together. The leaves and stalks of thyme, marjoram, rosemary,
and bay are also used after having been beaten in a mortar.

A dish very popular in France is one which is known as
fromage de cochon (pig cheese), and which somewhat resembles
brawn, as made in England. It is made as follows: After
having thoroughly cleaned the head, which has remained in
cold water for two hours, the bones are carefully removed, and
the whole of the flesh is pared from both skin and bones; this,
with the ears, and sometimes the tongue, being cut into small
pieces. The whole is then seasoned with finely-cut parsley,
salt, pepper, and spices. The skin of the head is then spread
open, the holes of the ears having been previously sewn up,
and the seasoned meat is placed upon it in alternate layers of
fat and lean, together with the edible portions of the ears and
some pieces of the fat of the intestines. The extremities of

the skin are next united and sewn up as nearly as possible into
a round form. Sometimes it is the custom to sew up the mixture in a cloth. The "cheese" is cooked in a boiler of water
to which a bottle of white wine, some parsley, cloves, and bay
leaves have been added. At the end of some hours it is taken
out and pressed (while still hot), in order to extract as much
of the water which has permeated it as possible. It is then
put into an iron mould or saucepan, together with some slices
of smoked bacon, and the whole is placed in an oven, where it
remains for from thirty to forty minutes for further cooking.
The "cheese" is not taken from the mould until it is cooled, but
its removal is assisted by plunging it for some minutes in boiling water. When served for table, it is garnished with slices
of lard, jelly, or raspings from the crust of bread.

Another favourite dish, known as *fromage d'Italie* (Italian
cheese), is reputed to have been a favourite with Louis XI. ;
but, although it is extremely delicious, it is somewhat indigestible. It is prepared as follows : About 5lb. of pig's liver is
finely chopped, together with about 1lb. of pork and ½lb. of
the finest fat from the interior of the pig. When well mixed,
the whole is seasoned with pepper, salt, shallots, thyme, and
nutmeg. An earthen or iron vessel is then lined with some
thin slices of bacon or the caul of the pig, and the "cheese"
placed within, and covered with some bacon or caul, as the case
may be. It is next cooked in an oven after a batch of bread
has been taken out. At the end of from two to three hours it
is finished, and, when cool, is removed by plunging the mould
or vessel for a few moments in boiling water. *Fromage d'Italie*
is eaten cold and garnished with jelly.

A delicious dish, known in Paris as *rillettes*, is made as
follows : 4lb. of lean meat is added to 6lb. of bacon or caul,
the whole being chopped fine and seasoned with salt, spices,
and bay leaves. The mixture is then cooked in a saucepan, care
being taken to stir it until it is finished, to prevent pieces
attaching to the bottom of the vessel. When the fat is
separated from the lean, the latter is finely chopped and put
into earthenware vessels, and the liquid poured upon the top
of it. It is then allowed to stand until cold, when it is ready
for the table. A *rillette*, well made, has a most agreeable
"nutty" flavour.

Y

Consumption of Pork in France.

It has been stated that pork is the principal food consumed by the country people of France. Whether this is absolutely the case or not, it is quite certain that the flesh of the pig is very extensively used, and that it is utilised in the formation of a very large number of most delicate dishes, which are highly appreciated by the poorer classes. There are in Paris alone between 7C0 and 800 *charcutiers*, or shopkeepers who sell pork in various forms. Pigs are very largely killed at Christmas time by those who keep them, chiefly for the purpose of providing hams and other preserved portions of the carcase for consumption at Easter. At that season a fair is held in Paris, which lasts three days, and at which pig-meat is generally sold to the extent of 500 tons.

The Pork Trade in America.

According to the report of the Department of Agriculture of the United States, the number of swine upon American farms in 1904 was 47,000,000, or some 3,000,000 more than was the case twenty years ago. Figures, however, have varied immensely, for they rose to 52,000,000 in 1902, and to 56,000,000 in 1901, and fell to 37,000,000 in 1900. The farm value of the swine in 1904 was estimated at 615 million dollars—a great increase. The greatest number of swine was in Iowa, where there were 7,000,000, valued at 47,000,000 dollars; next came Illinois and Missouri, followed by Nebraska, Indiana, Ohio, and Texas. The highest value of the swine of any State was recorded in the case of Rhode Island, where they reached 13 dollars, and in Connecticut and New Jersey, where they reached 11 to 12 dollars per head. The average price of live swine in Chicago in 1903 was 5.9 dollars per 100lb., against 6.78 in the preceding year; but the average varied from 4.4 dollars in December to 7.18 dollars in March. The swine are conveyed to Chicago and New York per rail at 30 cents per 100lb., while dressed pig-meat is charged at 45 cents per 100lb. in refrigerating cars. From Cincinnati to New York the average rate is 26 cents. The value of the pig-meat exported from the United States in the year ending June, 1903, reached $61\frac{1}{4}$

million dollars. Of these, 2 million dollars accounted for fresh pork, and 1⅓ million dollars for canned meat.

The following table will show the quantity of bacon, hams, canned pork, fresh and salted pork, and lard exported to the various countries of the world from the United States in the year 1903:

	BACON.	HAMS.	FRESH PORK.	LARD.
	lb.	lb.	lb.	lb.
United Kingdom	163,894,472	177,745,090	71,432,291	198,456,816
Belgium	11,340.581	4,818.973	4,961,973	28,525,453
France	930,215	96,365	133,437	7,349,167
Germany ..	15.260,392	892,543	3,130,886	174.877,447
Netherlands ..	2,337,661	1.154,151	2,472,272	53.603,949
Other Europe ..	7,777,669	2,444,581	11,791,314	21,486,947
British North America ..	5,300,412	8,683,054	12,132,967	1,724,912
Central American States	540,246	788,067	1,462,541	4,192,652
West Indies and Bermuda ..	3,716 392	6,796,930	11.802,874	23,889,324
South America..	1,838.206	830,694	3,723,799	17,345,373
Asia and Oceania	459,426	201,892	788.193	1,110,243
Other countries .	124,245	1,042,609	4,150,331	2,813,474
	213,519,917	205,494,949	127,982,878	535,375,757

Lard.

The importance of lard as an article of commerce can only be estimated by the figures showing the enormous imports into this country. The trade with America is simply astonishing in its vastness, and, wherever the pig is dealt with as an article of commerce, the lard which it furnishes is sufficient to create a great trade in itself. One reason why so much attention is paid to the preparation of lard is that it is preserved without the use of salt or any other antiseptic whatever. The cells, or globules of fat of which it consists, unlike those of butter, are exceedingly minute, and, when the membranes of which they are composed are destroyed while the fat is in its sweet state, its preservation is rendered perfect, inasmuch as it is these membranes that contribute, in other instances, to the process of putrefaction.

Formerly, lard was generally melted in an iron vessel, but steam is now used with greater advantage, the flavour of the fat being thus less liable to be affected. At the same time, it is found that it does not keep so sweet, owing, perhaps, to the fact that the heat of the steam is not so intense as is necessary. When lard is melted in large quantities, it is afterwards clarified and then cooled.

There are three qualities of lard : the first is sold for domestic use ; the second is utilised in the manufacture of oil ; and the third, generally prepared from inferior meat trimmings and tainted fat, is converted into one of the constituents of soap. In large factories there is a considerable quantity of waste or refuse left after the melted lard is drawn off. This, composed of bone, skin, and offal, together with the blood of the pig, is converted into manure.

In many parts of the world lard is largely used in the place of butter, whereas lard oil is purchased for mixing with, and consequently adulterating, olive, spermaceti, and other more valuable oils. In the manufacture of lard oil, quantities of the fat are placed in woollen bags, and subjected to high pressure for from twelve to fifteen hours. By this means the oil is thoroughly expressed. In making glycerine, another product, the lard is heated to 616deg. to 620deg. F. In this way the glycerine is divided from the other fats.

Lard is prepared in France as follows : After having cleaned the fatty parts of the pig which are intended for conversion into lard, and removed the offal and other useless parts, the fat is cut up into cubes about the size of a Barcelona nut. The whole is placed in a copper vessel, together with one quart of water to every 23lb. of fat. The copper is next placed upon a slow fire. When the fat is melted, the vessel is removed from the fire, and the lard which has risen to the top is left to cool slightly. Then, with the aid of a spoon, it is passed through a fine sieve or a straining cloth of extremely fire grain into earthenware pots, generally varnished on the interior.

The lard is usually considered to be good when the small

LARGE BLACK GILTS.

Mr. J. Robinson's Starkie Lady II. (Herd Book No. 4654) and Starkie Lady III. (4656), sister gilts bred by Mr. John Warne.

The head of the Large Black should be of medium length, with long and thin ears as shown; the back long and level; the sides very deep; the ribs well sprung; the belly and flank thick and well filled; the hams large and well filled to the hocks; the legs short and straight; and the tail set high.

scraps of offal which remain behind are slightly yellow. It is then generally clear, bright, and transparent. Lard of the first quality, which has been well made, should be a brilliant white in appearance. In some cases, those who prepare lard add 1oz. of fine salt to every 2lb. of the fat, in order to make the preservation more perfect. Lard which has been left exposed to the air frequently becomes rancid and yellow in colour, and care must, therefore, be taken in preserving it. It is believed to contain about 38 per cent. of stearine and 62 per cent. of oleine, and the French authorities have frequently discovered that the oils, more especially olive and spermaceti, to which we have referred above, are adulterated by one of these constituents.

The Chemistry of the Pig in Relation to Commerce.

One of the most careful and valuable investigations that have been made in connection with the pig was undertaken by Professor M'Murtrie, the chemist to the Illinois State Industrial University, and upon this he prepared an elaborate paper, which, unfortunately, has not been published in its entirety. The Professor's intention was to ascertain, as far as practicable, what were the actual component parts of the pigs of America as compared with those of England and other pork-producing countries, and the results which he has obtained will be found valuable not only to the breeder and dealer, but also to the general public, which consumes such an enormous quantity of American pork. Professor M'Murtrie had unusual opportunities of obtaining accurate information, as there is no field in connection with the pork trade which is so vast, and which is conducted with such accuracy, as the pork-packing trade of Chicago.

The first set of figures was obtained from a large packer of that city, who gives the following data as to the relative proportions of the pigs passing through his hands:

Per cent.

Offal, heart, lungs, livers, bones, entrails, except the small and anal guts, are made into dried fertiliser, containing about 7 per cent. ammonia, worth £4 7s. 6d. per ton 02.25

	Per cent.
Brought forward	02.25
Blood ditto, containing 15 per cent. ammonia, and worth £9 7s. 6d. per ton	00.75
Small guts, used for sausage casings, and worth 1s. 3d. per pound	00.12
Other guts ditto, worth 10d. per pound	00.10
Fat (lard, gut fat, and grease)	15.50
Flesh, (sides, hams, and shoulders)	57.00
Lean meat, trimmings, tender loins, hocks, &c., all taken from the sides, hams, and shoulders in cutting them into the shape necessary to offer for sale on the market	0?.50
Tongues, worth 3d. per pound	00.85
Total percentage	80.07

No data are given as to the percentage of weight of hair and bristles. These are sold by packers, at so much per hog, to hair manufacturers. The prices vary because the winter hogs yield more and stronger bristles, and longer hair, than is the case with summer hogs.

Another large packing firm stated that the yield of the several parts of the hog — the bones, flesh, &c. — depends not only on the weight, but upon many other things, such as the breed, feed, health, age, &c. It does not appear that winter or summer feeding makes any difference in the yield. It is generally true that hogs yield more in winter than in summer, but this seems to be due, not to the temperature, but to the feed—to the quantity and quality given. The distance from which hogs are brought to be slaughtered also affects the yield, other conditions being the same—the greater the distance the greater the percentage yield. For the American market, hog sides are cut in Chicago for the most part into what are called short ribs, or into mess pork.

From 4811 hogs cut, having an average weight of 215lb., there were made 6007lb. of dried blood, containing 16 per cent. of moisture, and 42,227lb. of pressed "tankage," containing 42 per cent. of moisture. When the shoulder and ham are cut off and trimmed, and a butcher's loin is made from the side, the loin will make about 9 per cent. of the weight of the hog. If the rest of the lard and fat trimmings were rendered, there would be a yield of lard of about 40 per cent.

of the live weight. These figures are taken from a test made upon 100 hogs, averaging 340lb. The word "tankage" refers to liquid and solid manure and waste.

In making an experiment for the purposes of his own investigation of this subject, Professor M'Murtrie selected a Poland-China sow, weighing 340lb. at the age of eleven months, and a Berkshire sow, weighing 245lb. at the age of nine months. The former was fasted for some hours before she was slaughtered, whereas the latter had been recently fed. After slaughtering, the various portions of the body were collected together as completely as possible, each part being weighed, and the weight recorded. The results were as follow:

| | POLAND-CHINA. | | BERKSHIRE. | |
CONSTITUENT PARTS.	Actual Weight.	Per Cent. of Live Weight.	Actual Weight.	Per Cent. of Live Weight.
	lb.		lb.	
Blood	11.50	3.38	6.00	2.44
Hair	2.25	0.66	1.50	0.61
Entrails and contents	17.25	5.07	15.00	6.11
Lungs, kidneys, spleen, and brains	5.75	1.07	4.75	1.95
Flesh, without fat or bones	101.00	29.70	80.00	32.61
Heart, liver, and tongue	6.50	1.91	6.00	2.44
Bones, crude	21.50	6.30	17.50	6.73
Side fat	104.50	30.73	70.00	28.57
Kidney fat	12.00	3.57	8.00	3.26
Fat on entrails ..	5.00	1.47	3.50	1.43
Skin	17.00	5.06	12.00	4.88
Loss	35.75	10.29	20.75	8.87
Total	340.00	99.21	245.00	99.90

In the Poland-China the percentage of fat was higher, and that of the flesh lower, than in the Berkshire in nearly all cases; while it was shown that, although the weight of the blood of the former exceeded that of the latter, the internal organs of the Berkshire, which, it should be remembered, had been recently fed, were nearly 3 per cent. heavier than those of the Poland-China. The unaccounted-for loss in the one animal was 10.29 per cent., and in the other, 8.87. The subjoined table shows the results of analyses:

PARTS.	MOISTURE.	FAT.	ASH.	PROTEIN.
Poland gut fat	9.630	85.523	1.910	2.937
„ side fat	5.000	92.331	0.001	2.668
„ kidney or leaf fat	4.118	94.566	0.066	1.250
„ flesh	60.530	13.505	0.800	25.165
„ bone	38.655	20.173	24.808	16 364
„ skin	53.320	3.382	0.344	42.954
Berkshire gut fat ..	19.350	78.608	0.002	2.040
„ side fat ..	8.130	90.847	0.042	0.981
„ kidney fat ..	1.730	96.426	0.044	2.800
„ flesh	67.300	15.084	0.779	16.837
„ bone	40.994	20.873	27.136	10.997
„ skin	49.380	4.265	0.640	45.715

If the reductions of percentages are made from the above tables, we get the following figures, showing the moisture, fat, ash, and protein in the dressed carcase of each animal:

POLAND-CHINA SOW.

PARTS.	PER CENT. OF CARCASE.	WATER.	FAT.	ASH.	PROTEIN.
Flesh	39.452	23.985	5.320	0.3156	9.6843
Bones	8.398	3.151	1.778	2.0832	1.4390
Side Fat	40.820	2.041	37.689	0.0061	1.0890
Kidney Fat	4.690	0.193	4.578	0.0030	0.0058
Skin	6.640	3.540	0.248	0.0228	2.8282
Total	100.000	32.910	49.613	2.4307	15.0463

BERKSHIRE SOW.

PARTS.	PER CENT. OF CARCASE.	WATER.	FAT.	ASH.	PROTEIN.
Flesh	42.666	28.714	6.639	0.3323	6.5155
Bones	9.333	3.882	1.949	2.5226	1.026
Side Fat	37.333	3.035	33.915	0.0159	0.367
Kidney Fat	4.266	0.074	4.070	0.0018	0.119
Skin	6.400	3.160	0.296	0·0409	3.325
Total	99.998	38.865	46.869	2.9135	11.3525

These figures afford the means of comparing these two animals, not only with each other as regards their composition, but, through the intervention of published analyses, with foreign animals as well. In the Berkshire, which had the lowest live weight, and was two months younger than the Poland-China, we find a higher percentage of flesh, and a generally lower proportion of side and other fat. We find also that its proportion of water is higher, and the total of pure fat lower, as perhaps might very naturally be expected. But it is interesting to note that in this animal, in which the proportion of flesh ranged higher than in the other, the proportion of protein substances is very considerably lower. This is an important point in the determination of the nutritive value of the product. If the standard of comparison be based upon the relative percentage of protein, it will appear that the Poland hog is of decidedly greater value for food supply, while the same remark will hold good if we base our estimate upon the proportion of fat. What changes the Berkshire hog might undergo in an additional two months of growth, feeding, and fattening, we have no means of knowing, and it would be manifestly unfair to base a definite comparison of the breeds upon these figures. Further analyses are necessary to confirm or to deny the relations here set forth. But it would seem doubtful if this animal could have gained the additional 100lb. in live weight during the period named, or change the proportion of protein from 12 to 15 per cent. Yet it is possible.

In order completely to understand the value of the figures arrived at with regard to American pigs, Professor M'Murtrie compared them with those of a similar character obtained by the analyses of animals in England and Germany, which were furnished by Drs. Lawes and Gilbert, and by Dr. Wolff respectively, and are as follow:

PARTS.	ENGLISH.		GERMAN.		AMERICAN.	
	Store Pig.	Fat Pig.	Well Fed.	Fat.	Poland.	Berkshire.
Water	58.1	43.0	57.9	43.9	32.8	38.8
Fat	24.6	43.9	24.2	42.3	49.5	46.3
Ash	2.8	1.7	2.9	1.9	2.4	2.9
Protein ..	14.5	11.4	15.0	11.9	15.3	12.0

It is generally believed that American pork suffers more from shrinkage in cooking than British. Bradley states that in cooking American pork loses 50 per cent., while Irish loses only 25 to 30 per cent. This greater shrinkage is to be ascribed to the higher proportion of fat it contains, as appears from the above figures. The bacon, however, is considered of inferior value on this account.

Davy gives the following as the composition of fat pork, and of green and dried bacon:

CONSTITUENTS.	FAT PORK.	DRIED BACON.	GREEN BACON.
Protein	9.8	8 8	7.1
Fat	48.9	73.3	66.8
Ash	2.3	2.9	2.1
Water	39.0	15.0	24.0

The difference is doubtless due to the system of feeding, or rather to the character of the food supplied. In America, maize is the staple, and, for the pork of commerce, almost the exclusive food; whereas in England, barley, peas, beans, rye, and other materials of a like character constitute the rations of the growing and the fattening hog. These foods contain higher proportions of protein than maize, and it is believed by the best authorities that an increased proportion of this substance in the animals fed on these materials may be due to this fact. Whether this be true or not, so high an authority as the late Professor Sanson, of the Agricultural School of Grignon, in France, believes that the addition of such nitrogenous foods as peas and beans to the rations of hogs greatly improves the quality of the pork. There can scarcely be a doubt that, if such material were added to the maize rations of American pigs, the pork would be enhanced in value.

The following table has been compiled for the use of students at the Academy of Sciences, in Paris, showing the nutritive value of domestic animals produced for food:

Food.	Matter Soluble in Ether.	Solid Matter.	Water.	Matter Insoluble in Ether.	Nutritive Order.
B ef	25.437	277.0	723.0	249.563	1
Fowl	14.070	263.5	763.5	248.930	2
Pork	59.743	294.5	705.5	242.577	3
Mutton ..	29.643	265.5	734.5	233.857	4
Veal	28.743	260.9	740.0	226.757	5

Weights and Measurements.

The following measurements of live pigs indicate the dead weight of pork that they will furnish, and a safe index to the profit of pig-feeding is the price of barley meal. When 1cwt. of the latter costs less than a score lb. of pork, and store pigs for fattening are at a moderate price, feeders may calculate on being remunerated for their outlay. 4ft. 1in. girth of a fat pig represents 10 score; 4ft. 4in., 12 score; 4ft. 7in., 14 score; 4ft. 11in., 16 score; 5ft. 2in., 18 score; 5ft. 7in., 20 score.

As regards corn, barley weighs 45lb. to 56lb. per bushel; oats 36lb. to 46lb.; maize and wheat 60lb. to 63lb.

CHAPTER XVI.

HERD BOOKS, RECORDS, AND SHOW CLASSIFICATION.

Herd Books.

THE first volume of the Herd Book of the National Pig Breeders' Association was published in 1885. There is more than ordinary necessity for referring to the question of herd books and records, inasmuch as, next to dairy cattle, the pig of Great Britain requires them most. It is argued by no person more strongly than by the British farmer that a carefully-bred animal, which can trace its descent back to a long line of meritorious ancestors, is of much greater value than any other; and it is therefore evident that, if a record of pedigree is necessary, it is equally essential that this should not be kept or handed down by word of mouth or tradition, but be carefully recorded in writing or print, so that it may be examined by those who are interested in pig breeding, and who can, if they choose, made additions on their own behalf.

The maintenance of a herd book depends chiefly upon the interest taken in the work by the members of the particular Association that has taken it in hand, and there is little doubt that the Herd Book of British pigs will last as long, and prove as successful, as those provided for stock of a more valuable nature. It is rather to the private breeder that our argument should apply, for in many cases he is either unwilling or unable to keep a record, which, simple as it is, he regards as a process similar to that of keeping a set of accounts. He should remember, however, that the

buyers of to-day—and they are increasing in number, as well
as in their demands—are seldom satisfied, in purchasing well-
bred stock, unless they are provided with some documentary
evidence as to their pedigree; and the man who is unable to
supply it from his own record certainly loses as much in a
single transaction as would pay for the trouble of keeping a
record for a whole year. It may fairly be argued that, if he
be too careless to keep a faithful account of his breeding
operations, he will be equally careless in his system of
breeding; and there are plenty of men who, failing to receive
satisfaction in their demand for recorded evidence, would
place little faith in what they were told, and, indeed, would
repudiate the transaction altogether. The successful breeder
is certain to meet with still greater success if he takes the
trouble to maintain a careful record of his own, and to enter
his breeding stock in the National Herd Book. There are
persons who claim to have improved, and in fact to have
made, certain families of stock, but who have no record of
any kind to support their statements. It is needless to
say that, whatever their efforts may have been, their reputation has not been enhanced by the work they claim to have
performed.

It is true there are men deficient in education, but who have
great capacity for stock-breeding, and who can carry in their
memories the names and pedigrees of all the principal animals
in their herd; but nothing should be trusted to mere chance
or recollection, which is almost certain to be at fault. A
breeder with an easy conscience may often be tempted to
claim blood for an animal which he wants to sell, but which
it does not possess; and in this remark we only suggest what
is believed to have frequently occurred. Knowing this fact,
the buyer very properly insists upon a record which is almost
certain to be trustworthy, and which is a very considerable
guarantee that the stock he is purchasing is as good as it
is stated to be. This is not all, however, for it is of the
greatest importance to the breeder himself, and is the only
means of enabling him to maintain the highest possible
degree of excellence in his herd.

The National Pig Breeders' Association issues separate forms
for boars and sows, for entry in its Herd Book. The

conditions of entry are : That the sire and dam are registered,
or that the pig proposed to be entered, if a boar, has five
crosses of registered blood, viz. : that the sire, sire of dam, sire
of granddam, sire of g. granddam, and sire of g.g. granddam,
are registered boars; and if a sow, has four crosses, *viz.* :
that the sire, sire of dam, sire of granddam, and sire of
g. granddam are registered boars. Also that the produce of
sows be accepted (which produce must be by a registered sire),
if the said sows have produce entered in any of the previous
volumes. The name, age, and breeder of the dam, granddam,
g. granddam, and g.g. granddam, when not registered, must
be given.

The fees are :—Members 2/6 each single entry. When more
than one boar of the same litter is entered by the owner, the
fee is 2/6 for the first entry, and 1/- for each other entry of the
litter. Non-members are charged double fees. The last day
of entry is January 31st, and the animals must have been
farrowed previous to the preceding September 1st.

There are one or two rules in connection with the
"American Berkshire Record" (for particulars of these we
are indebted to Mr. Phil. Springer, the secretary of the
American Berkshire Association), to which we will refer,
owing to their importance in this country. Rule 8 states
that "The breeder of an animal is the party owning the
dam at the time of service and dictating the cross." Rule 9
says, "In view of the fact that the value of a pedigree
depends largely on the character and standing of the party
or parties who bred and reared the animals represented therein,
it is a matter of no small importance that the name of the
breeder of every animal mentioned in each pedigree be given.
This, if known, should never be omitted." In the event of
a change taking place in the ownership of a recorded animal,
it is, of course, necessary that a Transfer Form should be
signed, and sent to the secretary for placing upon his file.

Private Record or Herd Book.

The following example is intended as a specimen page of
a private Herd Book, and will suggest how most minute
particulars may be kept relative to any herd of pigs bred

with care. It is quite possible that individual breeders may desire to add other columns, but it is impossible to arrange a herd book to suit everyone. Some breeders require a plain and simple record, while others, and particularly those with more time upon their hands, insist upon recording the most minute details. The date of service should not be entered until it is ascertained that the sow does not require a second service; it should, however, as in the case of other stock memoranda, be previously entered in a rough note book.

LADY DOLLY (343)—*Sire:* KING KOFFEE (221); *Dam:* SNOWFLAKE (84) Farrowed March 16th, 1904. Bred by Mr. JAMES THOMSON St. Oswald's.

Date of Service.	Date of Farrowing.	Sire.	No. in Litter.	Boars.	Gilts.	Dead (if any).	Remarks.	Result.
1905. Feb. 1.	May 24.	Brown's Jumbo (236).	11	6	4	1	Boars cut, June 29; weaned, July 25; fed on milk and barley-meal, and turned on the pastures.	8 sold to J. Smith, Aug. 3, at 18s. 6d. each, 2 gilts left for stock.

In the column for "Remarks," particulars of the growth of the pigs should be stated, together with any items as to their feeding, as this knowledge may be of considerable use in the future. It may also be here shown when they were cut. In the last column, dates of sale should be noted, together with the prices made. Pigs which are intended for stock should be entered accordingly, each animal being provided with a similar page in the book; but it would be well not to enter gilts until they have been served. Upon an ordinary farm, where pigs are bred without method, a record of this kind is scarcely necessary; but every person who desires to maintain a standard of excellence in his herd will find this or any similar system of the greatest possible assistance.

Service of Sows.

The model of a Service Ticket given below is one which, we think, will commend itself both to those breeders who keep stock of high quality and to those who find it necessary to employ other stock for the improvement of their own. A person selling young pigs which have been got by a celebrated boar should naturally have some tangible evidence of their paternity, and a certificate of this nature will not only satisfy a buyer, but also induce him to give a higher price than he would possibly otherwise do. The ticket should be accepted as a receipt for the service fee, and should never be supplied by the owner of a boar unless the fee is paid. It is not necessary that the amount of the fee should be stated upon the face of the certificate; but at the same time, for the information of the grantor, it should be inserted upon the counterfoil, together with the name of the herdsman who brought the sow. In case an animal is brought back for a second service, the owner of the boar, or his agent, should endorse the date, add his signature across the original ticket, and make a memorandum of the same upon his counterfoil.

A book of certificates and counterfoils, similar to the example given, would be supplied by any printer for a small amount—indeed, neither the trouble nor the expense, both of which are trivial, should be allowed to stand in the way of its use.

[COUNTERFOIL.]

No. 1.

June 24, 1905.
MR. BROWN'S
Sow Beauty (226)
to
Samson.

Herdsman,
JONES.

5s.

SERVICE TICKET.

No. 1.   THURLBY GRANGE.

This is to certify that MR. W. BROWN'S SOW (or GILT) (*   ) was this day Served by my MIDDLE WHITE Boar SAMSON (354).

JOHN SMITH,
June 24th, 1905.

P.S.—If Second Service is required, this Certificate must be brought with the Sow, for Endorsement.
* Here state the name and description in the Herd Book, if entered.

Classification at Shows.

It is questionable whether agricultural stock of any description has been subjected to such extraordinary classification as the pig. During recent years the principal classes provided have been those at the Royal Agricultural meeting, where the animals exhibited are expected to be suitable for breeding, although they have invariably been much too fat; at Smithfield, which is essentially a fat stock exhibition; and at Birmingham, where classes are provided for both fat and breeding stock. The Smithfield classes are now confined to two for each of the following breeds, viz., the Small White or Black, the Middle White, the Large White, the Tamworth, the Large Black, the Berkshire, and any other distinct breed or cross-breed. These classes have been for animals not exceeding nine months old and for those above nine and not exceeding twelve months, respectively. There are, however, also classes for single fat pigs of the White, the Black, the Berkshire, and the Tamworth breeds not exceeding one year.

Previous to the year 1883, the Smithfield Club omitted the Middle breed, but provided three classes for each of the other varieties, and, instead of two animals being shown in each class, as is the custom at present, exhibitors were required to enter three for the classes under nine months, two for the classes from nine to twelve months, and two for the class from twelve to eighteen months. At the Birmingham Show, in the fat pig classes, which are much more limited in extent than at Smithfield, the whole of the breeds are classified in two sections, the one for animals under nine months, and the other for pigs exceeding nine and not exceeding twelve months, with one exception—the Middle and Small Whites being placed together. There has, however, on some occasions been an extra class for the heaviest three fat pigs of any breed or age.

At the Exhibition of the Royal Agricultural Society, the classes are for the Large, Middle, and Small White breeds, the Large Black, the Berkshire, and the Tamworth, four classes being arranged for each variety (except the Small White, for which there are only two, and in which the entries in 1905 numbered only four and were all made by one exhibitor), viz., for old boars, for lots of three boars farrowed in the year, for old sows, and for pens of three sows of the same litter farrowed in the year of the exhibition. In the Large Black, however,

z

the second class is for single boars farrowed in the year of the
show. This classification is extremely liberal, and has resulted
in most successful displays of pigs.

We think there can be no better classification than the
following:

FOR FAT PIGS.

	Three Pigs under 8 months.	Three Pigs under 12 months.	Two Pigs under 18 months.
Large White breed			
Middle White breed	,,	,,	,,
Berkshire breed	,,	,,	,,
Large Black breed	,,	,,	,,
Tamworth breed	,,	,,	,,
Any other large breed, or cross between large breeds ..	,,	,,	,,
Any other small breed, or cross between small breeds ..	,,	,,	,,

FOR BREEDING PIGS.

BOARS.

	One not exceeding 18 months.	One not exceeding 6 months.
Large White Breed ..		
Middle White breed..	,,	,,
Berkshire breed ..	,,	,,
Large Black breed ..	,,	,,
Tamworth breed ..	,,	,,

SOWS.

	One of any Age over 12 months.	Three of the same Litter not exceeding 7 months.
Large White breed ..		
Middle White breed ..	,,	,,
Berkshire breed ..	,,	,,
Large Black breed.. ..	,,	,,
Tamworth breed	,,	,,

Since we pointed out the importance of carcase classes
such as those which were promoted at Chicago when this work
first appeared, and which we deal with below, the Smithfield
Club has instituted—if tardily—a carcase competition. There
are three clases, viz., for pigs up to 100lb., from 100lb.
to 220lb., and from 220lb. up to 300lb. live weight, and we
have noticed that in some cases the thickness of fat has

reached 3in. At the Fat Stock Show, at Chicago, in 1884, the classes referred to were described as follows:

DRESSED CARCASE.

Best carcase of barrow (*i.e.*, a cut pig), 12 and under 18 months.
Best carcase of barrow und r 1 year old.

GAIN PER DAY.

Barrow showing greatest average gain per day, including weight at birth. Entries to be accompanied by affidavit giving exact age.

COST OF PRODUCTION.

Barrow 12 and under 18 months, produced at the least cost per pound, live weight. Entries to be verified by affidavit.

We cannot protest too strongly against the custom which has so long prevailed, and which is still common, of employing as judges of porcine stock persons who have little or no practical knowledge of the subject. There may, in some instances, be financial reasons why judges of cattle or sheep, or even poultry, have to deal with the animals in the pig classes; but it is the duty of a committee, if it provides prizes for pigs, to do every justice to the exhibitors whom it invites to compete, and that justice cannot be rendered unless a judge is appointed who is thoroughly well qualified to perform the work required of him. Now, however, that the National Pig Breeders' Association is in a position of responsibility, the committees of shows cannot do better than seek its advice whenever any difficulty arises in the appointment and selection of a judge.

CHAPTER XVII.

DISEASES.

There is no greater proof, if any were needed, of the manner in which pigs have been neglected than the fact that there is no English work treating, in anything like a comprehensive form, of the diseases to which they are liable. We believe that the veterinary profession has given far less study to swine diseases than to those of any other of our domestic animals. Be this as it may, there are little or no data within the reach of the ordinary pig-keeper enabling him to treat successfully the simplest complaint, or to distinguish one disease from another. Unwritten codes of medicine and surgery, some of which are intelligent and others absolutely worthless, are common in different districts of England; but, as the pig becomes better understood and more scientifically cultivated, there is no doubt whatever that he will be far more easily managed, both in health and disease.

Diseases are commonly produced by improper or filthy food or drink, or by badly built, undrained, or dirty sties. If, therefore, the feeding and management are good, and the pigs are properly housed, there will be little or no disease to combat; but if the pigs are allowed to lie upon any manure-heap, and to drink the filthy liquid lying in pools in the farm-yard, or if they are compelled to consume offal (which many farmers consider quite good enough for them, instead of the grain which they grow), severe losses must sometimes be expected, and the disease which at such times usually appears is of such severity that neither medicine nor veterinary surgeon is of any use. There are, however, minor complaints, to which all live animals are liable; and it is to these, rather than to chronic or complicated diseases, that the attention of the pig-breeder is directed in the following remarks.

It should be remembered that the appetite of the pig is an unusual one: it is questionable whether there is any domestic animal which is always so ready to feed, and to feed so ravenously. In the first place, ravenous feeding is an incentive to indigestion, and, in the second place, a pig will, if he is permitted, invariably eat more than he can properly digest. This being so, it follows that anything abnormal in his management—as, for example, a very cold sty, a wet bed, or a serious change in the temperature—is liable to add the one straw that breaks the camel's back, and to bring about disease, which in a greedy hog of but moderate constitution is always ready to break out. As there is considerable similarity between the constitution of the pig and that of man, this argument will, by many readers, be the more easily understood. Thus it is that especial care should be taken to provide pigs with suitable food and quarters, to give them a run out on the green pasture in summer, and to furnish them, not only with warmth, but with suitable succulent food in winter.

The Pig Medicine Chest.

Most of the following remedies will be referred to in the remarks upon diseases, and as they are all in the nature of domestic medicine they may, with ordinary care, be used without fear.

The doses given are the minimum and maximum. The first would be for pigs that have just left the dam; about a middle dose should be given to half-grown animals; and the maximum to adults of large breeds. But this last must be regulated by the size of the breed. For instance, 90 drops of oil of male fern would be the dose for an adult of a large breed; but 60 would be quite sufficient for an adult of one of the small breeds. Except in special cases, medicine is not given to suckers, but is administered to the sow, and its action takes place through her.

ACONITE (TINCTURE OF).—Lessens fever.   Dose, 5 to 15 drops

ALOES.—This is a purgative, and should be preferably used in the form of a solution, previously dissolved in boiling water, when 10gr. to 60gr. may be administered to a full-grown pig.

ALUM.—Astringent.   As an injection, 60gr. to 1 pint of water.   A saturated solution is also used.

AMMONIA LINIMENT.—Professor Brown recommends this as an external stimulant.   It consists of liquor. ammoniæ fortis, 1 part; oil of turpentine, ⅛th part; soap liniment, 4 parts. It must be kept in a stoppered bottle, and applied by being rubbed well into the skin with the hand.

BELLADONNA.—Allays pain.   Dose of green extract, ¼ to 1gr.; dose of the tincture, 5 to 15 drops.   The compound liniment is used for pain in the joints and other parts.

BISMUTH.—Stomachic, sedative, and checks diarrhœa.   Dose, 10gr. to 60gr.; for suckers, 5gr.

BORACIC ACID.—Antiseptic and dressing for wounds.

CALCIUM PHOSPHATE.—Given for rickets.   Dose, 10gr. to 20gr.

CANTHARIDES.—Used externally with lard or vaseline as a blister.

CARBOLIC ACID.—This can be obtained of any chemist.   It is a very powerful antiseptic and disinfectant, and, whether used for disinfection or for application, should be diluted with water in the proportion of one part to fifty or sixty.   Carbolic acid should not be used as a skin dressing if a large surface has to be covered, as there is danger of absorption, and many deaths have occurred in consequence.   Oil of tar should be employed instead.

CARBOLISED OIL.—A valuable antiseptic and emollient, useful for wounds and at the time of parturition, when, if there is any fear of inflammation, the passage may be dressed with it. One or two parts of carbolic acid added to forty parts of olive oil forms the preparation known as carbolised oil.

CASTOR OIL.—Aperient.   Dose, ½oz. to 3oz.

CHALK (PREPARED).—This should always be at hand.   It is an important agent in checking diarrhœa.   Dose, 10gr. to 60gr.

CHARCOAL (WOOD).—An antiseptic absorbent and deodoriser, given in distension by intestinal gas, wind in the stomach, and abdominal pain, also in diarrhœa. It is an extremely useful and simple agent in restoring tone to the system, and assisting in the promotion of digestion. Dose, 1dr. to 3dr.

CHLORAL HYDRATE.—Relieves pain, especially when given in conjunction with opium. It may also be given in severe cough. Dose, 10gr. to 20gr.

CONDY'S FLUID.—A well-known and extremely valuable anti-septic and disinfectant.

EPSOM SALTS.—A well-known purgative. A dose should consist of from 1oz. to 2oz., or a little more, dissolved in warm water. A little ginger may be added with advantage.

GENTIAN (POWDERED).—Tonic and stomachic. Dose, 2dr. to 4dr.

IPECACUANHA. —Emetic and expectorant. Dose of the powdered root: as an expectorant, 1gr. to 2gr.; as an emetic, 15gr. to 30gr. Dose of the wine: as an expectorant, 10 to 30 drops; as an emetic, 4 to 6 drams.

IRON (REDUCED).—Blood tonic. Dose, 2gr. to 10gr.

IZAL.—A most useful disinfectant, deodorant, and insecticide.

JEYES' FLUID.—An excellent antiseptic and disinfectant.

LAUDANUM (Tincture of Opium).—Useful to relieve spasms and pain. Single dose, 20 to 60 drops; if often repeated, 20 to 60 drops.

LINSEED OIL.—A most valuable purgative. It should be given in doses of from 1oz. to 2oz., or a little more.

LUNAR CAUSTIC (Nitrate of Silver).—A powerful substance; it is used where actual cautery is required. The caustic pencil can be introduced into a wound with much greater facility than a hot iron.

MERCURIAL OINTMENT.—Useful as a dressing for skin disease, but must be used with care, as absorption may take place through the skin.

MORPHIA (HYDROCHLORATE OF).—Given in acute pain. Dose, $\frac{1}{8}$gr. to $\frac{1}{2}$gr.

NITRE, SALTPETRE, or NITRATE OF POTASH.—These are one and the same drug. Valuable as a diuretic in most fever cases. A teaspoonful may be given at a time. Sometimes nitrous ether is spoken of as "nitre."

NITROUS ETHER.—Diffusible stimulant. Dose, 2dr. to 1oz.

OIL OF MALE FERN.—Vermifuge. Dose, 20 to 90 drops. Capsules can be obtained each containing 15 drops of the oil, or it can be given in milk.

OIL OF TAR (*Oleum cadinum*).—Used in skin affections, especially in eczema and mange.

OIL OF TURPENTINE.—Valuable as a vermifuge (4 fluid drams with the same quantity of castor oil) and as a stimulating dressing. It should not be given to very young pigs. For adult animals, 4 fluid drams in the same quantity of castor oil is the proper dose.

OPIUM (POWDERED).—Sedative; relieves abdominal pain; also given in small doses with other astringents, and with chalk in diarrhœa. Dose, ½gr. to 2gr.

PERMANGANATE OF POTASH.—A most valuable antiseptic and disinfectant. The crystals can be purchased of any chemist, and dissolved in water. It must be kept well corked.

POTASH (BICARBONATE OF).—Lessens fever; diuretic. Dose, 30gr. to 60gr.

POTASSIUM (IODIDE OF).—Alterative; useful in rheumatism. Dose, 5gr. to 20gr.

SALICYLATE OF SODIUM.—Given in rheumatism. Dose, 10gr. to 30gr.

SALICYLIC ACID.—Of great use in foot-and-mouth disease, and as an antiseptic. Professor Brown recommends that 4 table-spoonfuls be dissolved in 1qt. of boiling water, in an earthen vessel, this being subsequently made up to 1gall. The solution may be syringed over the mouth and feet, or the dry powder itself may be sprinkled over the feet.

SANTONIN.—Vermifuge. Dose, 10gr. to 20gr.

SODA (BICARBONATE OF).—Antacid and diuretic; given for indigestion. Dose, ½dr. to 1dr.

SODA (HYPOSULPHITE OF).—An old-fashioned remedy, to purify the blood, and in cases of blood poisoning; also given for rheumatism. 5lb. of this salt dissolved in 100 gallons of water was recommended for the ordinary drink for cattle, as a preventative of cattle plague. Dose, 1dr. to 4dr.

SOFT SOAP.—Always valuable for use in the piggery.

SPIRITS OF TURPENTINE.—There is really no such preparation as spirits of turpentine, but the oil of turpentine is sometimes erroneously called spirits of turpentine.

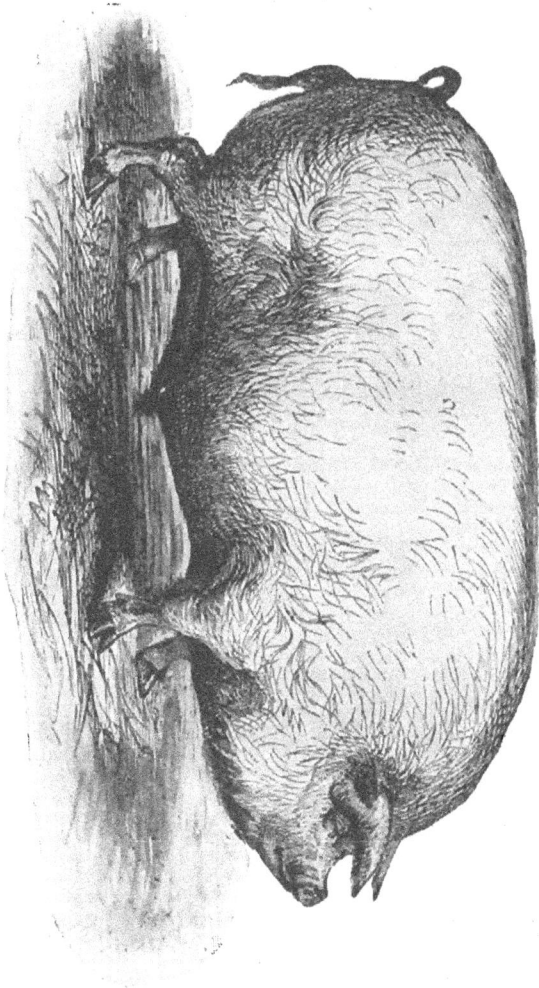

CHESTER WHITE BOAR.

This large American breed closely resembles the Poland-China, its thin, white, straight coat being the main distinction. It is usually very clean in its habits, docile, comes to early maturity, is an excellent variety for crossing on the common races, and is generally regarded by American breeders as one of the best of modern breeds of pigs.

SUBLIMED SULPHUR (Flowers of Sulphur).—An alterative medicine, which may occasionally be given with advantage in the food. Dose, ½dr. to 4dr. It also proves an effectual dressing, combined with oil or lard, for parasites or mange. If sulphur is given for a length of time or in overdoses, serious diarrhœa may be set up, which is most difficult to check; indeed, it sometimes will not yield to treatment, and death from exhaustion takes place.

SULPHUROUS ACID.—Antiseptic and disinfectant.

TAR.—Useful for the feet, whether for wounds or for foot-and-mouth disease.

ZINC (CHLORIDE OF).—A useful disinfectant and antiseptic. It should be diluted with sixty parts of water. As a lotion, 20gr. should be mixed with 1 fluid oz. of water.

Administering Medicine to Pigs.

In administering medicine to swine, a very common, and usually successful, method is to add it to their food; but it sometimes happens that, either from excessive sickness or from cunning, the animal refuses to eat it. In such a case, resort must be had to another plan. The animal is first caught, and roped as though the tusks were to be broken off or extracted (as explained on page 59). When firmly secured, the pig will, in his rage, continually open his mouth, when an old shoe or slipper, which has previously had the toe cut off, should be at once thrust in. This he will commence to chew, and an opportunity should be taken to pour the medicine from a long-necked bottle into the shoe, when it will successfully pass down his throat without any fear of his being choked.

Another method is that described by the late Mr. G. T. Turner, which is the best we have ever heard of. It is the plan adopted by a village blacksmith who used to ring his pigs. The animal is secured as for ringing (as described in Chapter VII.), and the attendant, standing with the pig's head between his knees, pours the medicine, which must be at least as thin as oil, first into one nostril and then into the other, from a feeding can, similar to those used for oiling machinery, until the whole is administered.

Diseases and Their Treatment.

ABORTION.—In the pig this is of such rare occurrence that it hardly merits a paragraph. Premature birth occasionally occurs. It may be caused by a plethoric condition, a debilitated condition, injuries, strong purgatives, or obstinate constipation.

ABSCESS.—Abscesses may be divided into two categories—acute and chronic. The former develop themselves under the influence of a sharp local inflammation. Generally speaking, they are found under the throat, at the bottom of the neck, in the groin, or upon the limbs. They are red in appearance, extremely painful, and frequently exhibit a considerable tension of the skin under which they develop. At the same time, infiltration is going on, and the pain increases. In a short time, the hand, if placed upon one of the inflammatory spots, will detect a certain fluctuation, which indicates the presence of pus in the thicker part of the tissues. At the commencement of the treatment, the abscesses should be well poulticed with linseed meal; but if, from the difficulty of their application to the pig, a poultice cannot very well be used, the formation of the abscess should be hastened by the application of ointment (1 part of lard or vaseline to 12 parts of powdered cantharides). When the tumour has come to a "head," it should be cut with a bistoury (a surgical knife) or any similar instrument.

The chronic abscess is usually the local expression of a general morbid state in the pig, and frequently arises from bad food or unhealthy sties. The remedy employed may be the same as that above mentioned; but it is of the highest importance that, at the same time, there should be a change in the management, by the adoption of a proper system of hygiene and alimentation.

ANÆMIA.—This complaint is most generally found amongst animals which are in low condition, resulting from bad feeding; it indicates that the blood is poor in both quantity and quality, there is a diminution in the number of red corpuscles, and consequently the blood is thin. Animals affected have a feeble but rapid pulse, they become easily and dangerously excited, and there is an emaciated and generally weakly condition. The appetites of animals suffering from the complaint are very changeable, and indigestion and pains are frequently present. The extremities are always cold, and constipation often sets in.

It is necessary to convert the blood into a healthy condition, and, after the cause of the complaint has been removed, mineral and vegetable tonics, such as 5gr. to 10gr. of reduced iron in food twice or thrice a day, or 2 to 4 drams of powdered gentian flavoured with a little fenugreek powder. Cod liver oil may also be given. A diet of the most nutritious character should be given in such proportions as the animals can easily digest and assimilate; the other natural hygienic conditions being, of course, attended to.

ANTHRAX.—A specific contagious disease, that may spread from one animal to another, and is *communicable to man.* The disease is due to a bacillus known as *Bacillus anthracis,* which is, under the microscope, a rod-like organism; the rods split up in the blood and multiply very rapidly, blocking up the blood-vessels, and often causing death in a few hours after their introduction into the system. The disease in pigs is generally introduced into the system by eating parts of the carcases of anthrax-affected animals, or through abrasion of the skin, especially in the mouth. When man is affected it is through some abrasion, usually on the hand.

Anthrax is a most fatal disease. The symptoms appear suddenly, the animal becomes dull and dejected, and unwilling to move, the head hangs down, and the ears droop. If the animal is forced to move, partial paralysis shows itself in the hind-quarters, and this, later, becomes complete. The fore-limbs may not be affected. Then the characteristic symptoms of a swelled throat and neck appear, and ultimately the whole face and head is involved, and the whole appearance is altered. Breathing becomes more and more difficult, the mouth is held open, the tongue often protrudes; the skin is affected, red patches appear, and these may form into abscesses.

The following explicit instructions and regulations with regard to anthrax-infected animals have been issued by the Board of Agriculture and Fisheries:

1. Anthrax has long been known as a generally fatal disease, communicable to cattle, sheep, swine, and also to man. Prior to the discovery of its true cause it was regarded as a disease affecting cattle only, attributed to feeding them on highly nutritious or artificial foods, which induced an attack of apoplexy or enlargement of the spleen, resulting in sudden death. This view as to the cause of anthrax appears still to exist in many parts of the country, for it is a common practice amongst owners of stock, being unaware of its dangerous character, to slaughter their cattle as soon as they present symptoms of serious illness, in order that the carcase and hide may be utilised. The blood of the diseased animal is thus in many cases distributed over the floors of the sheds, or upon the mangers, or is carried upon the boots of the attendants, by which means other parts of the farm or premises become infected.

2. It is important that it should be widely known that the disease is solely due to the introduction into the blood of an animal or of man of the minute spores or germs contained within the anthrax bacilli, which are to be found in the blood of animals recently dead of anthrax soon after it has been exposed to the air. On the other hand the bacilli of anthrax and the spores therein die speedily if kept within the intact carcase.

3. It will thus be recognised that in order to prevent the extension of anthrax from diseased to healthy animals or to persons, it is essential that the carcase of the affected animal should not be opened, and that none of the blood or the excretions which may contain blood should escape, as the spores contained within the blood will, when exposed to the air, multiply with rapidity, and may become the means of infecting other animals.

4. In most instances, the first intimation of an outbreak of anthrax is the discovery of a dead animal in the pasture or byre. The animal may have been left a few hours earlier in apparent good health; at least, there may have been nothing to attract attention, or give warning of the near approach of death. Occasionally there are, however, premonitory symptoms of an attack of anthrax which can be recognised by an expert. The affected animal is dull, and disinclined to move. If the

case occurs in a herd at pasture the fact is sometimes indicated by the separation of the sick animal from the rest. The affected animal will occasionally cease to feed, and stand with its head bent towards the ground, and sometimes a little blood is discharged from the nostrils and also with the fæces. Close attention will enable the observer to detect an occasional shiver and trembling of the limbs, which passes rapidly over the body, and then ceases. The shivering fits may then become more frequent, and perhaps, while these signs are being noted, the animal will suddenly roll over on its side, and, after a few violent struggles, expire. On close inspection, especially in the case of swine, it will often be found that there is a good deal of swelling under the throat extending down the neck; and the swollen part will at first be hot and tender to the touch, but as the disease progresses it becomes insensitive and cold.

5. Although anthrax is, as already indicated, a communicable disease, it is doubtful whether it is often transmitted from the living diseased animal to the healthy by association, as in the case of cattle plague, foot and mouth disease, or other animal diseases of a contagious nature; experience goes to show that it is usually transmitted to the healthy animal through the medium of food or water containing the spores of the disease. These spores may also find their way into the circulation through a cut or abrasion. The disease may be introduced through the spreading of infected manure on the pastures, and occasionally outbreaks have been traced to the distribution of manure containing the cuttings or scrapings of hides removed from diseased animals.

6. For the reason above given it is evident that, in their own interest, it would be better for owners of stock to allow their animals when affected with anthrax to die, rather than slaughter them and thus incur a serious risk not only of infecting their sheds, stock-yards, and other parts of their farms and premises, but also of causing the death of those persons who may be engaged in slaughtering them.

7. It will be gathered from the preceding remarks that, since the means by which anthrax may be spread are somewhat different from those in the case of other contagious diseases of stock, the measures to be adopted for preventing extension should also be dissimilar.

8. Whenever an animal dies suddenly from some unaccountable cause, the fact should be at once reported to the Local Authority, and in the meantime the owner should forthwith plug the nostrils and all the natural openings with hay or tow saturated with a strong solution of carbolic acid, in order to prevent the oozing of any blood therefrom. The Veterinary Inspector should forthwith inquire into the cause of death, and determine by careful investigation whether anthrax exists or not. This can be done by examining with a microscope a few drops of blood taken from one of the superficial veins soon after the death of the animal.

9. So soon as it has been decided that the disease is anthrax, the owner should cause all the cattle, sheep or swine which have been in association with the dead animal, pronounced by the Veterinary Inspector to be apparently healthy, to be moved from the shed or field or other place where the disease has originated, to some other place on the farm or premises, where they can be isolated and kept under observation. The isolated animals should be given an entire change of food and water, and as the period of incubation of anthrax is usually very short, seven days will as a rule suffice to enable the Veterinary Inspector to determine whether any further animals have become infected or not.

10. The carcase of every animal which has died of anthrax should be buried in some part of the farm remote from any watercourse, and to which animals cannot or do not ordinarily have access, such as a wood or enclosure. The burial and disinfection of the carcase should be carried out under the supervision of an inspector of the Local Authority.

11. The inspector of the Local Authority should carry out or supervise a rigid system of disinfection of the place or premises where the diseased animal has been detained or has died; also of all manure and broken fodder remaining therein.

12. There is no doubt that the cause of the periodic recurrence and persistence of anthrax on many farms in this country has been due to the skinning of the carcases of animals which have died of that disease and to the neglect of proper precautions for their burial and disinfection. The most effectual manner of destroying the germs of anthrax is by burning the carcase, or destroying it with chemical agents,

and where facilities exist for carrying out either of these methods a licence of the Board must be previously obtained. In cases where burial is adopted every facility should be afforded by the owner to the inspector of the Local Authority in order that this duty may be effectually carried out.

13. It has been found by experience that where all the above-named precautions have been scrupulously adhered to, the disease frequently ceases after the death of one animal on the farm.

ANUS (IMPERFORATION OF THE).—This is dealt with under "Malformation of the Anus."

APHTHOUS FEVER.—The scientific name for "Foot-and-Mouth Disease" (to which heading the reader is referred).

APOPLEXY.—This malady is almost entirely confined to young pigs in close confinement and to pigs that have been fatted too much, and, unless it is energetically treated, the probability is that the animals will die. The attack is sudden. The afflicted pig at first exhibits a most stupid movement; it then falls down, stretches itself considerably, and becomes apparently lifeless—indeed, it might very easily be supposed to be dead, but for the fact of its breathing, which, upon close examination, may be observed. The snout and tongue become dark-red, the eyes are wide open, and may be touched without causing the eyelids to close or to respond. Urine and excreta pass involuntarily, and the breathing is heavy. The causes of apoplexy are a sudden change of food from a farinaceous diet to butcher's offal, exertion on the part of a fat animal, also heart disease and over-loading the stomach.

Bleeding is not practised now; the disease being so quickly developed, there is little hope of successful treatment, and consequently, from an economic view, the animal should be killed and bled at once, as the meat is perfectly good for consumption.

BLACK JACK.—A popular name for "Swine Fever" (to which heading the reader is referred).

BLIND STAGGERS.—A common name for "Epilepsy" (under which heading it is dealt with).

CANKERS are ulcerous sores, which usually appear in the

mouth and at the extremity of the ears. They are sometimes hereditary, or the result of an eruptive malady.

The only treatment that can be effectually applied is cauterisation by fire or by nitrate of silver. The parts of the ears affected are sometimes cut completely off.

CATARRH (NASAL).—See " Nasal Catarrh."

CHOLERA (HOG).—A common name for " Swine Fever."

CONSTIPATION indicates that a change of diet is necessary. Sows after parturition, and young pigs that have been too highly fed. are very often constipated. Other causes are insufficient water, feeding on astringent articles such as acorns or chestnuts, consumption of dry and indigestible food, and want of exercise. The animals affected eat little, their excrement is dry and hard, and they drink a great deal.

Withhold all solid food for twenty-four hours and substitute mash. Then give 4 drams of sulphur with 4 tablespoonfuls of Epsom salts in a little tempting food; or 3oz. of castor oil can be administered. (These doses are for a large pig; half quantities will suffice for a half-grown animal.) Soap-and-water enemas are also useful. For a time afterwards mix a little linseed oil with the food, and give some roots or grass if the latter can be obtained.

COUGH is a symptom of many diseases, but is not a disease of itself, and it is quite impossible to treat it without ascertaining the cause; on removing the latter the cough will cease. Cough often arises from indigestion or an irregularity in the system, caused by bad feeding or want of method. " Rising of the Lights " is a common name for cough.

CUTS.—For treatment of these the reader is referred to " Wounds."

DIARRHŒA, OR SCOURS.—Pigs which are suckling or have been lately weaned are most afflicted with this complaint; indeed, it is seldom that older pigs are attacked. If the little ones are not taken vigorously in hand, the probable result will be most serious. The principal causes of diarrhœa are an insufficient supply, or poverty, of the sow's milk, improper food, irregular feeding, cold and damp sties, and sudden variations in the temperature. Green food given to sows with litters, when they have been long deprived of it, not infrequently causes the complaint. It may arise from intestinal worms, and is caused by putrid animal or vegetable matter, and by a vitiated state of the blood; it is also present in constitutional affections, such as diseases of the liver, pulmonary tuberculosis, and rickets. In bad cases, intestinal inflammation sometimes sets in, and this is followed by severe prostration, and ultimately by death. The excrement of the afflicted animal is generally of a whitish colour, and if the pig is, at the time of the attack, suffering from a cold or cough, the chances of recovery are considerably diminished.

Sows, when attacked, should be shut up and fed upon dry maize (although this food, given alone for any length of time, has been stated to cause the disease frequently) or sweet skim-milk. If the little pigs are large enough to eat, they may be given a quantity of dry, raw flour mixed with arrowroot. In grown animals the first treatment should be a dose of castor oil (1oz. to 3oz.), to clear the bowels of any undigested matter. Then give ½dram of subnitrate of bismuth. (Suckers may be given 5gr. three times a day.) If there be any pain ½gr. to 2gr. of powdered opium, with 5gr. to 15gr. of bicarbonate of potash, and 10gr. to 60gr. of prepared chalk may be given in the food two or three times a day. It is a good plan to change the diet of the sow when diarrhœa is prevalent, and to allow the animal a fair amount of outdoor exercise; but her young ones should be kept in the sty, which at all times should be clean, warm, comfortable, and well ventilated. A quantity of earth sprinkled about the pen will be found advantageous, as this will tend to absorb the various gases which are generated. The feeding-pen, if separated from the sleeping apartment, should also be strewn with dry earth, and, after being thoroughly cleansed, and disinfected with lime or carbolic acid, a small quantity of straw may be added. The troughs, too, should be scalded with hot water, so as to make them perfectly sweet.

Beans have been recommended by some authorities, but, in practice, they will be found both heating and forcing, and liable to intensify the illness of the pig. We have frequently and most successfully used soot, which, mixed with the food of young pigs that can feed, will, in almost every case, effect a cure.

Eczema.—In its initial stages, at least, eczema cannot be classed as a true skin disease. If neglected, changes take place in the skin; these local changes may probably be considered a skin disease, but be this as it may, eczema is due to a morbid condition of the blood—a plethoric state of the system, commonly called "over-rich blood"—and even the opposite condition may cause eczema. A sudden change from a farinaceous to a flesh diet may be responsible for the eruption, as also may intestinal worms, and in old animals liver or kidney trouble may occasion the malady. One of the symptoms is an eruption of the skin, which presents a number of red, more or less rounded patches, raised from the surface; there is a discharge of a watery nature which eventually dries up and forms scales or scabs, and these fall off later, leaving the skin red. Itching accompanies these symptoms, and the animal by rubbing the parts may cause bleeding.

The treatment of eczema must be consistent with the cause. If the animal is plethoric give from 1oz. to 3oz. of Epsom salts at once, and repeat this once or twice at a few days' interval. The diet must be less stimulative and possibly less in quantity. If the animal is "poor," give a more generous diet, and if possible turn it out in an orchard or grass meadow for a little time during the day. In all cases a blood tonic is desirable. Give 5gr. to 10gr. of reduced iron in food twice or thrice a day—it is best given in a little gruel. If the skin is very irritable, rub with a preparation of one part of flowers of sulphur and eight parts of olive or linseed oil every third day until the irritation is overcome. If worms are present expel them in the manner prescribed in the article on "Worms." If sucking pigs are affected, they must be treated through the mother, as the latter, or rather her milk, is the cause, and she must be accorded the treatment suggested above, having regard to the cause. In all cases give a clean bed, and let the sties generally be clean with good ventilation.

Enteritis.—Pigs affected with this disease exhibit a careless disposition to feed; they drink much, have a dull and spiritless appearance about the eyes, the mouth is dry and purplish-red, and the skin is also red. Generally, the animals are weak, costive, and flatulent, making plaintive groanings, and remaining in their sties. Among the causes of the disease are want of cleanliness in the sties, and impure drink or food, or food which contains too little nutriment, either of which may affect the intestines and bring about inflammation. As the disease is nearly always fatal, the animal should be slaughtered upon the first symptoms showing themselves.

EPILEPSY, OR BLIND STAGGERS.—This complaint has often been confounded with apoplexy, but it is not exactly the same thing. Both, however, are diseases of the nervous system, and may occur from time to time at intervals of a week or months, and continue for years, and not seriously endanger the health. The pig, when afflicted, is very restless both by day and by night; it has red and inflamed eyes, the pulse is quick, the bowels are often constipated, and frequent gyrations are made, nearly always in the same particular direction. It walks as though it were blind and ascending a number of steps—hence, we presume, the origin of the term "blind staggers." As in apoplexy, the animal falls to the ground, but the limbs are convulsed, and are moved violently backwards and forwards, and the jaws are clenched. These symptoms gradually disappear; the jaws relax, consciousness returns, and after a time the animal often walks off as if nothing had occurred. At other times, when consciousness returns, the animal becomes frenzied, and jumps about and behaves in a wild maner. The most frequent causes of epilepsy are intestinal worms, teething, and engorgement of the stomach, particularly after fasting.

In treating the animal the cause must be ascertained. If worms are present, give 10gr. to 20gr. of santonin in 1oz. of castor oil, and in four-and-twenty hours give 1oz. of castor oil. Or 20 to 60 drops (the maximum dose for a large pig is 90 drops) of oil of male fern may be given shaken up in a little milk. If the disease occurs in sucking pigs, owing to teething, the dam's bowels should be opened by giving 1oz. to 2oz. of Epsom salts; and this will operate on the suckers through the dam. In place of either of the above remedies, the following recipe may be adopted with advantage: Dash a quantity of cold water well over the

2 A 2

entire body of the pig, after placing a purgative injection
within the rectum. The injection may consist of 10oz. of
water, 6oz. of sulphate of soda, and 1 or 2 teaspoonfuls
of spirits of turpentine. A plentiful supply of water should
always be within reach of the animals. Professor Low is of
opinion that bleeding should not be practised for epilepsy.

ERYSIPELAS.—This is referred to under "Swine Erysipelas."

FEVER (SWINE). See "Swine Fever."

FOOT-AND-MOUTH DISEASE, OR APHTHOUS FEVER.—A specific
contagious fever. This disease is eruptive, and appears in
the mouth, or in the feet between the, toes, in the form of
blisters; these ultimately may burst and form ulcers, which are
either isolated or confluent. It is contagious, and communicates itself to almost all classes of domestic animals. The
contagion in this disease is believed to be due to minute
organisms, but the latter have not been entirely separated.
The virus may be conveyed by rats or on men's boots, but the
most frequent cause is giving milk from cows that are affected
with the disease.

In treating the patient, place on dry litter, thoroughly
cleanse the feet, and dress the latter with a saturated solution
of alum, and with a little carbolic acid or strong solution
of Jeyes' added to it. Repeat the dressing night and
morning. Perhaps the most convenient way to dress the feet
is to drive the animals through a trough containing the liquid.
The snout and mouth may be dressed with a 5 per cent.
solution of salicylic acid. But precaution should not end
here; for, as there is much to fear from the manure, this
should be thoroughly mixed with lime, and removed to where
it cannot possibly come into contact with any other stock.
Some persons prefer to use a solution of carbolic acid, in the
proportion of 1 part to 50 of water in which soft soap has
been dissolved. This is also a good remedy, but the smell is
objectionable, and for this reason we prefer the salicylic acid.

Another remedy is to use a mixture of sulphurous acid and
water, mixed in the proportion of 2oz. of the former to 1qt.
of the latter. The animals should be given a teaspoonful of
this three times a day, and, in addition, the feet may be
moistened with the solution, applied with a cloth or sponge.

For further safety, a quantity of sulphur may be burned in the piggery daily. The Rev. F. H. Brett, of Carsington, says he has proved conclusively that sulphurous acid, given to cattle affected in this way, has a most beneficial effect; and, as it can be purchased cheaply, it is worth remembering.

Instead of the preceding remedy, tar and hyposulphite of soda may be used. The tar should be so placed that the animals will be compelled to walk in it two or three times a week, to a depth of some 2in. or 3in. For this purpose, it might be put in a trough, stood upon one side of a partition in which there is a hole through which the pigs would have to pass in order to get at their food. The hyposulphite of soda should be given as a dose, in a mild form (1dr. to 2dr. daily for a week or ten days), and, as it is almost tasteless, little difficulty will be found in making the animals swallow it.

It should not be forgotten that, inasmuch as foot-and-mouth disease is acquired, in almost every case, by carelessness, it may, in a similar way, *be perpetuated, and carried from animal to animal, and from farm to farm.* The attendant, by the exhibition of the least carelessness, may carry the disease from the piggery to the cowshed; or he may, even by walking across a neighbour's field, or upon the high road, in the boots that he has been wearing in the piggery, communicate it indirectly to animals which may travel over either. A person attending an animal afflicted with any disease of this nature should invariably change his clothes before leaving the piggery, and wash his hands in water into which a disinfectant has been poured.

At the present moment, the law upon the question is clear. It is impossible to move an animal in an infected district without considerable risk of catching or spreading the disease, as well as of being heavily fined. Much annoyance is experienced by a person owning healthy stock, and who happens to be living in an unhealthy district, for he cannot deal with his own animals with impunity. It is, however, for the general good, and the restrictions placed upon the removal of pigs should materially assist in bringing about a healthier state of things. It may be well to remark here that, in buying or selling an animal, it should always be clearly understood whether the district in which either party lives is in an

infected circle, in order that the proper official permits may
be obtained for removal, or that the bargain may be declared
"off" in case removal is impossible. When pigs are sent by
rail in a truck, care should always be taken to see that it
has been freshly lime-washed.

FOOT-ROT.—This disease, according to Heuzé, does not exist
in the pig. He says it is the same as aphthous fever (foot-
and-mouth disease). A number of small vesicles appear
between the toes, and these are filled with a pale yellow
matter, which escapes by the bursting of the membrane enclosing it. When these vesicles are once opened they leave a
red and bleeding sore. The inflammation extends from place
to place, weakens the corneous secretion, and may ruin the
hoof. The disease is discussed fully under the heading "Foot-
and-Mouth Disease."

GASTRITIS, OR INFLAMMATION OF THE STOMACH, is of frequent
occurrence in the pig, and this is not to be wondered at,
considering the voracious appetite of the animal, and its
habit, especially when hungry, of consuming irritating and
even poisonous substances. The commonest cause of gastritis
is the practice of adding strong brine to pig-wash; it is not
added with any beneficial object, but merely because it happens
to be at hand and increases the bulk of the wash given. Again,
over-feeding, after long fasting, and also the consumption of
large quantities of putrid flesh, will produce the same result.
Inflammation of the stomach is a common symptom of swine
fever.

The symptoms are first a disinclination for food and a
tucked-up appearance, and the animal isolates itself. Later
symptoms are a hot and red skin, and extreme thirst; while
if food is forced upon the animal, or it takes any solids, pain
ensues, vomiting occurs, and diarrhœa sets in; but we
have known cases where constipation was a prominent symptom. Gas may accumulate in the stomach, and cause
distension of that organ, which is very noticeable.

In treating a case, all solids should be withheld, and gruel
or linseed-tea should be given. If constipation exists, give
1oz. to 2oz. of Epsom salts. If diarrhœa is present, remove
the undigested matter in the stomach by giving 2oz. or 3oz.

of castor oil; and if pain is present, add to this 10gr. to 20gr. of chloral hydrate; or 10 to 30 drops of tincture of opium and 15gr. of chloral together will often allay pain where either of the drugs given separately fails to do so (this may be repeated every two hours until the pain subsides). When the animal is recovering, care must be taken not to give solid food hurriedly or in large quantities. Gradually bring the animal to it, by first offering some liquid food and some solid, and lessening the former until it is entirely withheld.

GRAPES.—The butcher's name for "Tuberculosis."

GRAVEL.—The urine of pigs, in common with all animals, contains salt, in solution. Chemically the salts vary in different animals, and naturally their composition is regulated in a great measure by the food taken. In the pig the salts present are magnesia phosphate and ammonia phosphate, with oxalate of lime. In the ordinary course of nature, in a healthy animal, these salts are passed with the urine, but at other times they are deposited in the bladder. Retention of urine may cause the deposit, and the retention may be due to inability to pass the urine owing to obstruction at the neck of the bladder or in the penis, and distension may cause paralysis of the muscular coat of the bladder. In fat animals sometimes the urine is retained from sheer unwillingness upon the part of the animal to rise and urinate.

The symptoms are very variable. Sometimes the urine is passed frequently and in small quantities; it may flow involuntarily, or not at all, and the character of the urine may alter and become whitish. Pigs in this condition should be fattened and slaughtered. Treatment is not likely to be successful, at least from an economic point of view.

HERNIA, OR RUPTURE.—This condition is caused by an escape, usually of the bowel, through an opening in the abdominal muscles, and the skin is pressed out into a sac or bag, which is filled with a portion of the intestine; but the organ protruding is not always the bowel—it may be another organ, or part of one. The two usual forms of rupture seen in the pig are known as *scrotal* and *umbilical*. In the former, some portion of the abdominal viscera passes into the scrotum with the testicle. In umbilical hernia, a portion of the bowel, or omentum, escapes through the navel opening, due to imperfect closure of the latter. A blow that injures the abdominal wall may cause rupture, so will straining, as in parturition, or obstinate constipation accompanied by straining. But rupture mostly occurs in young animals, and is hereditary, or at least there is a predisposition to it.

In umbilical hernia—the commonest form—there is a swell-ing in the abdominal wall, more or less round and fluctuating; it may appear quite suddenly, or may slowly develop. The most characteristic symptom, and the one that usually helps the amateur to diagnose it correctly, is the difference in size of the swelling at various periods of the day. When the animal is fasting the swelling is comparatively small and soft, but after a full meal it is much larger and denser. By a careful manipulation of the part, the escaped viscera or part of viscera can be returned into the abdomen; this is easily accomplished if the animal is turned on its back.

Treatment in these cases is not profitable. The animal should be fattened and killed. From the foregoing remarks it will scarcely be necessary to say that in sows so affected they should never be bred from, or boars used where one or two litters have shown symptoms of this complaint, and there is reason to believe that the boar is responsible.

HOG CHOLERA.—Another name for " Swine Fever " (to which heading the reader is referred).

IMPERFORATION OF THE ANUS.—This is dealt with under " Malformation of the Anus and Rectum."

INFLAMMATION OF THE LUNGS.—The common name for " Pneumonia " (under which heading it is described).

INFLAMMATION OF 'THE STOMACH. — The common name for " Gastritis."

INFLAMMATION OF THE UDDER OR TEATS.—Sows affected with this malady usually become so during the first few days after farrowing, those which are fed highly, give a great quantity of milk, or have large pendent udders, generally being the most subject to it. An unclean sty will also cause the disease, which may be either local or general, and which varies in its intensity. Generally speaking, the complaint is unnoticed until the pig displays a want of appetite, or an abnormal heat of the skin, although such is only the case when the disease has extended throughout the glands. The udders and nipples become hard, warm, tender, and a brilliant red, and the sows do not care to be troubled with their young.

The sties should be kept in a perfectly clean condition in cases where the sow is affected with this complaint; the animal herself should be subjected to a cooling *régime*. Give the following: Epsom salts, 4oz.; bicarbonate of soda, 2oz.; nitrate of potash, 2oz.; water, 1½ pints. Two table-spoonfuls in food night and morning. Dress the teats daily with compound liniment of belladonna, using a little friction.

INVERSION OF THE RECTUM. — This complaint generally affects pigs that are badly fed and housed, although it sometimes arises from farrowing, constipation, or some undetermined disease. The best thing to do is to empty and cleanse the protruding part, using warm water for the purpose, then to push it back into the anus. If the animal strains, mix ½gr. of hydrochlorate of morphia in a little treacle and put this in the mouth. The animal must have no solid food for a day or two. If constipation is the cause, give 1oz. to 2oz. of castor oil. Soap-and-water enemas are also useful. Good feeding and management are the chief preventives. We have occasionally seen bad cases of this kind that have been allowed to exist without treatment, the suffering animals being permitted to rough it with the rest. In such cases, they must have been exposed to great pain, which proper attention and humane feeding would easily have modified.

ITCH.—This is treated under "Scabies."

LICE.—Of all pests, properly so called, few are a greater torment to the pig-keeper or breeder than lice. The particular louse infesting the pig is *Hæmatopinus urius;* it is of goodly size, and young pigs in particular suffer severely from it. Where pigs are at all badly housed or kept, lice are sure to appear, and, if they are allowed to increase, the owner must not be surprised to find that his animals do not thrive. There is no difficulty connected with the extermination of lice. An application of a mixture of 1 part of Jeyes' fluid to 60 of warm water will kill the lice, or any of the sheep-dips commonly advertised will serve the same purpose. More than one application will be necessary on account of the unhatched generations. The sties, and particularly the sleeping quarters, must also be thoroughly treated with a hot solution of Jeyes' fluid. Lousy pigs have the skin marked with red pimples, and the itching set up is intense, especially at night.

LIGHTS (RISING OF THE).—A popular name for "Cough."

LIVER (DISEASES OF THE).—Notwithstanding the somewhat gluttonous habits of the pig and its varied provender, it rarely suffers from liver disease. The organ occasionally becomes congested, causing sickness and irregular bowels; but a couple of doses of salts ($\frac{1}{2}$oz. to 2oz. given dissolved in water and repeated in twenty-four hours) usually relieve the symptoms, and a light diet given for a day or two completes the cure. In old sows and boars structural changes of the liver have been noticed, but such cases are rare.

LUNGS (INFLAMMATION OF THE).—The common name for " Pneumonia."

MALFORMATION OF THE ANUS AND RECTUM.—This complaint is common to both pigs and calves, especially that form of it which is termed "Imperforation of the Anus and Rectum." The symptoms, according to Professor Hill, are constipation during the first two days, nausea, restlessness, pains, distension of the abdomen, and repeated expulsive efforts, which are attended, where the anus only is affected, with expansion of the fundament.

If the anus alone is to be dealt with, an incision should be made through the membrane, and, upon the escape of the excrement, it should be noticed whether there is any other malformation. The aperture thus made should be kept open by the insertion of a piece of oiled lint or linen, and the fæces removed from time to time. So soon as cicatrisation takes place, the lint may be removed entirely, and all will go well with the animal. In addition to the same measures being indicated in deficiency of the rectum, Professor Hill recommends the stitching of the bowel to the edges of the external wound, after it has been drawn out and opened.

MANGE.—The disease known to pig-keepers as mange is really " Scabies," and to that heading the reader is referred.

MEASLES.—This is a parasitic disease, due to the presence of bladder worms (or "measles") in the muscles. The scientific name given to these worms or measles is *Cysticercus cellulosæ*. They are really the larval form of the Tapeworm, *Tænia solium* (*T. armata*) infesting man. Pigs readily become infested with measles by picking up the ripe ova of *Tænia solium*. Once in the digestive organs, these eggs are dissolved, and the embryo Cysticerci penetrate the walls of the stomach and disperse themselves all over the body. If meat that is measly is thoroughly cooked the parasites are destroyed, and no harm comes of its consumption. Unfortunately for man, in most cases, the presence of measles is not detected in an animal during life, and the animals may, and often do, fatten and thrive notwithstanding the presence of large numbers of measles. But if the muscles of the pig are extensively invaded, there may be stiffness of the limbs and wasting of the frame. Vesicles on the tongue also point to the existence of measles. It has, too, often been noted that pigs afflicted with measles have a high carriage of the shoulders.

No treatment is of any avail. If the worms do not apparently interfere with the animal's general health, they eventually die, and a little white gritty substance is left in their place, and may be easily discovered in the flesh. From the foregoing remarks it will be seen that pigs should not have access to human excrement.

NASAL CATARRH is very similar to the disease of horses known by the name of glanders—that is, so far as symptoms are concerned—but it has no pathological similarity. All nasal catarrhs are communicable from one animal to another. The opinion of many eminent breeders is, that pigs affected with it should be destroyed, in order to prevent the complaint from spreading to other members of the herd. The symptoms are, a considerable discharge from the nose, with inflammation, which extends gradually to the pharynx, gullet, and larynx. The pig coughs, the mucous membrane enlarges, and the nose becomes thick and ill-shaped, the discharge also becoming more and more like blood. Although the animal may consume its food with apparently as much relish as before, yet it fails to put on any flesh, and the chances are that, in the end, it will pine away and die. Little hope can be entertained as to the curing of a pig affected with this disease.

NETTLE RASH.—The popular name for "Urticaria."

PARALYSIS OF THE MUSCLES OF THE LOINS.—This is described under the heading "Paraplegia."

PARAPLEGIA, OR PARALYSIS OF THE MUSCLES OF THE LOINS, does not seem to have a very detrimental effect upon either the appetite or the health of the pig. It is generally caused by concussion of the spinal column, but also by intestinal worms and constipation.

Apply mustard and water along the spine from the root of the tail to the middle of the back, and repeat the treatment at an interval of three or four days, desisting if the skin becomes too tender. The diet should preferably consist of sour milk and soft foods, and the sties should be of the most comfortable description. If there is constipation, warm water injections should be freely employed.

PARASITES.—These are dealt with under the headings "Lice," "Measles," "Scabies," "Trichinosis," and "Worms."

PLETHORA.—This complaint, frequently found amongst stock which are fed highly and upon a stimulating diet, indicates that there is an excess of blood in the system. The pulse of an affected animal may be observed to be very full and bounding, the bowels to be constipated, and the skin of an excessively high temperature.

The food in all such cases should be diminished, both in quantity and quality, this treatment being accompanied by the administration of aperients.

PNEUMONIA, OR INFLAMMATION OF THE LUNGS.—This disease, when contracted by a pig, is most likely to prove fatal, especially if it is allowed to exist for any length of time before being attended to. The general symptoms are disinclination to feed, difficulty in breathing, coughing, and shivering.

Pigs suffering from pneumonia should be housed in warm and comfortable apartments, provided with a suitable diet and clean water, and a dose consisting of ½oz. of hyposulphite of soda and 15 drops of tincture of aconite given night and morning in gruel or mash. Or the following may be given as a drench night and morning: Spirits of nitrous ether, ½oz. to 1oz.; nitrate of potash, 1dr. to 2dr.; water or gruel, ½ pint. Too much dry food is deleterious.

In ordinary cases of pneumonia, it is a common practice to rub the sides of the pig with a mixture of turpentine, ammonia, and olive oil, in equal parts. Mustard water will also answer the purpose. The animal should be removed to a warm sty, kept thoroughly clean, and fed upon an even, nourishing diet. Pneumonia is speedily fatal, and if it is not vigorously attacked there will be little hope of saving the animals affected.

QUINSY.—This is described under "Strangles."

RASH (NETTLE).—The common name for "Urticaria."

RECTUM (INVERSION AND MALFORMATION OF THE).—These complaints are dealt with under "Inversion of the Rectum" and "Malformation of the Anus and Rectum."

RED SOLDIER.—A popular name for "Swine Fever" (to which heading the reader is referred).

RHEUMATISM.—Pigs suffering from rheumatism exhibit lameness, and this is more or less noticeable according to the extent to which they are afflicted. There is, moreover, an indisposition to move about, with a general lack of activity, the complaint sometimes extending from limb to limb, and causing an elevated temperature, with tenderness in the joints or muscles affected. Rheumatism may or may not be attended with inflammation and swelling of the affected parts, and the characteristic symptom is that the pain or lameness changes from one limb to another. In many cases the cause of rheumatism is hereditary (stock born of rheumatic parents). It may arise from low and damp situations, long exposure to cold, damp sties, and injuries may also cause an attack in animals predisposed to it.

Place the patient in a warm and dry sty, with plenty of bedding. Give 1oz. or 1½oz. of Epsom salts, and repeat this at an interval of three or four days. If the symptoms continue, give 10gr. to 20gr. of iodide of potassium in the morning, and 30gr. to 60gr. of bicarbonate of potassium in the evening. If this is not successful, give 20gr. to 30gr. of salicylate of sodium night and morning; if the affected parts are sore or tender, rub into the parts daily compound liniment of belladonna.

RICKETS is a disease usually affecting the young animal, and in many cases is hereditary. Other causes are breeding from old or debilitated sows, exposure to wet, improper nourishment, badly-ventilated sties, and general insanitary conditions. Many of the early symptoms of rickets would be overlooked by an amateur. The first noticeable is a stiffness of the joints, later perhaps there is lameness; then the joints become enlarged, and are tender to the touch, and the limbs have a bowed appearance. The spine may be affected, giving the pig a "roached back"; or the opposite condition may exist, when the spine appears to drop, i.e., there is a depression in it.

In treating cases of rickets, all conditions that are likely to foster the disease, such as wet, badly-ventilated sties, must be overcome by removing the animal to better quarters. If the food is poor or insufficient, this must be altered. Give plenty of clean litter. Medicinally, give 10gr. to 20gr. of phosphate of calcium to each pig, night and morning; it can be mixed with some boiled potatoes and a little barley meal. Rickety animals, although they may eventually do fairly well to fatten, never attain the same size as healthy ones, and they should not be used for breeding purposes.

RISING OF THE LIGHTS.—Another name for "Cough."

RUPTURE.—The common name for "Hernia."

SCABIES, OR ITCH.—This disease, which is often popularly referred to as "Mange," may be mentioned, although, fortunately for pig-owners in this country, it is seldom seen. It is due to a parasite, *Sarcoptes scabiei*, var. *suis*. The symptoms are as follow: Violent itching, the skin presents small red points; these develop into pimples, and a watery fluid exudes which dries into a scab and eventually falls off. The skin cracks and becomes thickened, and unless the disease is checked the whole body may become a mass of scabs and cracks. Usually scabies is first noticed on the head—the ears and eyes—but the parasite quickly spreads. Inasmuch as one disease is *communicable to other domestic animals and to man,* it needs to be dealt with carefully.

A good dressing is flowers of sulphur, 4oz.; linseed oil, 1qt.; and oil of tar (*oleum cadinum,* P.B.), 3oz.; or a dressing of Jeyes' fluid (say, 1 in 60 of water) may be tried. Spratt's mange lotion, as used for dogs, may also be tried. A mixture of oil of turpentine, 8 parts, and sulphur 1 part, has been known to effect a cure in very bad cases. Whatever dressing is used, it must be repeated at intervals of a few days. Three or four dressings should prove sufficient. Thorough disinfection of the sties by scrubbing or spraying, or both, is all-important. A 5 per cent. solution of carbolic acid in water is possibly the best and surest preparation to use, but strong Jeyes' also answers well. In all cases of scabies it will be advantageous first to wash the affected animal with soap and warm water.

Pigs are sometimes affected with an outbreak upon the skin which resembles scabies, but which is not the work of a parasite. This is caused by an impoverished state of the blood. It will be found treated under the heading of " Eczema."

SCOURS.—A popular name for " Diarrhœa " (to which the reader is referred).

SCROTAL HERNIA.—This is described under " Hernia."

SORE TEATS become very troublesome in certain sows. Sometimes this is due to eczema, or to the sharp teeth (tusks) of the young. If the latter are the cause, nip off the teeth with pincers. If soreness is due to chafing, apply a little glycerine once or twice a day. If eczema is the cause, give a few doses of castor oil, and follow with 5gr. of reduced iron night and morning; this is tasteless, and can be given shaken up in a little milk or gruel.

STAGGERS (BLIND).—A common name for " Epilepsy " (under which heading it is described).

STOMACH (INFLAMMATION OF THE). — A common name for " Gastritis."

STRANGLES, OR QUINSY.—The symptoms of this disease are inflammation of the throat and tonsils, difficulty in breathing

and swallowing, a protruding tongue, which is more or less
covered with saliva, and a swollen and sometimes gangrenous
neck. Moreover, the animal refuses to eat, and its voice is
hoarse. Upon opening the jaws, a red or scarlet tint is
noticed in the mouth, and then a greyish tint, while some parts
of a false membrane appear upon the palate and the supports
of the tongue. The presence of this membrane induces suffo-
cation, the interior obstruction extends considerably, and becomes covered with spots of different colours. There is also
a serous fluid which exudes to the surface, and, unless the
disease is checked in a very short time, the animal dies. By
immediate attention it is possible to conquer the disease, and to
prevent a fatal result. Pigs that are fed too well, or are too
fat, are those which are generally attacked.

On the first appearance of the disease a strong emetic should
be given, which may consist of 30gr. of powdered ipecacuanha.
This may be either given by itself, as a dose, or mixed with
the food. A little castor oil in addition will also prove
beneficial. In cases where there is very great difficulty in
breathing, a blister should be placed over the throat; and,
provided the animal can swallow, ½ dram of nitrate of potash,
mixed in water, should be given. The mouth should be washed
out with the following gargle: Boracic acid, 1 dram; water to
6oz. Put the neck of the bottle in one side of the mouth, and
let the liquid run out of the other side. If a little is swallowed
it will do no harm.

SWINE ERYSIPELAS.—Under this name has been known a very
virulent disease that on more than one occasion has caused
considerable trouble amongst pig-owners in this country and
abroad. A quite recent outbreak in Cambridgeshire has
directed attention to the disease; but as long ago as 1896 and
1897 there was drawn up a report, the result of a departmental
inquiry, and this was published. The presence of the disease is
first manifested by a redness behind the ear; but so quickly does
it spread that the whole body is soon involved and the surface
becomes purple. Abroad it assumes a very virulent form, and
it is recorded that death ensues in from twelve to thirty-
six hours. In France it is known by the name of "Rouget du
porc," and in Germany by that of Swine "Roth-lauf." Though

often confused with swine fever, it appears to be perfectly distinct therefrom.

Besides the external symptoms that characterise the disease in its earlier stages it may be well to state what the *post-mortem* examinations have revealed. In the acute form the valves of the heart are involved, and fibrous growths are found thereon, these harbouring large numbers of the bacilli responsible for the disease. The blood, however, is the chief vehicle by which the bacillus is propagated. Apart, too, from the redness and swelling of the skin, there is slight inflammation of the lining membrane of the stomach and intestines, as well as a swelling of the spleen. Those pig owners having reason to suspect the disease in their stock should at once communicate with the proper local authority.

SWINE FEVER.—Swine plague is a specific fever, which was for many years regarded as a form of anthrax. It is communicable through the atmosphere, by means of food and by inoculation, and by means of persons who have been in contact with diseased pigs elsewhere, and the virus is contained in the excrement, saliva, blood, and discharges from the skin. This disease, which has been the cause of such immense loss to pig-breeders of all classes, and which is known under the names of "Red Soldier," "Black Jack," "Hog Cholera," &c., may generally be recognised by anyone who has once seen it. It sometimes, however, exists in an obscure form, and in such cases it is not difficult to understand that animals which are removed from one part of the country to another, unsuspected, may be the means of spreading the disease.

The following embodies the latest information drawn up by the Ministry of Agriculture and Fisheries for the guidance of those whose pigs are suffering from this virulent disease.

The onset is rapid, especially in young pigs. In the acute form death usually occurs in about three days. In chronic cases the symptoms are less definite; breathing is quick, temperature high, the pigs seem to have lost control of their hind quarters and stagger about in attempting to walk. A red rash appears on the skin at the base of the tail, under the belly, the inside of the thighs, and on the ears. More usually

2 B

the symptoms come on slowly; the pigs appear dull; they lie
under cover and are disinclined to move; the appetite is lost and
frequently animals vomit. A mucous discharge may be present
around the eyes. Red patches are also observed, and the
temperature rises to 104-106 F. Not unusually symptoms of lung
trouble are manifested, and in such cases sick animals suffer from
a short cough and laboured breathing. Lung disturbance is not
necessarily due to swine fever, but frequently accompanies it, and
if it appears in a number of pigs this is a suspicious sign of the
presence of swine fever. In the less acute form of the disease
animals die in from one to three weeks. They may, however,
recover or drag on for two to three months or longer in an emaciated
condition.

Owners of pigs affected with or suspected of swine fever are
required by law to give notice at once to the police. Strict watch
must be kept of the suspected animals, which should be separated
from all others until the nature of the malady has been determined.
Pigs so isolated should be attended by special persons, who must
not in any circumstances go among other pigs either on the home
premises or elsewhere. All unauthorised persons should be
excluded wherever there are symptoms of illness the precise nature
of which is obscure, as it is well to consider the possibility of swine
fever. Continued unthriftiness among young pigs should of
itself be looked upon with suspicion. Wherever premises have
been declared by a notice to be a swine fever infected place the
owner or occupier should read carefully the rules printed in the
notice. These must be strictly observed, and failure to do so
renders the offender liable to a penalty.

From time to time the slaughter of pigs on infected premises has
been tried as a means of stamping out swine fever. This measure
is combined with general restrictions on the movement of pigs from
district to district. While the restrictions, particularly those
controlling the movement of pigs from markets, have proved most
useful in checking the spread of the disease from an infected to a
clean area, no success has attended the policy of slaughtering swine
on infected premises.

During the years preceding the War experiments were made in
treating pigs by the injection of serum as a preventive against

attacks of swine fever, and the success of this experiment led in 1915 to the introduction of the practice by the Ministry of Agriculture. The injection is employed wherever a pig-owner consents to its adoption. It does not cure swine fever in pigs already ailing, but it will protect from any damaging attack those not infected, making them safe from risk of infection by contact for a period of ten days after the injection of the serum. If the pigs so treated mix with pigs actually suffering from swine fever they will probably catch the disease in a mild form, and if they do so before the effect of the serum has worn off they will remain permanently immune. If, however, they are not mixed with infected pigs their immunity will last only for ten days.

It is therefore to the advantage of pig-owners that they should immediately report suspected cases to the police, so that prompt measures may be taken for the treatment of other pigs on the premises. In this way losses can be minimised. The Ministry no longer orders slaughter of pigs on infected premises, because that policy is very expensive to the community and has not been found effective in eradicating disease. Slaughter is practised by the Ministry of Agriculture only for the purpose of determining definitely whether or not swine fever is actually present in suspected cases. If the serum treatment is used it is possible to allow owners of infected premises to restock earlier than formerly, as the new pigs can be treated with serum and so protected.

Teats (Sore and Inflamed).—These complaints are dealt with under "Inflammation of the Udder or Teats" and under "Sore Teats."

Tetanus.—Recent research has established the fact that this disease is due to a microbe, and is contagious. The microbe has been found to exist in the earth, and it has been proved that this may spread from one animal to another through the agency of wounds. The virus remains near the wound; it does not attack remote organs. This disease occurs mostly after castration, and is due to the virus gaining access through the wounds caused thereby. The period of dentition is also a favourable time for the virus to gain access. The first symptom is a stiffness of the limbs, then the eyes stare; the breathing becomes short; these symptoms become accentuated,

2 B 2

and finally the animal is scarcely able to turn round, the
limbs are rigid, and exhaustion is followed by death.

Pigs rarely recover from this malady. They should be killed
upon the first symptoms showing themselves. Authorities
declare that if the flesh around the wound is removed, the meat
is quite wholesome, and fit for consumption.

TRICHINOSIS, OR TRICHINIASIS.—This disease is common to
most other mammals, and is due to the presence in the system
of a very minute Nematode worm (*Trichina spiralis*). The
parasite was first discovered in 1828, and the portion of flesh
containing the trichinæ is still preserved at Guy's Hospital,
London. Four years later it was again observed by Mr. J.
Hilton in the muscles of an old man who died at Guy's Hospital.
He referred it to Professor Owen, who called it *Trichina spiralis*,
a name that is very appropriate on account of the hair-like
nature of the worms, and the way they are spirally coiled up
in their cyst. They may be seen by the unaided eye. In 1856
and 1857 Leuckart worked at the life-history of these remarkable worms, and a year later also made some important
investigations in the same direction. To Zenker, however,
belongs the chief credit for working out the remarkable life-
history of these parasites. In 1859, the German naturalist,
Virchow, explained how it developed and transformed itself
in the body of the pig and the human being. According
to Leuckart, an ounce of the flesh of the cat which
he examined contained no less than 300,000 of these parasites.
The body of the pest is round and thin, and the head narrow
and pointed. The trichina lives and multiplies in the intestines of the pig. When a human being consumes pork
affected with it, the larvæ enter the intestines, and remain
there for some little time. They subsequently pass into the
veins and muscles, and, reaching the centre of the latter,
enclose themselves in a sort of cyst, or cell, and multiply
with great rapidity, altering the entire muscular system, and
producing the greatest disorders, which often end in death.

Pigs which are allowed to run at large, and to consume
decaying animal food, can never be depended upon to supply
meat free from trichinæ. In two days after the worm is
taken into the stomach of the pig, it reaches its adult stage,
and in five more days it has so greatly multiplied that little

time is required for it to pass from the intestines to the muscles, causing, from its assiduous boring, no small amount of pain. An attack lasts some four to eight weeks, but if the pig lives over the sixth week, its restoration to health may be anticipated, although it has been stated by some authorities that they have no knowledge of an instance where swine have been lost by being infested with trichinæ. It is estimated that *T. spiralis* gives birth to a hundred young worms at the end of six or seven days, and a pig that had swallowed, say, a pound of trichinæ-infested flesh might readily contain after five days 250 millions.

The usual symptoms of the disease are a wasting of the flesh, and suppuration of the muscular fibres, great pain and considerable weakness, and, besides internal disorders, there are sometimes a swelling of the tongue and abundant perspiration.

So far as we are aware, the first epidemic attack of trichinosis occurred at Dresden, towards the end of the year 1859, and, according to Dr. Zenker, it was introduced by a single pig. Other epidemics have since occurred, and especially in Germany; but we believe it has never, either in an epidemic or any other form, existed in France, where, contrary to the German custom, the meat is well cooked before being eaten.

To prevent an epidemic among human beings, the meat should always be thoroughly cooked before consumption, its temperature being raised to not less than 200 to 212deg. Fahr. throughout its whole body. In this way the parasites will be completely destroyed. Prolonged salting, however, or a "hot fumigation," will have the same effect.

TUBERCULOSIS.—This disease is common among cattle, and pigs are not exempt from it, although they are rarely affected. The symptoms are a cough, wasting of the body, and frequently a fluid discharge from the bowels. The chief signs are usually found in the lungs, but other organs, especially the glandular structures of the bowels, may be affected. Upon examination of the carcase, more or less rounded growths are noticeable, ranging in size from that of a pea to that of a pigeon's egg, and these, especially when observed in the chest, may be in clusters—hence butchers often call the disease "Grapes."

A question arises in this disease as to whether or not affected animals are fit for food. Some contend that if the disease is in the early stages, and especially when confined to the lungs, the flesh is quite fit for human consumption : others assert that the carcases of all animals, no matter to what extent they are affected, should be condemned. Careful experiments have been made with animals that are susceptible to the disease, and it has been proved that the disease can be communicated by feeding them with tubercular meat. This being so, and as the same experiments cannot be tried upon man, to make the matter conclusive, it is hardly expedient to run the risk by using tubercular meat, and it would be safer to condemn it.

UDDER (INFLAMMATION OF THE).—This is dealt with under " Inflammation of the Udder or Teats."

UMBILICAL HERNIA.—This is described under " Hernia."

URTICARIA, OR NETTLE-RASH.—This condition is often met with in pigs, and in many there is a predisposition and a hypersensitive skin, but the usual causes are disarrangement of the stomach and bowels, and a congested liver, due to improper diet. Stimulating foods, such as peas or new beans, given in too great quantities, may cause it, so also may food of an inferior quality. The skin in urticaria becomes congested in patches, the fluid part of the blood oozes from the vessels, and the skin exhibits the appearance of having been stung with nettles—hence the common name, " nettle-rash." There are little raised white patches, surrounded by a congested ring or band, and this may be accompanied by intense itching, and the scratching against objects increases the symptoms; but extreme itchiness is not always present. The rash may come on quite suddenly, and disappear in the same way, leaving no trace behind. The whole skin may be attacked in patches, or there may only be one or two patches in all.

As above mentioned, the disease is usually due to deranged digestive organs, so these should have rest. Abstention from food, at least of a solid nature, for eighteen hours is beneficial, and a purgative should be given— 1oz. to 2oz. of Epsom salts, or 2oz. or 3oz. of castor oil; whichever is given, it should be repeated once or twice at intervals of a day or two. Afterwards give a few powders each consisting of bicarbonate of potish, 1 dram; carbonate of bismuth, 15gr.—one night and morning in food, change the diet, and regulate the quantity carefully.

WORMS.—The commonest species of worm infesting the pig is *Ascaris suilla,* a close ally, if not a variety, of the Round Worm affecting man; still, it is not an abundant species. It is several inches in length, white, thick in the middle, and tapering to both ends. Usually the pig, when attacked, has a highly ravenous appetite, but fails to put on any flesh. The reason of this is that the *Ascaris suilla* lives on the alimentary principles of the bowels of its host, and starves the latter by reducing its supply of nourishment. Moreover, the pest causes colic and bowel-obstruction. There are others that affect the liver.

For the removal of the Round Worm there is nothing more effective than santonin. This is a vegetable extract, obtained from *Artemisia maritima,* which is largely used in the composition of vermifuges. The dose is from 10gr. to 20gr., given in 1oz. of castor oil, and followed in twelve hours by ½oz. to 1½oz. of Epsom salts. It can be given for three alternate days. This treatment is effectual in most cases for all worms affecting the pig. Should it fail, 20 to 90 drops of oil of male fern may be given in 1oz. to 1½oz. of castor oil; three doses can be administered, allowing four days between each; three hours after each dose give a dose of Epsom salts, as suggested above. Thorley's Worm Powders are another effectual remedy for worms. See also "Measles."

WOUNDS.—Wounds in the pig are not infrequent, but they are extremely difficult to deal with, inasmuch as a bandage or any solid application is certain to be rubbed off. When the animal is healthy, the wound will generally heal of itself, there being nothing to fear, unless it be septic influence from without. To prevent this, some diluted Condy's Fluid, or a solution of permanganate of potash, or a 5 per cent. solution of carbolic acid, or a solution of Jeyes' fluid, or a saturated solution of boracic acid, should be syringed over the affected part. If the wound is in the foot, there is nothing better than an occasional dressing of tar.

INDEX.

A.

Abortion, 346
  cause of, 65
Abscess, 346
Acorns, 272
Administering medicine, 345
Age attained, 4
  determining, 17-28
  for breeding, 29, 51, 62, 103
  for breeding gilts, 103
  influence of, on litter, 53
  registration of, 82
Albuminoids, 294
Alençonnaise pig, 183
Allen, Hon. A. B., on the Berk-
    shire, 148
Allender, Mr. G. M., on the
    Tamworth, 169
Alsacienne pig, 184
American breeds, 7
  Berkshire record, 334
  Berkshires, 147
  curing, 307
  exports, 2, 323
  piggeries, 224
  pork trade, 322
  slaughtering establishment, 301
Anæmia, 346
Ancient authorities, 3
Angevine pig, 183
Animal food, 279
Anthrax, 347
Anus, imperforation of, 362
Aphthous fever, 356
Apoplexy, 351
Ardennaise pig, 184
Arnold, Mr. Levi, on the
    Poland-China, 177
Artésienne pig, 184
Artichokes, 277
Ascaris suilla, 375

Ashcroft, Mr. W., on the Berk-
    shire, 142
Auberjonois' (M. Gustave)
    Swiss piggery, 209
Augeronne pig, 182, 183
Augoumoise pig, 182

B.

Bacillus anthracis, 347
Bacon, American exports of, 323
  analysis of, 330
  beetle infesting, 315
  composition of, 330
  curing, 306, 313
  feeding for, 260
  keeping, 312, 315
  pests, insect, 315
  pigs, 11, 12
  salting, 306, 310
  smoking, 306, 314
Baconers, feeding, 260
"Band-box" hogs, 173
Barker, Mr., quoted, 115, 126
Barley, 257, 270, 331
Barrow, definition of, 339
Baths, 29, 30, 205-238, 253
Bayonne pig, 185
Beans, 257, 271
Beau Cèdre piggery, 195, 209
Bedding or benches, 72, 103,
    191, 192
Beechnuts, 272
Beetle, bacon, 315
Beginner, pig for, 97
Belgian pigs, 187
Benches or bedding, 72, 103,
    191, 192
Benjafield, Mr. N., quoted,
    103-106
  on feeding, 254-258
  on the Berkshire, 146

Bennett, Mr. Thomas, quoted, 103-107, 172
Berkshire breed, 135
  Allen, Hon. A. B., on, 148
  American Association, 149
  American imports, first, 151
  analyses of, 327, 329
  as a bacon pig, 141, 152
  Ashcroft, Mr. W., on, 142
  at shows, 138
  Benjafield, Mr. N., on, 146
  Camperdown's (Countess of) herd, 144, 145
  celebrated specimens, 136, 145, 150
  characteristics of, 137, 140, 147
  Coburn, Hon. F. D., on, 147
  colour and marking of, 136, 140, 143, 148, 152, 154
  crosses, 136, 137, 139, 142, 144, 148, 149
  deterioration of, 137
  faking, 153
  fattening, 144, 145
  Fisher, Mr. J., on, 139
  Fowler, Mr. R., on, 145
  Gibson, Mr. A. S., on, 140
  Humfrey, Mr. Heber, on modern points, 153
  in America, 135, 147
  judging, 140
  modern, 137, 153
  objections to, 139, 140
  original, 135, 149
  points of, 141, 143, 146, 153, 154
  popularity of, 135
  scale of points, 145, 149
  Siamese cross, 149
  Sidney on, 136
  Smith, Mr. J., on, 143
  snout of, 151
  Society, 111, 126, 129, 140, 166
  standard of excellence, 148
  weights of, 136, 141, 142, 144, 145, 146, 149, 150
  Wykes, Mr. W. H., on, 144
Berrichonne pig, 182

Bins for food, 205-248
Birth, weight at, 29
Black Dorset breed, 162
  as mothers, 164
  characteristics of, 163
  Coate, Mr., and, 163-165
  crosses, 163, 164
  early maturity of, 163
  improved, 164
  local breeds, 165
  weights of, 164
Black Jack, 369
Black, Large, 155. (See Large Black)
Black puddings, 319
Black, Small, 157. (See Small Black)
  Suffolk, 157
Blind staggers, 355
Blood, 301
  composition of, 29, 325
Boar, the, 55
  age of, 54, 58
  castration of, 60, 87
  catching, 59
  classification at shows, 336, 338
  cost of keeping, 60
  definition of term, 14
  fatting, 60
  feeding, 254, 264, 265
  for cottager's pig, 99
  house for, 57
  importance of, 8, 40, 54
  management of, 57
  on the farm, 60
  points of a good, 56
  rations for, 264, 265
  rearing, 56
  removing tusks of, 59
  second serice of, 58
  selecting, 56
  utilising old, 60
  young, 81
Board of Agriculture's regulations, 347, 369
Bones, 15, 16
  proportion of, 325
Bourbonnaise pig, 182, 183, 185
Boussingault's experiments with food, 252, 276

Brawn or boar, 14
Breed, choice of, 33
Breeders, bad, 83
Breeding, 31, 63
  age for, 29
  art of, 35
  for the butcher, 10
  gilts, age for, 103
  good *v.* bad stock, 32
  herd book, value of, in, 39
  inconsistent in different countries, 7
  points to be obtained in, 44
  sire, importance of, 40
  stock, changes in, 10
  sows, rations for, 263-265
  what to avoid in, 42
Breeds, principal, 6. (See under their names)
Bressane breed, 185
Bresse pig, 185
Bretonne pig, 182
Brett, Rev. F. H., on foot-and-mouth disease, 357
Brewers' grains, 279
Brick floors, 190
Brine, 306
Bristles, 15, 296, 297
British Berkshire Society, 111, 126, 129, 140, 166
British pigs, 1, 9, 187
  superiority of, 188
Brown's pig-ring, 93
Buckwheat, 271
Bugey pig, 185
Building materials for piggeries, 190, 205-248

C.

Cabbages, 275
Camperdown's (Countess of) Berkshires, 144, 145
Canada, pigs in, 12
Cankers, 351
Canton, Illinois, piggeries at, 227
Carbo-hydrates, 294
Carcase, chemical constituents of, 325
  classes at shows, 47, 338
  cutting-up, 299

Carrots, 261, 277
Castration, 60, 76, 86
  of adult boars, 60, 87
  of young pigs, 76, 86
Catarrh, nasal, 363
Catching-gate, 90, 91
Catching pigs, 59, 89, 90
Cauchoise pig, 183
Chambers' pig-catching gate, 90, 91
Champion ring and pig-holder, 94, 95
Characteristics, general, 4
Charcutier, the French, 184, 316
Charollaise pig, 183, 185
Chase, Dr., on the Poland-China, 176
Cheese, pig, 320, 321
Chemistry of the pig in relation to commerce, 325
Cheshire breed, so-called, 118, 165, 180
Chester White breed, 179
  Coburn, Hon. F. D., on, 180
  colours of, 180
  crosses, 180
  objections to, 179
  origin of, 179
  points of, 180
  weights of, 180
Chicago slaughtering establishment, 301
  show, carcase classes at, 338
Chinese breed, 6, 131. (See Poland-China)
Chitterlings, 318, 319
Cholera, hog, 369
Cinders for pigs, 84, 193
Clapham Park piggeries, 208
Classics, ancient, quoted, 3
Classification at shows, 138, 336
  model, 338
  scientific, 14
Clay, Mr. S. H., on cooked food, 282, 283
Clover, 273
  compared with other foods, 273-275
  Sullivant, Mr., on, 274
Coal for pigs, 84, 193

Coate, **Mr.** Frederick, quoted, 103-106, 163-165
  on feeding, 254-259
Coburn, Hon. F. D., author of "Swine Husbandry," quoted, 60, 115, 133, 147, 160, 174, 180, 230, 266, 273, 277, 280, 282, 286, 309
Coche, 14
Cochon, 14
Cohoon's (**Mr. A. R.**) piggery, 233
Coleshill breed, 116
Colour in pigs, 15
  choice of, 47
Commerce of the pig, 296
Composition of pig foods, 292
  of pork, chemical, 325
Concrete floors, 180
Conover, Mr., on the Poland-China, 176
Constipation, 352
Constituents of food consumed by pigs, 293
Continental pigs, 182-188
Cooked food, 257, 260, 275, 280
  apparatus for, 202, 283
  Clay, Mr. S. H., on, 282, 283
  Coburn, Hon. F. D., on, 280, 282
  experiments with, 280, 285
  Sisson, Messrs., on, 281
  Sullivant, Mr. J., on, 284
  value of, 280
Cooling-rooms, 300
Corn, composition of, 292. (See Foods)
Corse breed, 185
Cost of boar-keeping, 60
  of pig-keeping, 32, 97
Cotentine pig, 183
Cottager's pig, 97
Cough, 352
Courts of piggeries, 201
Craonnaise breed, 182-184
Crépinettes, 316
Cumberland breed, 116, 118, 132, 133
Curing, 306
  American systems of, 307

Curing, Coburn, Hon. F. D., on, 309
  hams, 311
  "home," 309
Curtis, Col. F. D., quoted, 117, 118, 171, 172, 181
Cuts, 352, 375
Cutting (castration), 60, 76, 86
Cutting-up Carcase, 299
Cuvier quoted, 14
Cysticercus cellulosæ, 363

D.

Danish pigs, 2, 187, 188
Debatable points, 102
Demole, M., opinions of, 186
Denmark, pigs in, 2, 187, 188
Dentition, 17
Dermestes lardarius, 316
Detmers, Dr., quoted, 250
Diarrhœa, 352
Diseases, 340
  abortion, 346
  abscess, 346
  anæmia, 346
  anthrax, 347
  anus, imperforation of, 362
  aphthous fever, 356
  apoplexy, 351
  black jack, 369
  blind staggers, 355
  cankers, 351
  catarrh, nasal, 363
  causes of, 340
  cholera, hog, 369
  constipation, 352
  cough, 352
  cuts, 352, 375
  diarrhœa, 352
  eczema, 354
  enteritis, 354
  epilepsy, 355
  erysipelas, 368
  fever, swine, 369
  foot-and-mouth disease, 356
  foot-rot, 358
  gastritis, 358
  grapes, 374
  gravel, 359
  hernia, 359

Diseases, hog cholera, 369
imperforation of anus, 362
inflammation of lungs, 364
inflammation of stomach, 358
inflammation of teats, 360
inflammation of udder, 360
inversion of rectum, 361
itch, 366
lice, 361
lights, rising of, 352
liver diseases, 362
lungs, inflammation of, 364
malformation of anus and
 rectum, 362
mange, 366
measles, 363
nasal catarrh, 363
nettle rash, 374   [364
paralysis of muscles of loins,
paraplegia, 364
parasites, 363, 367, 372, 375
plethora, 364
pneumonia, 364
premature birth, 346
quinsy, 367
rash, nettle, 374
rectum, imperforation of, 362
rectum, inversion of, 361
rectum, malformation of, 362
red soldier, 369
rheumatism, 365
rickets, 366
rising of lights, 352
round worms, 375
rupture, 359
scabies, 367
scours, 352
scrotal rupture, 359
sore teats, 367
staggers, blind, 355
stomach, inflammation of, 358
strangles, 367
swine erysipelas, 368
swine fever, 369
tapeworm, 363
teats, inflamed, 360
teats, sore, 367
tetanus, 371
trichiniasis, 372
trichinosis, 372

Diseases, tuberculosis, 373
udder, inflammation of, 360
umbilical rupture, 359
urticaria, 374
worms, 375
wounds, 375
"Dished" snout, 17
Distillery waste, 279
Domestic pig compared with
 wild boar, 4, 5, 8, 14, 27, 46
Doors of piggeries, 192, 205-248
Dorset breed, 7, 162. (See Black
 Dorset)
Doses of medicine for pigs at
 various ages, 341
Drainage of piggeries, 190-192,
 205-248
Drugs, 341
Dry foods, 251, 253, 265. (See
 Foods)
Drying the sow, 78
Ducie's Lord, Small Whites, 133
Duroc-Jersey breed, 170
Bennett, Mr. Thos., on, 172
colour of, 171, 172
Curtis, Col., on, 171, 172
feeding, 172
Illinois breeder on, 172
Lyman, Mr. J. B., and, 172
origin of, 171
points of, 171
Red Pig Club, 171
standard of excellence, 171
weights of, 171, 172
Dutch pigs, 187

E.

Early breeders, 6
importations, 5, 6
writers, 3
Ears, 16
Eber, 14
Economic feeding, 13
Economy of the carcase, 5
Eczema, 354
Ellesmere's (Lord) piggeries, 207
Ellsworth, Mr. W. W., on feed-
 ing, 254-259
on the Poland-China, 177
piggery owned by, 233

England, pigs in, 1, 10
English-French cross, 186
English pigs in France, 186
    pigs in Italy, 187
    Small White breed, 132, 133
Ensilage, 98
Enteritis, 354
Epilepsy, 355
Erysipelas, 368
Essex breed, 7, 136, 158, 162
    Neapolitans, 158
Exhibitions, classification at,
    138, 336, 338. (See Shows)
Expense of keeping boar, 60
    of pig-keeping, 32
Exports from America, 2, 323
Eyes, 16

F.

Factories, large, 298, 301
Farrow gilt, feeding, 64
Farrowing, 66, 67
    after, 83
    attention during, 66, 68
    bedding or benches for, 72
    before, 69
    excitement to be avoided
        during, 66
    indications of, 65, 68
Farrow sow, feeding, 65, 253,
    263-265
Fat pigs, classification of, at
    shows, 336, 338
    proportion and composition
        of, 46, 325
Fattening, effects of temperature
    on, 285
    foods for, 256
    for exhibition, 258
    on purchased foods, 262
Feeding, 249. (See also Foods)
    apartment for, 202, 205-238
    appliances, 202
    Benjafield, Mr. N., on, 254-
        258
    boars, 254, 264, 265
    Boussingault's experiments in,
        252
    changes necessary in, 250
    Coate, Mr. Fredk., on, 254-259

Feeding, Dettmers, Dr., on, 250
    economic, 13
    Ellsworth, Mr. W. W., on,
        254-259
    exhibition pigs, 258
    experiments in, 251
    farrow gilt, 64
    farrow sows, 65, 253, 263-265
    fatting pigs, 256, 258
    hand, 79
    Howard, Mr. James, on, 253-
        258
    in-pig sow, 65, 253, 263-265
    Jones, Col. Walker, on, 254-258
    Knapp, Prof., on, 251
    litters, 254
    object in, 250
    on cooked food, 257
    opinions on, 253
    Platt, Col., on, 254-258
    principles to be observed in,
        252
    rations, Heuzé's, 263
    Robertson, Mr. James, on,
        254-258
    Sanson, Prof., on, 331
    sows, 65, 69, 254, 263-265
    Spencer, Mr. S., on, 254-258
    systems of various breeders,
        253
    " Thrifton," Mr. Phil, on,
        255-258
    troughs for, 193
    weaners, 77, 255
    young pigs, 59, 254, 255, 259,
        263-265
Fees for stockman, 85
Fever, swine, 369
Fidgeon, Mr. F. C., on the
    Tamworth, 168
First-prize American piggery,
    231
Fisher, Mr., quoted, 111, 112,
    117, 120, 130, 139
Flamand pig, 184
Flesh, composition and propor-
    tion of, 15, 29, 325
    food, 279
Floors of piggeries, 103, 190,
    205-248

Foods, 249, 265. (See also Feeding)
acorns, 272
albuminoids, 294
animal, 279
artichokes, 277
barley, 257, 270, 331
beans, 257, 271
beechnuts, 272
bins for, 205-248
brewers' grains, 279
buckwheat, 271
cabbages, 275
carbo-hydrates, 294
carrots, 261, 277
change of, necessary, 250
cheap, for cottager's pig, 98
clover, 273
composition of, 292
constituents of, consumed by pigs, 293
cooked, 257, 260, 275
corn, composition of, 292
distillery waste, 279
dry, 251, 263, 265
flesh, 279
for baconers, 260
for boars, 254, 264
for farrow gilts, 64
for farrow sows, 65, 253, 263, 265
for fatting pigs, 256, 258
for porkers, 260
for sow and young, 69, 254
for weaners, 77, 255
for young pigs, 255
garden refuse, 98, 99, 101
grains, 279
grass, 272
green, 272, 293
green, dry, 98
hotel waste, 280
Jerusalem artichokes, 277
maize, 251, 265, 331
mangels, 258, 277
meals, composition of, 292
milk, 278
oats, 270, 331
parsnips, 277
peas, 257, 271

Foods, potatoes, 258, 260, 275
pumpkins, 278
roots, composition of, 292
seed, composition of, 292
troughs for, 193
turnips, 258, 277
wash, 203
water in, 294
weeds, 101
wheat, 257, 269, 331
whey, 278
Foot-and-mouth disease, 356
Foot-rot, 358
Fowler, Mr. R., on the Berkshire, 145
France, consumption of pork in, 322
names used in, 14
pigs of, 182-188
French dishes from the pig, 316
piggeries, 215
pigs, 182-187
sausages, 316
terms, 14
French-English cross, 186
Fresh blood, importance of, 8
Fromage de cochon, 320
d'Italie, 321
Furstenberg on the teeth, 18

G.

Garden, food from, 98, 99
Gascogne pig, 185
Gastritis, 358
Gate for catching pigs, 90, 91
German names, 14
pigs, 2, 7, 187, 298
sausages, 318
slaughtering establishment, 298
Gestation, 65
table, 67
Gibson, Mr. A. S., on the Berkshire, 140
Gilbert, Dr., on composition of the pig, 330
Gilt, age for breeding, 103
definition of, 14, 62
feeding the farrow, 64
ringing, 64

Gilt, young, 81
Goret, 14
Grains, 279
Grapes, 373
Grass as a food, 272
    runs, 64, 205-236, 272
Gravel, 359
Grazer, pig as a, 250, 272, 274
Great Britain, pigs in, 1
Green foods, 272, 293
    foods, dry, 98

H.

Habits, 4, 5
Hæmatopinus urius, 361
Hair, 5, 15, 296, 297
Half-blood, 40
Hallo's (M.) piggeries, 216
Hamburg hams, 311
    sausages, 318
Hampshire breed, so-called, 165
Hams, American exports of, 323
    curing and smoking, 311, 314
    Hamburg, 311
    keeping, 312
    Nessler, Prof., on, 312
    prices of imported, 14
    Westphalian, 311
York, 313
Hand-feeding, 79
" Harry," the, 50
Head, 16
Heart, 28
Heat, period of, 63, 99
Hebron piggeries, 224
Henry, Prof., quoted, 70
Herd-books, 332
    private, 334
    value of, 39
Heredity, 36
Hernia, 359
Heuzé, M. Gustave, quoted, 28,
    29, 182, 184, 186, 200, 215,
    314, 320, 358
    rations drawn up by, 263
Hill, Prof., quoted, 362
Hilt, definition of, 14, 62
Hilton, Mr. J., and trichinosis,
    372

Hobbs, Mr. Fisher, and the
    Essex pig, 136, 158
Hog Cholera, 369
    definition of, 14
Holland, pigs of, 187
Holloway, Mr. Cephas, quoted,
    175
Home-curing, 309
Hoppers or jumpers in bacon,
    310, 316
Hotel waste as food, 280
Houses, 189. (See Piggeries)
Howard, Mr. James, quoted, 71,
    103-105, 112
    on feeding, 253-258
    piggery of, 208
Humfrey, Mr. Heber, on Berk-
    shire points, 153

I.

Ice-rooms, 300
Imperforation of anus, 362
Importations, early, 5, 6
Improvements in production
    methods, 9
Improving breeds, 8
In-breeding, evils of, 42
Incisors, 18-28
Inflammation of lungs, 364
    stomach, 358
    teats, 360
    udder, 360
In-pig period, 62, 65, 253, 263-
    265
Insect pests infesting bacon and
    hams, 315
Intelligence of the pig, 4, 30
Internal system, the, 28
Intestines, 28
    preparing, 318, 319
Inversion of rectum, 361
Ireland, pigs in, 1
Irish Grazier pig, 175
Italian cheese, 321
    pigs, 187
Itch, 366

J.

Jaws, 17-28
Jefferson County breed, 180

Jersey Red breed, 170
Jerusalem artichokes, 277
Johnson, Mr., on feeding, 267
Jones, Col. Walker, quoted, 103-106
on feeding, 254-258
Judging at shows, 339
Jumpers or hoppers in bacon, 310, 316

K.

Keeping bacon and hams, 312, 315
Kennedy Mr., quoted, 126
Killing, 299, 301
establishments, great, 298, 301
Knapp, Prof., quoted, 251
Koopman's, Mr. J. C., establishment, 298

L.

Labourer's pig, 97
Lancashire breed, 116, 118
Landaise pig, 185
Lard, 323
American exports of, 323
qualities of, 324
salting, 323
Large Black breed, 155
disposition of, 156
disqualifications of, 157
Herd-book, 157
points of, 156
registration, 157
society, 155, 156
trade mark, 156
weights of, 155
Large v. small breed, 116, 118
Large White breed, 109
American opinions of, 115, 117
Barker, Mr., on, 115
Bennett Mr. Thos., on. 173
breeders and mothers, 113
celebrated specimens, 110
Coburn, Hon. F. D., on, 115
colour of, 112
crosses, 109
Curtis, Col., on, 117
for bacon, 113
Howard, Mr. James, on, 112

Large White breed, Jones, Prof., on, 115
Kennedy, Mr., on, 115
origin of, 110
points of, 112, 113, 114
popularity of, 109
prices realised, 111, 113, 117
purity of, 116
Spencer, Mr., on, 115
standard of excellence, 111
Tuley's (Joseph) strain, 110
weights of, 111, 113, 115
Lawes, Sir J. B., on composition of the pig, 29, 330
on food increase and manure, 234
on foods, 268
Lean, composition and proportion of, 46, 325
and fat meat compared, 10
Léouzon's (M. Louis) piggery, 219
Leuckart on trichinosis, 372
Lice, 361
Lights, rising of, 352
Limousine pig, 185
Lincolnshire breed, 7, 118, 165
Litter (straw), 71, 72, 192
Litters, dividing, 78
feeding, 254
growing, treatment of, 74
influence of age on, 53
number of, 29, 79, 84
sale of, by cottager, 100
size of, 29, 74, 84, 124, 169
unequal, 72
Liver, 28
diseases, 362
Lorraine breed, 184
Low, Prof., quoted, 150
Lowe, Mr. John, on the Tamworth, 167
Lungs, 28
inflammation of, 364
Lying upon young, 66, 73, 104

M.

M'Murtrie, Prof., on the chemistry of the pig, 325

Maize, 251, 265, 331
  Coburn, Hon. F. D., on, 265
  Johnson, Mr., on, 267
  Lawes, Sir J. B., on, 268
  Miles, Prof., on, 266, 268
  Moore and Son, Messrs., on, 267
  Sullivant, Mr. Joseph, on, 268
  Voelcker, Dr., on, 266
Malformation of anus and rectum, 362
Mancelle pig, 183
Manchester breed, 116
Mange, 366
Mangels, 258, 277
Manure from sties, 190
  table, by Sir J. B. Lawes, 294
Markham, quoted, 3, 51
Mastication, imperfect, 249
Materials for piggeries, 190, 205-248
Mating, 63
Meal foods, composition of, 292
Measles, 363
Measurements and weights, 331
Meat, composition of, 15, 29
Mechi, Mr., quoted, 104, 190
Medicine, administering, 345
  chest, 341
  doses of, for pigs at various ages, 341
Memory, 30
Methods of production, improvements in, 9
Middle Ages, wild boar of, 4
Middlesex breed, 116
Middle White breed, 119
  American opionion of, 126
  breeders and mothers, 120
  celebrated specimens, 119, 120, 121
  characteristics of, 123
  colour of, 120, 126
  crosses, 124, 125
  Fisher on, 120
  head of, 121, 122
  origin of, 119, 120
  points of, 126
  prices realised, 125
  quietness of, 123

Middle White breed, shows, poor entries at, 120
  Sidney on, 123
  size of, 120
  size of litters, 124
  standard of excellence, 126
  Tuley's (Joseph) strain, 119
  weights of, 127
Middle York breed, 123, 126
Miles, Prof., experiments by, 266, 268
Milk, 278
  for young pigs, 79
  sows with poor supply of, 83
Mitchell Bros., Messrs., on the Tamworth, 170
Molars, 18-28
Mongrels, 40
Moore, Messrs. A. C., and Son quoted, 178, 262, 267
  piggeries owned by, 227
Moreton's (Lord) piggery, 205
  Small Whites, 133
Mouth, 17-28
Movable sties, 247

N.

Nasal catarrh, 363
National Pig-breeders' Association, 39, 111, 126, 129, 166, 332, 339
Navarine pig, 185
Neapolitan breed, 6, 158, 187
Nessler, Prof., on hams, 312
Nettle rash, 374
Nightcap part of chitterlings, 319
Nippers, 18-28
Nivernaise pig, 182, 183
Nomenclature, 14
Nonant pig, 183
Norfolk breed, 7, 118
Normandy, pigs of, 182
Norwegian pigs, 187
Number in litter, 29, 74, 84, 124, 169

O.

Oats, 270, 331
Offal, 5
  composition and proportion of, 325

Omnivorous feeding, 4, 29
Open gilt, 62
  sow. 62
Ovaries, removing, 76, 88
Overcrowding, 193
Owen, Prof., on trichinosis, 372

P.

Paaren, Dr., on the teeth, 18
Packing pork, 308
Paralysis of muscles of loins, 364
Paraplegia, 364
Parasites, 363, 367, 372, 375
Parsnips, 277
Parturition, 66, 67
Pay, pig-man's, 85
Peas, 257, 271
Pedigree, importance of, 8, 37, 39
Pens, 189, 205-248. (See Piggeries)
Perche pig, 183
Périgord pig, 185
Périgordine breed, 185
Period of gestation, 65, 67
Permanent teeth, 19-28
Pests infesting bacon and hams, 315
Physiology, 14
  internal system, 28
  teeth, 17-28
Picarde pig, 184
Pickled pork, 306
Pig catchers, 89, 90
  cheese, 320, 321
  medicine chest, the, 321
  puddings, 319
  skin, identification and uses of, 294
Piggeries and sties, 189-248
  American, 224
  aspect of, 190, 205-238
  Auberjonois' (M. Gustave), 209
  Beau Cèdre, 195, 209
  beds and benches, 72, 103, 191, 192
  breeding-sties, 191
  Canton, 227

Piggeries and sties, Clapham Park, 208
  Cohoon's (Mr. A. R.), 233
  cooking apparatus for, 202
  courts of, 201, 205-248
  division of, 200
  doors of, 192, 205-248
  double, 191
  drainage of, 190-192, 205-248
  Ellesmere's (Lord), 207
  Ellsworth's (Mr. W. W.), 233
  features of, essential, 190
  feeding apartment and appliances in, 202, 205-248
  first-prize American, 231
  flooring of, 190, 191, 205-248
  for pork, 111
  French, 215
  grass runs in, 205-238
  Grignon, 215
  Hallo's, M., 216
  Hebron, 224
  Heuzé, M. Gustave, on, 200
  Howard's (Mr. James), 208
  large, 205-238
  large, objection to, 193
  Léouzon's (M. Louis), 219
  litter in, 192
  manure from, 191, 192
  market types of, 11
  materials for, 190, 205-248
  Moore and Sons' (Messrs.), 227
  Moreton's (Lord), 205
  movable, 247
  overcrowding in, 193
  points necessary in, 200
  Poule (La), 219
  roofing of, 190, 205-248
  St. Maurice (Colonie Agricole de), 216
  sawdust for, 193
Piggeries, sizes of, 200, 205-248
  small, 238
  straw in, 192, 193
  Street and Sons' (Messrs.), 224
  Swiss, 209
  Tortworth Court, 205
  tramways in, 203, 205, 238

Piggeries, troughs for, 193, 205-
    248
    ventilators in, 205-248
    walls of, 201, 205-248
    warmth in, 189, 192
    water in, 202, 205-236
    Worsley, 207
    yards of, 193, 205-248
Platt, Col., quoted, 103-105
    on feeding, 254-258
Plethora, 364
Pneumonia, 364
Points to be obtained in breed-
    ing, 44
Poitevine pig, 182
Poitou pig, 183
Poland-China breed, 174
    American-bred, 177
    analyses of, 327, 329
    Arnold, Mr. Levi, on, 177
    characteristics of, 177
    Chase, Dr., on, 176
    Coburn, Hon. F. D., on, 174,
        176
    colour of, 177
    Conover, Mr., on, 176
    crosses, 174-176
    Ellsworth, Mr., on, 177
    Holloway, Mr. Cephas, on, 175
    Moore, Mr. A. C., on, 178
    origin of, 174, 177
    points of, 177, 178
    Warren County pigs, 175
    weights of, 176-178
Poor man's pig, 97
Porcelet, 14
Pork, American trade in, 322
    analysis of, 325
    composition of, 325
    consumption of, in France,
        322
    curing, 306
    derivation of name, 14
    exports, American, 323
    packing, 308
    pickled, 306
    preparations of, 316
    salting, 306
    trade, 298, 322
    type of pig for, 11

Porkers, feeding, 260
Potatoes, 258, 260, 275
    Boussingault's experiments
        with, 276
    cleaner for, 204
Poule, La, piggery, 219
Precocity, 51
Premature birth, 346
Premolars, 18-28
Preparations of pork, 316
Prepotency, 44
Prices of imported bacon, &c.,
    13
Prince Consort's small whites,
    129
Production methods, improve-
    ments in, 9
Profit, 32
Prolificacy, 8, 44
Puddings, black, 319
    pig, 319, 320
Pumpkins, 278
Purchased food, fattening on,
    262
Purchasing at shows, 80
Pyrenean pigs, 185

                Q.

Quinsy, 367

                R.

Rash, nettle, 374
Rations for pigs, 263
Records, 332, 334
Rectum, imperforation of, 362
    inversion of, 361
    malformation of, 362
Red pigs, 165, 170
Red Pig Club, American, 171
Red soldier, 369
Refrigerating chambers, 300
Registration book, 82
Rheumatism, 365
Rickets, 366
Rillettes, 321
Ringing, 92
    appliances for, 93
    effects of, 92
    gilts, 64
    time for, 96

Rings, 93
Rising of lights, 352
Robertson, Mr. James, quoted. 103-105
on feeding, 254-258
Root cleaner, 204
Roots, composition of, 292
Roth-lauf, 368
Rouget du porc, 368
Round worms, 375
Runs, grass, 64
Rupture, 359
in young pigs, 87

S.

St. Maurice, Colonie Agricole de, piggeries at, 216
Salting pork, 306, 310
Sanders, Mr., on the Cheshire or Jefferson County breed, 180
Sanson, Prof., on feeding, 331
Sarcoptes scabiei suis, 367
Sau, 14
Sausages, 316
crépinettes, 316
French, 316
German, 318
Hamburg, 317
type of pig for, 11
Sawdust as litter, 193
Scabies, 367
Scalding, 299, 303
Scandinavian pigs, 187
Scent, power of, 4, 17
Schwein, 14
Scientific classification, 14
Scotland, pigs in, 7
Scours, 352
Scrotal rupture, 359
Sebright, Sir John, quoted, 35
Second service, 58
Seed foods, composition of, 292
Selection of stock, 49
Service, certificates of, 336
of sows, 63, 99, 100, 336
second, 58
ticket, model of, 336
Seven-eighths blood, 40

Shelton's (Professor) experiments in fattening, 285
Shows, Birmingham, 337
carcase classes at, 47, 338
classification at, 138, 336, 338
fatting for, 258
judges at, 339
purchasing at, 80
Royal Agricultural, 337
Smithfield, 337
Siamese cross with the Berkshire, 149
Sidney quoted, 27, 116, 123, 132, 133, 136, 165
Silo, food, preserved in, 98
Simonds, Professor, on teeth, 26
Singeing, 305
Sire, importance of, 40. (See Boar)
Sisson, Messrs., on cooked food, 281
Size, 9, 51
Skeleton, 15, 16
Skin, 15
identification and uses of, 296
Skull, 18-28
Slaughtering, 299, 302
American system of, 301, 303
Coburn, Hon. F. D., on, 303
English system of, 302
establishments, great, 298, 301
French system of, 303-305
Sleeping apartments, 205-248
Small Black breed, 157
American error, 160
crosses, 157-159
early maturity of, 161
Hobbes, Mr. Fisher, and, 157
objections to, 159, 161
origin of, 157
points of, 159, 162
popularity declining, 159
Smith, Mr. J. A., on, 161
Western, Lord, and, 157
Small piggeries, 238
Small White breed, 127
American opinion of, 133
breeding, 134
Carhead herd, 131, 132
celebrated specimens, 129, 131

Small White breed, Coburn.
    Hon. F. D., on, 133
  colour of, 129, 134
  crosses, 131, 132
  deterioration of, 128
  disqualification of, 130
  Ducie's (Lord) herd, 133
  early breeders of, 130
  economical feeders, 128
  English, 132, 133
  " fancy " breed, 127
  Fisher on, 130, 132
  head of, 129, 130
  origin of, 129, 130, 132, 133
  Moreton's (Lord) herd, 133
  points of, 129, 134
  Prince Consort's herd, 129
  Sidney on, 132, 133
  Small York breed, 132
  Solway breed, 131, 139
  standard of excellence, 129, 134
  Suffolk breed, 131-133
  Victoria's (Queen) herd, 129
  Wentworth, Hon. John, on,
    133
Small v. large breed, 116, 118
Small York breed, 132
Smell, sense of, 4, 17
Smith, Mr. Joseph, on the
    Berkshire, 143
  Mr. J. A., on the Small
    Black breed, 161
Smithfield Show, 47
Smoking, 306, 314
  Heuzé, M., on, 314
  houses, 314
Snout, 17
  in the Berkshire, 151
Solway breed, 131, 139
Sore teats, 367
Sow after farrowing, treatment
    of, 83
  and young, treatment of, 62,
    69
  classification at shows, 336,
    338
  clumsy, 68
  cottager's, 97
  definition of, 14, 62
  devouring young, 68, 105. 107

Sow after farrowing, drying,
    78
  exercise for, 69
  feeding, 65, 69, 253, 254, 263-
    265
  lying upon young, 66, 73, 104
  number of teats of, 71
  parturition of, 66, 67
  points of a good, 45
  rations for, 263
  rearing young, 99
  ringing, 96
  selecting, 49, 73
  service of, 63, 99, 100, 336
  short of milk, 83
  spaying, 76, 88
  vicious, 68, 104
  young, not thriving, 85
Sparred floors, 190, 240
Spaying, 76, 88
Spencer, Mr. Sanders, quoted,
    106, 115
  on feeding, 254-258·
Springer, Mr. Phil, quoted, 74,
    334
Staffordshire breed, 165
Staggers, blind, 355
Sties, 189-248. (See Piggeries)
Stock, selection of, 49
Stockman's fees, 85
Stomach, 28
  inflammation of, 358
Stone floors, 190
Strangles, 367
Straw, 71, 72, 192, 193
Street and Sons' (Messrs.) pig-
    gery, 224
Suckling sow, treatment of, 69
Suffolk breed, Black, 157
  White, 116, 125, 131, 160, 162
Sullivant, Mr. J., on foods, 268,
    274, 284
Sunshine, advantages of, 192,
    222
Sus aper, 14
  larvatus, 14
  scrofa, 14
Sussex breed, so-called, 165
Swedish pigs, 187
Swimming, 29

Swine, classification of, 14
    erysipelas, 368
    fever, 369
Swiss pigs, 187
    piggery, 209
System, internal, 28

T.

Tænia solium, 363
Tail, 15, 16
Tamworth breed, 165
    Allender, Mr. G. M., on, 169
    antiquity of, 167
    as baconers, 169
    at shows, 166
    Bennett, Mr. Thos., on, 173
    butcher's opinion of, 170
    characteristics of, 165, 168
    colour of, 166-168
    crosses, 166, 167, 170
    Fidgeon, Mr. C. F., on, 168
    Lowe, Mr. John, on, 167
    Mitchell Bros., Messrs., on, 170
    number in litter, 169
    objectionable features, 166
    precocity of, 169
    Sidney on, 165
    Staffordshire breed, 165
    standard of excellence, 166
    weights of, 170
Tapeworm, 363
Teats, development of, 66
    inflamed, 360
    number of, 71
    sore, 367
Teeth, Development of, 17-28
    Furstenberg on, 18-28
    of young pigs, nipping, 71
    Paaren, Dr., on, 18
    Sidney on, 27
    Simonds, Prof., on, 26
Temperature, effects of, on fattening pigs, 285
Testicles, removing, 60, 76, 86
Tetanus, 371
Thin-Rind, 115
Three-quarter blood, 40
"Thrifton," Mr. Phil., on feeding, 255-258

Tongue, 17, 27
Tortworth Court Piggeries, 205
Tramways in piggeries, 203, 205-238
Trichina spiralis, 372
Trichiniasis, 372
Trichinosis, 372
Troughs, 193, 205, 248
Truffles, hunting for, 4
Truie, 14
Tuberculosis, 373
Tuley, Joseph, the weaver, 110, 119
Turnips, 258, 277
Tushes, 27
Tusks, 18-28
    removing, 59
Twain, Mark, on the Chicago trade, 299

U.

Udder, development of, 66
    inflammation of, 360
Umbilical rupture, 359
Unequal litters, 73
Uniformity in breed, 50
United Kingdom, pigs in, 1
Urine, 29
Urticaria, 374

V.

Vegetable foods, 249. (See Foods)
Vendeënne pig, 182, 183
Ventilators, 205-248
Verrat, 14
Viborg, quoted, 28
Vicious sows, 104
Victoria breed, 181
Victoria's (Queen) Small Whites, 129
Vigour, age, and size, 51
Virchow on trichinosis, 372
Viscera, 28
Voelcker, Dr., quoted, 266, 279
Voice, modifications of, 30
Vosgienne pig, 184

W.

Wainman's (Mr.) Large Whites,
    111, 112
Walker Jones, Col., quoted,
    103-106
Walls of piggeries, 201, 205-248
Warmth, effects of, on fattening
    pigs, 285
    Coburn, Hon. F. D., on, 286
    Shelton, Prof., on, 285
Warren County pigs, 175
Wash, 203
Water in piggeries, 202, 205-248
    253
    quantities of, in various foods,
    294
Weak pigs bad breeders, 85
Weaners, feeding, 77, 255
    value of, 75
Weaning, 77, 255
Weight at birth, 29
Weights and measurements, 331,
    great, 5
Wentworth, Hon. John, on the
    Suffolk, 133
Western's, Lord, breed, 157
Westmorland breed, 7, 118
Westphalian hams, 311
Wheat, 257, 269, 331
Whey, 278
    Voelcker, Dr., on, 279
White breeds, 109-134. (See
    under separate headings)
    Chester, 179
    Large, 109
    Middle, 119
    Small, 127
White, Gilbert, quoted, 5
White puddings, 320
Wild boar compared with
    domestic pig, 4, 5, 8, 14, 27,
    46
Windsor breed, 116, 129
Wolf teeth, 18, 28

Wolff, Dr. Emil, quoted, 293,
    330
Wood unsuitable for piggeries,
    190
Woods for smoking, 314
Worms, 375
Worsley Piggeries, 207
Working-man's pig, 97
Wounds, 375
Wykes, Mr. W. H., on the
    Berkshire, 144

Y.

Yards of piggeries, 192, 205-248
Yelt, definition of, 14, 62
York hams, 313
Yorkshire breed, 116
    allied breeds, 118
    Bennett, Mr. Thos., on, 173
    Cumberland breed, 116
    origin of, 116
    prices realised, 117
Young, bedding or benches for,
    72
    castration of, 76, 86
    dead, 85
    disposal of, by cottager, 100
    exercise for, 76
    feeding, 79, 254, 255, 259, 263-
    265
    hand-feeding, 79
    management of, 69, 81
    nipping teeth of, 71
    number of, 74
    rations for, 263
    ringing, 96, 107
    ruptured, 87
    sow eating, 68, 105, 107
    spaying, 76, 88
    value of, 75, 100
    weaning, 255

Z.

Zenker on trichinosis, 372, 373